AF789842

ENTOMOLOGY

ENTOMOLOGY

Editor

Rochus Vogts

scitus
academics

Entomology

Edited by **Rochus Vogts**

This Book is Printed on Acid-Free Paper. ♾

This book contains information obtained from highly regarded resources. Copyright for individual articles remains with the authors as indicated. All chapters are distributed under the terms of the Creative Commons Attribution License, which permits unrestricted use, distribution, and reproduction in any medium, provided the original author and source are credited.

ISBN: 9781681179100

Library of Congress Control Number: 2017948938

© 2018 by

SCITUS Academics LLC,
616, Corporate Way, Suite 2, 4766,

Valley Cottage, NY 10989

www.scitusacademics.com

Notice

Reasonable efforts have been made to publish reliable data and views articulated in the chapters are those of the individual contributors, and not necessarily those of the editors or publishers. Editors or publishers are not responsible for the accuracy of the information in the published chapters or consequences of their use. The publisher believes no responsibility for any damage or grievance to the persons or property arising out of the use of any materials, instructions, methods or thoughts in the book. The editors and the publisher have attempted to trace the copyright holders of all material reproduced in this publication and apologize to copyright holders if permission has not been obtained. If any copyright holder has not been acknowledged, please write to us so we may rectify.

Printed in United States of America.

Preface

Agricultural crops are of considerable importance to the nation economy and also vital to other industries, such as poultry, the leading agricultural enterprise, which relies upon certain field crops for feed and a place for waste management. Controlling insects and other arthropods has become an increasingly complex issue over the past two decades. Minimizing losses due to insect pests and insect vectors of important plant, animal, and human diseases remains an essential component of the Agricultural Entomology. Nonetheless, the deleterious effects associated with the use of chemicals, such as development of insect resistance to insecticides, poisonous residues on foodstuffs, human illnesses associated with pesticide applications, contamination of soil and water, and diminution of biotic diversity must be reduced or eliminated.

This volume, Entomology, presents environmentally sound pest management practices under the umbrella of integrated pest management (IPM). It consists of contributions from the renowned authors providing information on pest management approaches as chemical alternatives. It presents emphasis on pest management strategies for use in corn, soybeans, small grains, forage crops, vegetables, and beekeeping. As a result, considerable emphasis has been placed on tactics other than chemical controls, including biological, cultural, and genetic methods and the deployment of varieties resistant to pests. A re-evaluation of the economic thresholds of insect pests and vectors and the concurrent development of models capable of generating management strategies are a direct result of the need to reduce our reliance on chemical controls. To ensure profitable production it addresses both short- and long-term solutions to the most important arthropod-related problems facing agriculture. The goal of this book is to provide the most up-to-date review on information available around chemical alternatives in IPM. Therefore, this book will provide academia and industry with adequate basic concepts and applications of IPM as eco-friendly pest management option.

Table of Contents

CHAPTER 1

Reproductive Interference and Niche Partitioning in Aphidophagous Insects

Suzuki Noriyuki[1,2] and Naoya Osawa[3]

[1]*Department of Environmental Science, Policy, and Management, University of California, Berkeley, 201 Wellman Hall, Berkeley, CA 94720, USA*
[2]*Center for Geo-Environmental Science, Rissho University, Tokyo, Japan*
[3]*Laboratory of Forest Ecology, Graduate School of Agriculture, Kyoto University, Kyoto, Japan*

ABSTRACT

The range and quality of prey species differ greatly among closely related species of predators. However, the factors responsible for this diversified niche utilization are unclear. This is because the predation and resource competition do not always prevent species coexistence. In this paper, we present evidence in support of reproductive interference as a driver of niche partitioning, focusing on aphidophagous insect. Firstly, we present closely related generalist and specialist species pairs in aphidophagous lacewings to compare the reproductive interference hypothesis with two other hypotheses that have been proposed to explain niche partitioning in lacewings and sympatric speciation through host race formation and sexual selection. Secondly, we present a case study that shows how reproductive interference can drive niche partitioning in sibling ladybird species. Thirdly, we show that many ladybird genera include species inhabiting the same region but having different food and habitat preferences, raising the possibility that reproductive interference might occur in these groups. Finally, we show that intraguild predation cannot always explain the niche partitioning in aphidophagous insects including hoverflies and parasitoids. On the basis of the evidence presented, we urge that future studies investigating predator communities should take account of the role of reproductive interference.

1. INTRODUCTION

In nature, closely related species often occupy niches that diverge with respect to both type and breadth, and such niche divergence is observed not only in herbivores but also in many predators [1–4]. A frequently proposed explanation of niche differentiation is the trade-off hypothesis, according to which adaptations of a species that allow it to exploit one resource decrease its fitness on other resources, and this trade-off leads to different niche utilization by different species [5]. In predators, the main driving force of food specialization may be morphological and behavioral adaptations that enhance a species' prey capture performance against one prey species but simultaneously reduce it against another [6]. However, the trade-off hypothesis is inadequate to explain differences in food choice among some closely related predator species. In particular, some species that possess traits that make them highly efficient foragers specialize in a highly defended prey species, even though they could potentially utilize a wide variety of food items, including less well-defended prey (e.g., [7–9]). For example, large-mouthed marine cottid fish species have been shown in a laboratory setting to have a high feeding performance on both slow-moving prey (e.g., crabs, isopods, and gastropods) and more mobile, and thus more elusive, prey (e.g., fishes and shrimp), but in nature they feed predominantly on elusive prey [10]. Such examples raise the possibility that negative interspecific interaction, rather than a species' own resource-use traits, can restrict its host range in nature.

Examples of negative interspecific interactions that might promote niche partitioning among predators include intraguild predation and parasitism, where species competing for a shared resource also prey on or parasitize one another [11, 12]. Predation by predators on predators has been widely reported under both laboratory and natural conditions [13, 14]. Moreover, multiple species involved in intraguild predation have been observed to persist in the same locality in field studies (see the "Aphidophagous Guilds"). Therefore, the idea that intraguild predation, which combines predation with resource competition, is sufficient to drive niche partitioning in predators is still controversial. Thus, we would propose that it should be fruitful to consider alternative interspecific interactions that might explain the divergent food choices of carnivorous species.

One reported type of interspecific interaction between closely related predator species is heterospecific mating interactions (e.g., [15–17]). For example, in damselflies (Nehalennia), males frequently attempt to form a tandem pair with a heterospecific female, a behavior that may waste the time and energy of both species even if hybridization does not occur [18]. Moreover, in a molecular study, Fitzpatrick et al. [19] detected genetic introgression between specialist and generalist garter snakes, which is indirect evidence that interspecific copulation occurs in nature. In fact, such reproductive interference is a general mechanism that is applicable to various species assemblages with overlapping mating signals, regardless of trophic level, if at least one species engages in sexual reproduction [20, 21]. However, we think that the potential effect of reproductive interference on the spatial distribution and resource use by predator species has been underestimated, possibly because researchers have focused on mechanisms that are specific to predator-prey systems, such as intraguild predation and strong resource competition [22]. Therefore, we assert that reproductive interference should be given greater consideration in the search for a mechanism to explain ecological specialization in predators.

In this paper, we focus on aphidophagous insects. First, we describe aphidophagous lacewing genera that include generalist and specialist predatory species pairs. In lacewings, it has been hypothesized that sympatric speciation has occurred through host race formation or sexual selection. We argue instead that the observed niche partitioning between closely related species may actually be a consequence of niche diversification through reproductive interference after allopatric speciation. Next, we present as a case study of niche partitioning by reproductive interference our own research on two species of ladybirds (Harmonia) in Japan. Then, we consider whether reproductive interference-driven niche partitioning can explain niche partitioning in other predatory ladybirds. Finally, we describe some aphidophagous insect communities in which intraguild predation is known to occur and discuss reproductive interference as a possible mechanism that can drive niche partitioning of intraguild predators. We hope that the arguments made in this paper will encourage researchers to pay more attention to interspecific mating behaviors and their ecological consequences in predator communities.

2. SPECIALIST AND GENERALIST LACEWINGS IN NORTH AMERICA

In this section we first examine the possible role of reproductive interference in the ecology of two sibling species (i.e., reproductively isolated species that are nearly identical in their appearance [25]) of the lacewings (Neuroptera: Chrysopidae) in North America, one in genus Chrysopa and another in Chrysoperla. TheChrysopa pair includes a strict specialist that uses a strange woolly aphid, and the Chrysoperla pair includes a conifer specialist. Then we compare the reproductive interference model with models of sympatric speciation based on host race formation and sexual selection, which have been considered the main drivers of genetic and habitat diversification in sympatric lacewings.

2.1. Chrysopa and Chrysoperla Species Complexes

Chrysopa quadripunctata is a generalist predator that preys upon a wide variety of aphids and other soft-bodied arthropods that live on various plant species, including apple, elm, goldenrod, hickory, maple, oak, and rose. In contrast, Chrysopa slossonae is a specialist predator that exclusively utilizes the woolly alder aphid, Prociphilus tessellatus [31]. Interestingly, the woolly alder aphid covers its body with secreted wax, which it uses to attract the ants that protect the aphid colonies from predators [31]. The wax may also serve as a physical defense against attacking predators [32]. However, C. slossonae larvae exhibit a sophisticated behavior in which they scrape the wax from their prey and attach it to their own back, thus preventing their detection by the attending ants [31]. In addition to this camouflage strategy, the specialized morphology (long legs and a large mandible) of C. slossonae larvae may enable them to consume the woolly alder aphid efficiently [33].

We argue that reproductive interference by C. quadripunctata may be responsible for the restricted food range of C. slossonae, for the following reasons. First, C. slossonae larvae can develop on less elusive prey that they do not utilize in nature, such as Myzus persicae and A. pisum [7, 34]. These observations are evidence that the fundamental niche of C. slossonae is wider than its realized niche. Moreover, interspecific copulation and hybridization occur between C. slossonae and C. quadripunctata, at least under laboratory conditions [34]. Although interspecific pairs of C. slossonae and C. quadripunctata can produce viable hybrids, the ecological traits of such hybrids show the intermediate values between their parent species [34]. This indicates that hybridization may hamper the local adaptation to each habitat. Therefore, it is possible that C. slossonae is forced to specialize on a highly defended prey to avoid heterospecific mating interactions with C. quadripunctata, which utilizes various preferred prey species in nature.

Chrysoperla carnea is a habitat generalist that lives in fields and meadows during the summer and migrates to deciduous trees in autumn. In contrast, its sibling species, Chrysoperla downesi, is a habitat specialist, living all year only in coniferous trees, in the region of eastern North America where the two species are sympatric [38]. It can be inferred that C. downesi specializes on conifer-associated prey items, but the information about the variety and quality of prey utilized by these species is still scarce. Each species, however, exhibits phenotypic traits that suit it to its preferred habitat. First, both species display cryptic adult body coloration that is specific to their habitats: C. carnea is bright green in spring and summer and reddish brown in autumn and winter, whereas C. downesi is dark green all year [39]. Second, each species has life-history traits, such as the critical day length for diapause induction (e.g., [40]), that suit it to its voltinism. Chrysoperla carnea can utilize a variety of food sources depending on the season and produces several generations each year, whereas C. downesi is restricted to a single generation in the spring, probably because the availability of their prey in coniferous trees is more seasonal. As in the Chrysopa species complex, hybridization can occur between C. carnea and C. downesi, at least under laboratory conditions [39], which suggests that reproductive interference might occur in nature.

At present, the available laboratory findings on interspecific mating behavior are insufficient for the role of reproductive interference in these Chrysopa and Chrysoperla species to be evaluated. The hybrids were experimentally produced to clarify the existence of postmating reproductive isolation [34, 39] and to examine the genetic basis for the phenotypic traits [39, 40]. However, all possible behavioral mechanisms that can reduce adult reproductive success should be considered in investigations of the population dynamics of these two species pairs. For example, interspecific sexual harassment prior to interspecific copulation can reduce a female's oviposition rate, longevity [41], and mating success [42], which might have population dynamics consequences and lead to subsequent niche separation. Thus, future studies of these sibling lacewing species should not only examine hybrid production but also quantify lifetime female reproductive success to evaluate the possibility that reproductive interference has driven the niche separation.

2.2. Sympatric Speciation through Host Race Formation

The difference in niche width between sympatric lacewings has been attributed to sympatric speciation [43, 44]. Sympatric speciation occurs when ecological adaptations to a different host (habitat) also produce nonrandom mating in a conspecific population, which eventually leads to a phenotypically and genetically divergent subpopulation, or species, even without geographical isolation [45]. In lacewings, courtship and copulation generally take place on plants that also serve as the feeding sites for both the adults and their offspring [46], indicating that adaptation and fidelity to a different habitat by each species may contribute to the development of reproductive isolation.

Tauber et al. [44] have argued that the comparative evidence from this Chrysopa species pair demonstrates for the first time a sympatric speciation mechanism in predatory insects. They proposed that a generalist ancestor of C. slossonae adopted a specialized foraging behavior and morphology as well as cryptic adult coloration that were well suited to a specific habitat. At the same time, a seasonal reproductive cycle, different from that of the ancestor, became established in the specialized subpopulation. This subpopulation then became reproductively isolated, both spatially and temporally, from the original population, which eventually led to its becoming a new specialist species by sympatric speciation.

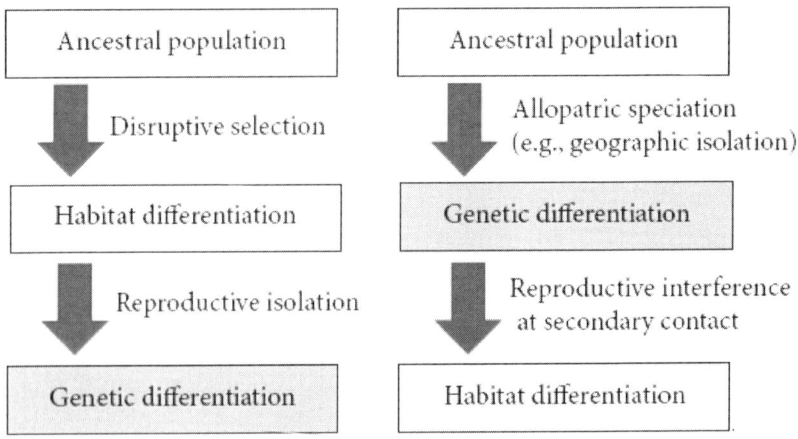

(a) Sympatric speciation by habitat specialization

(b) Character displacement by reproductive interference

Figure 1: Two models of ecological specialization and niche differentiation in sibling lacewing species. (a) Sympatric speciation model. Disruptive selection within an ancestral population (i.e., ecological adaptation to different habitats) drives reproductive isolation, which leads to genetic differentiation and, subsequently, complete speciation. (b) Reproductive interference model. Reproductive interference drives habitat differentiation between the two species when secondary contact occurs after allopatric speciation.

Here we compare the two evolutionary scenarios, sympatric speciation and ecological specialization via reproductive interference, that have been proposed to explain the current niche partitioning between sibling lacewing species. The assumed order of ecological adaptation and genetic divergence differs between these two models [47] (Figure 1). In the sympatric speciation model, habitat differentiation is the primary impetus of genetic divergence in an originally uniform population. In this model, therefore, reproductive isolation is a by-product of adaptation to a particular habitat [48] (Figure 1(a)). In contrast, in the ecological specialization through reproductive interference model, genetic divergence (speciation) occurs allopatrically and reproductive interference promotes habitat differentiation when the two already distinct species come into secondary contact [49, 50]. In the reproductive interference mechanism of the second model, therefore, genetic incompatibility between the species is the cause of niche diversification (Figure 1(b)). Thus, the two models predict similar ecological consequences, namely, utilization of different habitats by two genetically close species, which prevents interspecific sexual interactions [47]. Accordingly, it is difficult to judge which model is applicable to a particular study system by examining current phenotypic traits and fitness.

The model of sympatric speciation in Chrysopa and Chrysoperla lacewings has some as yet unresolved problems. Most importantly, the generalist predator has been observed to visit the habitat on which its sibling species specializes, even though that habitat is less suitable for the generalist species. For example, C. quadripunctata is sometimes found on alder trees, where C. slossonae preys on the woolly alder aphid [34]. In addition, C. carnea visits not only meadows and deciduous trees but also coniferous trees, where C. downesi is found year-round [39]. This flexible behavior on the part of generalist predators may allow them to add a less suitable prey to their menu when the availability of more suitable prey is limited for some reason (e.g., [51]). Moreover, Henry [38] pointed out that the breeding periods of the sibling lacewing species greatly overlap, even though C. A. Tauber and M. J. Tauber [39] considered the seasonal mismatch to be an important mechanism of temporal reproductive isolation. Therefore, it is possible that observed niche overlap in space and time might be insufficient for sympatric genetic divergence to have occurred in the Chrysopa and Chrysoperla systems. Nevertheless, because genetic divergence sometimes occurs despite copulation and gene flow between taxa [52–54], the apparent incomplete habitat isolation does not rule out sympatric speciation in these lacewings. At this time, therefore, it is not possible to rule out any particular view of speciation in the lacewing system.

2.3. Speciation through Sexual Selection

Henry [38] proposed that divergence in the courtship song, rather than habitat specialization, might be a primary driver of speciation in Chrysoperla lacewing species, whether allopatric or sympatric. In Chrysoperla, males produce substrate-transmitted calls on plants by vibrating their abdomen, and females respond to a potential partner by reciprocal signaling according to their own mating preference [55]. Henry [38] proposed that, based on the diversified and complex songs among close relatives, differentiation of courtship signals among populations should create ethological barriers to hybridization, thus accelerating the rate at which complete speciation can be achieved.

However, there is at least one problem with speciation through sexual selection. Chrysoperla females sometimes respond to heterospecific mating calls [55], which indicates that negative interspecific mating interactions can occur. In fact, under laboratory conditions copulation and hybridization can occur among populations (species) having different courtship songs [39]. These findings suggest that an ethological barrier alone might not ensure premating reproductive isolation.

This weak point of the speciation through sexual selection model (i.e., inevitable heterospecific mating interactions), however, supports the reproductive interference model. In general, the maintenance of incomplete species recognition abilities can be interpreted as a consequence of adaptive decision-making in reproductive animals. When the signals of high-quality conspecifics resemble those of heterospecifics, a trade-off is likely to exist between species recognition and mate-quality recognition within a species [56]. In this situation, strict species discrimination skill is unlikely to be maintained because individuals may also lose opportunities for acquiring high-quality conspecific mates. As a result, promiscuous mating occurs between the species. By this logic, reproductive interference is likely to be common in sibling species (because their mating signals are likely to be similar), and the reproductive interference might drive niche partitioning.

3. A CASE STUDY IN HARMONIA

3.1. A Brief History of Studies of Harmonia Sibling Species

The multicolored Asian ladybird Harmonia axyridis is a common aphidophagous predator originally distributed in Russia, China, the Korean Peninsula, and Japan [24]. Because of its extreme polymorphism with regard to elytral color, H. axyridis has been extensively used for genetic studies. Theodosius Dobzhansky famously discovered geographic variation in the frequency of elytral color morphs across Siberia and East Asia [57]. Subsequently, extensive surveys established that elytral color morph frequencies change along a geographic (latitudinal) cline in Japan, where red (nonmelanistic) morphs at higher latitudes are gradually displaced by black (melanistic) morphs as latitude decreases [58]. Then, Hosino, who worked at an agricultural high school in Sanage, central Japan, discovered that elytral color morph frequencies were slightly different between Harmonia ladybirds collected from Japanese red pine (Pinus densiflora) and those collected from agricultural crops such as wheat, pear, and peach [59]. In addition, he reported that a ridge at the tip of the elytron, which is a genetic character exhibited by approximately 40% of H. axyridis individuals in the Sanage population, was seldom observed in individuals collected from pine trees [59]. For a few decades, Japanese entomologists were unable to account for these strange patterns.

Eventually, Hiroyuki Sasaji demonstrated the existence of a cryptic Harmonia species. He did so by experimentally demonstrating that a clear postmating reproductive barrier existed between H. axyridisindividuals and those found on pine trees, even though they were observed to copulate together and produce eggs [60]. He also showed that larval coloration was distinct between these groups of individuals [61] (Figure 2), even though the adults were difficult to distinguish by their external and genital morphology [60], and, furthermore, he showed that the characteristic ridge at the tip of the elytra of H. axyridis adults never emerged in the pine-associated individuals [60]. On the basis of these findings, he concluded that the pine-associated individuals belonged to a cryptic species that could be clearly distinguished from H. axyridis by its postreproductive isolation and some phenotypic traits.

However, when Sasaji proposed denominating the new, pine-associated cryptic species, he discovered that Takizawa [62], at the dawn of insect taxonomy in Japan, had already described a species of ladybird collected from a pine tree and had named it Ptychanatis yedoensis (Ptychanatis is a currently unused synonym for genusHarmonia). Takizawa's illustration of a larva of his new species (Figure 2(c)) clearly shows the larval characteristics of the cryptic species "discovered" by Sasaji. Sasaji [63], therefore, identified the cryptic species as Harmonia yedoensis Takizawa. In addition, deeply moved by Takizawa's insight, Sasaji proposed Kurisaki-tento as the Japanese name of this species, because Takizawa had adopted the name "Kurisaki" when he became a Buddhist priest, and tento is the Japanese word for ladybird.

Figure 2: (a) A final instar larva of H. axyridis on a Spiraea thunbergii leaf. (b) A final instar larva of H. yedoensis on Japanese red pine, Pinus densiflora. (c) Illustration of a final instar larva of H. yedoensis by Kurisaki [23].

3.2. Host Specialization in H. yedoensis

The existence of these two species, H. yedoensis and H. axyridis, living in close proximity raises some questions; namely, what determines their habitat ranges? And how did two such phylogenetically close species with such a close morphological resemblance, at least in adults, come to occupy dramatically different niches? Harmonia axyridis is a truly generalist predator found in many different habitats, including scrub, orchards, grasslands, and coniferous and deciduous woodlands (e.g., [51]). In contrast, H. yedoensis is a specialist in central Japan that breeds exclusively on pine trees, where its larvae are regarded to prey mainly on the giant pine aphid(Cinara pini) [24] (Figure 3(a)). Because aphid colonies are spatially heterogeneous and temporally variable in terms of both quality and quantity, a generalist strategy that allows the ladybird to utilize multiple food patches has obvious advantages for maximizing lifetime fitness [51], as it does for other generalist insect species (e.g., [64–66]). Therefore, how to explain host specialization, a strategy that appears to forego potentially available resources, has been an interesting problem in ecology and evolutionary biology.

(a) (b)

Figure 3: (a) A giant pine aphid, Cinara pini, on Japanese red pine, Pinus densiflora. (b)Harmonia yedoensis hatchling capturing a C. pini individual, observed in a laboratory experiment. Because C. pini has long legs and is highly mobile, H. yedoensis larvae are often unsuccessful in their attempts to capture this aphid.

It is clear from field surveys that H. yedoensis is much less abundant than H. axyridis. One result of this situation is that much more effort is required to collect enough H. yedoensis individuals to carry out research projects, compared with that required for H. axyridis. The rarity of H. yedoensis might be at least partly due to the low abundance of its prey item on pine trees. Although the yellow egg batches of H. yedoensis are easy to locate on dead or live pine tree needles, it is hard to locate nearby colonies of the giant pine aphid, which are usually very small (S. Noriyuki personal observation), unlike those of other aphid species, which tend to congregate on young shoots of their host plants. Syunsuke Shimamoto, a junior high school student studying the ecology of H. yedoensis, once remarked to one of the authors (S. Noriyuki), "I wonder why H. yedoensislarvae stay where there is so little food to eat, while H. axyridis larvae are on plants with lots of aphids." We, too, ask this simple question.

Not only are giant pine aphids rare, they are also so mobile that ladybirds seem to have difficulty capturing them. In contrast to most arboreal aphids, which walk very slowly at best, the giant pine aphid has long legs and can walk fast (Figure 3(b)). We therefore experimentally evaluated the prey capture performance of H. yedoensis and H. axyridis hatchlings against the giant pine aphid and other aphid species [9]. As predicted, it was hard for H. axyridis hatchlings to capture the giant pine aphids. Interestingly, however, although H. yedoensis apparently specializes on the giant pine aphid, this species is not the easiest prey for H. yedoensishatchlings to catch. Rather, we found that H. yedoensis hatchlings could capture other aphid species more easily, even though they never encounter them or prey upon them in nature. These results led us to hypothesize thatH. yedoensis mothers deliberately choose a host that is less suitable for the prey capture activities of their offspring even though other suitable hosts are available in the region.

Harmonia yedoensis mothers help their offspring cope with their elusive aphid prey by investing more maternal resources in each hatchling. In aphidophagous ladybirds, the hatch rate of an egg batch is often less than 100% and the newly hatched larvae consume the unhatched sibling eggs. These unhatched eggs are not just an unavoidable developmental side effect; they are an adaptive maternal strategy known as trophic egg provisioning, because ladybird mothers are able to control the proportion of unhatched eggs in a clutch according to food availability [67]. In fact, trophic egg provisioning is more intense in H. yedoensis than in H. axyridis [26, 28]. Moreover, artificial manipulation of the trophic egg number provided to each hatchling has revealed that trophic egg consumption enhances the prey capture performance of H. yedoensis hatchlings against the giant pine aphid [9]. In addition, H. yedoensis eggs are significantly larger than H. axyridis eggs [26, 28]. Consequently, the amount of maternal resource invested per offspring is much higher in H. yedoensis than in H. axyridis. Importantly, however, the number of offspring that a mother can produce in her lifetime is lower in H. yedoensis than in H. axyridis, because of the trade-off between offspring size and number in the similar-sized species [26]. Therefore, H. yedoensis mothers apparently sacrifice reproductive success in order to supply more resources to each of her larvae, which are obligated to prey on the elusive aphid.

Furthermore, we found by experiment that the giant pine aphid is nutritionally less suitable for the larval development of ladybirds. We froze some aphids, to exclude the effects of prey mobility on larval performance, and then fed them to larvae of H. yedoensis and of H. axyridis to evaluate their intrinsic suitability as food [27]. The result showed that in both ladybird species larval developmental performance was lower in larvae fed the giant pine aphid than in those fed the other prey species. Although we have not yet examined the proximate mechanism for the lower developmental performance, secondary compounds derived from the pine trees and stored in the aphid's body (see [68, 69]), or simply an insufficient nutrient content, might be responsible.

Taken together, these findings show that the giant pine aphid is poor prey for both H. yedoensis and H. axyridislarvae with respect to abundance, capture difficulty, and its intrinsic suitability as food. Therefore, we conclude that food quality per se does not determine the food range of H. yedoensis. In contrast to the rarity and niche specialization of H. yedoensis, H. axyridis is abundant and utilizes various preferred prey species. These differences suggest that negative interactions between H. yedoensis and H. axyridis might greatly influence the food utilization of H. yedoensis in nature.

3.3. Reproductive Interference between the Two Harmonia Species

The results summarized in Section 3.2 led us to hypothesize that reproductive interference by H. axyridisagainst H. yedoensis is responsible for the specialization on less suitable prey by H. yedoensis. When H. yedoensis was first identified as a cryptic Harmonia species, it was already known that although interspecific copulation occurred, a postmating reproductive barrier existed between H. yedoensis and H. axyridis [60, 70]. Sasaji [60] also showed that the male genitalia were very similar between the two species with respect to both size and morphology, so there was no structural barrier to interspecific copulation. To examine the possible influence of reproductive interference on niche partitioning, it was necessary to quantify the effects of interspecific mating behaviors on reproductive success, which might affect the population dynamics of the two species.

Therefore, we performed some laboratory experiments to examine interspecific mating behaviors and their subsequent effect on reproductive success [29]. Both H. yedoensis and H. axyridis males attempted to mate with both conspecific and heterospecific females, and these attempts often resulted in interspecific copulation lasting a few hours or more. Females of both species can lay eggs even after interspecific copulation, but no viable offspring hatched. However, H. axyridis males tended to choose conspecific over heterospecific females, whereas H. yedoensis males promiscuously copulated with both conspecific and heterospecific females. We are unable to account for this difference in discrimination skill between H. yedoensis and H. axyridis, although chemical signals based on cuticular hydrocarbons or visual cues related to body size and color might be important for the interspecific communications. But, probably owing to this difference in discrimination skill, most H. yedoensis individuals failed to copulate with conspecifics, especially when the density of H. axyridiswas high in the experimental arena. By contrast, most H. axyridis individuals successfully copulated with conspecifics regardless of whether H. yedoensis was present [29]. Here, it is important to note that the fitness reduction at the mating stage is asymmetric between these two species;

that is, H. yedoensis is inferior to H. axyridis in terms of reproductive interference.

These results support the hypothesis that reproductive interference from H. axyridis has led to host specialization in H. yedoensis. In nature, H. yedoensis might suffer from a considerable fitness reduction if they utilize less elusive aphids in a patch occupied by H. axyridis. By specializing to the habitat of the giant pine aphid, which is especially elusive prey for H. axyridis, H. yedoensis can mitigate the cost of reproductive interference. Therefore, we argue that H. yedoensis mothers are obligated to choose lower quality prey for offspring performance where the ranges of these sibling species overlap.

(a) (b)

Figure 4: (a) Natural distribution of H. yedoensis (red solid line) and H. axyridis (blue dotted line) around the Japanese archipelago (modified from Sasaji [24]). The two species are sympatric in central Japan (Honshu), whereas only H. yedoensis is found in the Ryukyu Islands and only H. axyridis in Hokkaido. (b) Harmonia yedoensis on a lead tree, Leucaena leucocephala, on Miyako Island, southern Ryukyus. In areas where H. axyridis does not occur, H. yedoensis adopts a generalist strategy.

Biogeographical distribution patterns also suggest that reproductive interference plays a nonnegligible role in determining the food range of H. yedoensis. In the Ryukyu Islands, southern Japan, where H. axyridis is not found (Figure 4), H. yedoensis often visits various deciduous tree species

[71, 72] (S. Noriyuki, personal observation). In contrast, on Hokkaido Island, northern Japan, where H. yedoensis does not occur, H. axyridisis a generalist predator, just as it is in regions of central Japan where the two species are sympatric. These patterns suggest that H. axyridis utilizes various preferred prey species without regard to the presence of H. yedoensis, whereas the negative impact of reproductive interference of H. axyridis on H. yedoensis forces the latter to become a specialist predator on less preferred prey where the two species coexist.

Comparison of Harmonia (ladybirds) and Chrysopa (lacewings) (Table 1) reveals many similarities in the relationships between these sympatric sibling species pairs. Both specialist species can potentially develop on a variety of food sources that are in nature utilized exclusively by their generalist counterpart. Moreover, the specialist species have morphological, behavioral, and physiological adaptations that enable them to handle their less suitable prey. Furthermore, interspecific copulation can occur between members of each pair, at least under laboratory conditions. Given these observations, it would be interesting to explore whether the mechanism of ecological specialization/niche differentiation is the same in these two aphidophagous taxa.

Of course, alternative interspecific interactions, not necessarily mutually exclusive with reproductive interference, must also be considered as possible mechanisms of the ecological specialization of H. yedoensis. In particular, intraguild predation may drive host specialization in H. yedoensis, because H. axyridis is known to be a strong intraguild predator against other ladybird species. However, the results of a laboratory experiment suggest that at the larval stage H. axyridis might not be dominant over H. yedoensis as an intraguild predator [73]. It is probable that the body size of the individual larvae, rather than their species identity, mainly determines the winner in interactions between these two predatory larvae, because the final instar larvae of both species feed on earlier instar larvae of the other species when they are put together in a Petri dish [73]. We think it is likely that intraguild predation and reproductive interference jointly contribute to niche partitioning between these two ladybird species, because both mechanisms can destabilize species coexistence locally. However, the direction of the food specialization, with H. yedoensis feeding on more elusive and H. axyridisfeeding on less elusive prey, might be determined by the asymmetry of the reproductive interference. Nevertheless, the combined effect of predation and competition on individual fitness and species coexistence is predicted by mathematical models to produce complex outcomes [74, 75]. We therefore think that it would be very interesting to see how the incorporation of reproductive interference in a trophic module along with intraguild predation would alter population dynamics and community structure.

Table 1: Comparison of the niche, life-history traits, and interspecific mating interactions in sibling species between Harmonia ladybirds and Chrysopa lacewings.

	Harmonia ladybirds		*Chrysopa* lacewings	
	H. axyridis (generalist)	*H. yedoensis* (specialist)	*C. quadripunctata* (generalist)	*C. slossonae* (specialist)
Niche				
Habitat	Various deciduous trees including pine trees	Pine trees	Various deciduous trees including alder trees	Alder trees
Prey	Various aphid species	The giant pine aphid	Various aphid species	The woolly alder aphid (with attended ants)
Alternative prey	Various aphid species and artificial diet	Various aphid species and artificial diet	Various aphid species	Various aphid species
Life-history traits				
Adult body size	Similar		Small	Large
Egg size	Small	Large	Small	Large
Rate of oviposition	High	Low	High	Low
Sibling cannibalism within clutch	Low	High	NA	
Hatchling head size	Small	Large	Small	Large
Larval leg length	Short	Long	Short	Long
Larval behavior	Low walking ability	High walking ability	Camouflage	Camouflage with wax
Seasonality	Bivoltine (central Japan)	Univoltine	Bivoltine and a partial third generation per year	Univoltine
Critical photoperiod for diapause induction	NA		Short	Long
Interspecific reproduction				
Interspecific mating attempt	Yes		Yes	
Interspecific copulation	Yes		Yes	
Viable hybrid offspring	No		Yes	
Reproductive interference	Superior	Inferior	NA	

Information on *Harmonia* is from Sasaji [24], Osawa and Ohashi [26], Noriyuki and Osawa [27], and Noriyuki et al. [9, 28–30]; information on *Chrysopa* is from Albuquerque et al. [7], Eisner et al. [31], Milbrath et al. [32], Tauber et al. [33], and C. A. Tauber and M. J. Tauber [34].

Now, about 100 years after Takizawa first described a new Harmonia ladybird species and several decades since its rediscovery by Sasaji, it is becoming clear that the mechanism of ecological specialization and generalization in H. yedoensis and H. axyridis involves negative interspecific interactions. This case study illustrates how the integration of taxonomy, ecology, and evolution can deepen our understanding of natural history and the great principles that underpin biological communities.

4. OTHER SPECIALIST LADYBIRDS

Reproductive interference may have led to niche partitioning not only in Harmonia species but also in predatory ladybirds belonging to other genera. Therefore, we picked congeneric species pairs known to show different food (habitat) ranges or types in the same geographical region and examined the incidence of interspecific mating interactions in these congeneric pairs. Then we investigated conifer- (pine) associated specialist ladybirds in Britain and Japan to assess the generality of our hypothesis that reproductive interference can lead to niche partitioning in closely related species.

4.1. Niche Partitioning between Congeneric Ladybird Species Pairs

To examine niche partitioning in other ladybirds, we selected genera that contain at least two species with different food (habitat) types in the same region of Britain or Japan (Table 2). In both these countries, the natural history of ladybirds has been well studied by professional and amateur entomologists. Because, in this paper, our focus is on aphidophagous insects, we exclude phytophagous and mycophagous ladybirds from the list in the table, but this exclusion does not mean that reproductive interference might not be important in these functional groups [47]. We identified numerous closely related species in various ladybird genera that exhibited niche partitioning (Table 2). For example, Pseudoscymnus sylvaticus mothers lay their eggs on galls of certain aphid species (Tuberocephalus sasakii, which forms galls on cherry tree leaves, or Ceratovacuna nekoashi, which forms galls on fruits of Japanese snowbell, Styrax japonica), and the hatched larvae forage exclusively on the aphids in the gall. In contrast, a sympatric congener, Pseudoscymnus pilicrepus, utilizes the eusocial aphid Colophina arma on the subshrub Clematis stans [24]. At present, the drivers of this niche partitioning in regionally coexisting predatory ladybirds are not well understood. Here, we focus on Adalia andCoccinella species pairs, because each of these genera contains generalist and specialist species pairs distributed sympatrically in Britain and other parts of Europe and considerable information is available about their ecology.

Adalia bipunctata is one of the most common ladybird species in Europe and it occupies a great variety of habitats, including woodlands, scrub, and grassland. By contrast, its sibling species, Adalia decempunctata, is more habitat-specific; it is found mainly on deciduous trees and in hedgerows [76–78]. Because in A. decempunctata larvae can develop and forage on various aphid species [79] and mothers produce more eggs when fed on an aphid that is not regularly used in the field [80], this habitat restriction cannot be explained solely by food suitability. Moreover, Sloggett and Majerus [78] have shown that the prevalence of parasitoids is similar between A. bipunctata and A. decempunctata, suggesting that the habitat specialization of the latter is not driven by a need for a habitat with low predation risk (i.e., an enemy-free space [81]).

Table 2: Comparison of niche utilization among sympatrically distributed congeneric ladybird species of the United Kingdom (UK) and Japan.

Genus	Species	Food, host plants, and habitats
UK		
Adalia	*A. bipunctata*	Various herbaceous and arboreal habitats
	A. decempunctata	Various arboreal habitats, but more specialized than *A. bipunctata*
Chilocorus	*C. renipustulatus*	Coccids; broad-leaved deciduous trees
	C. bipustulatus	Coccids; *Calluna*, Leyland cypress, and other trees; heathland
Coccidula	*C. rufa*	Reeds, reed-mace, rushes, and wetland grasses
	C. scutellata	Reeds, reed-mace, and rushes
Coccinella	*C. undecimpunctata*	Aphids; herbaceous habitats, especially in coastal areas
	C. quinquepunctata	Aphids; low-growing herbaceous plants such as nettle, thistles, bitter-cress, and angelica; unstable river shingle
	C. septempunctata	A variety of aphid species on an extensive range of low-growing herbaceous host plants; habitats including agroecosystem, grassland, heathland, and coniferous and deciduous woodland
	C. magnifica	Ant-attended aphids; Scots pine and other plants close to ant nests of genus *Formica*
	C. hieroglyphica	Larvae of the heather leaf beetle *Lochmaea suturalis* and the heather aphid *Aphis callunae*
Nephus	*N. bisignatus*	Low-growing vegetation in coastal regions
	N. quadrimaculatus	Coniferous and deciduous woodlands
	N. redtenbacheri	Various low-growing vegetation in both inland and coastal regions
Scymnus	*S. nigrinus*	Needleleaf conifers
	S. frontalis	Low-growing vegetation in dry habitats and on coastal dunes
	S. femoralis	Low-growing vegetation on well-drained soils
	S. schmidti	Various types of low-growing vegetation
	S. haemorrhoidalis	Low-growing vegetation and small shrubs, particularly in damp areas
	S. auritus	Oak trees
	S. limbatus	Willow, sallow, and poplar trees
	S. suturalis	Needleleaf conifers, occasionally deciduous trees
	S. interruptus	Pseudococcids and diaspidids in diverse habitats
Rhyzobius	*R. chrysomeloides*	Pine trees, deciduous trees, and ivy
	R. litura	Low-growing vegetation, especially grasses and thistles
	R. lophanthae	Coccids and diaspidids on trees
Japan		
Calvia	*C. quindecimguttata*	Reeds
	C. muiri	Various habitats, especially bamboo grasses
Chilocorus	*C. kuwanae*	Coccids such as *Pseudaulacaspis pentagona*
	C. rubidus	*Kermococcus* coccids on plum, cherry, chestnut, and oak
	C. mikado	*Quercus gilva*
Harmonia	*H. axyridis*	Various habitats, mainly arboreal
	H. yedoensis	Pine trees
Oenopia	*O. scalaris*	Pine trees
	O. hirayamai	Various arboreal habitats
Pseudoscymnus	*P. sylvaticus*	Inside galls of *Tuberocephalus sasakii* on cherry and of *Ceratovacuna nekoashi* on storax
	P. pilicrepus	Eusocial aphid *Colophina arma* on the subshrub *Clematis stans*
Scymnus	*S. posticalis*	Various herbaceous and arboreal habitats
	S. yamato	Wetland, mainly on reeds
	S. babai	Wetland
	S. ohtai	Wetland
	S. hoffmanni	Wetland
	S. nakaikemensis	Wetland
	S. otohime	Chestnut; prey is *Moritziella castaneivora*, which infects chestnut cases

Information on UK ladybirds is from Roy et al. [35] and Majerus [36]. Information on Japanese ladybirds is from Sasaji [24] and Shiyake [37].

Under laboratory conditions, interspecific copulation occurs between A. bipunctata and A. decempunctata, and the resulting sterile hybrids occasionally develop into adults [82]. Thus, interspecific mating interactions should reduce their reproductive success when the two species occupy the same local patch. During fieldwork in the Netherlands, Brakefield [76] importantly observed that the principal spring mating habitats of A. bipunctatawere shrubs, whereas he found almost all mating A. decempunctata on trees. This use of different mating habitats suggests that past negative mating interactions might have caused small-spatial-scale divergence of reproductive sites in these species. It would be interesting to explore this possibility empirically in the future.

Coccinella septempunctata is also very common in gardens and agroecosystems, but it is also found in heathland, scrub, and coniferous and deciduous woodlands. In contrast, each of the four other Coccinellaspecies in Britain specializes in a different niche (Table 2). In particular, Coccinella magnifica is a myrmecophile that lives only near ant nests and it preys on aphids that are attended by the ants [83]. Obviously, foraging on ant-attended aphids is costly for C. magnifica adults and larvae, because the ants often attack the foraging ladybirds [84, 85], suggesting that the ant attendance itself cannot explain the niche differentiation between C. septempunctata and C. magnifica. Importantly, C. magnifica can be successfully reared in the laboratory from adult to adult by feeding nonattended aphid species such as Acyrthosiphon pisum and Aphis fabae (i.e., alternative foods). Thus, its potential dietary breadth appears to be similar to that of its generalist congener C. septempunctata [83]. Even though interspecific mating behaviors have not been thoroughly studied in these two ladybird species, the adult body size and color pattern are very similar between them. It might be illuminating, therefore, to consider interspecific mating interactions in future investigations of niche partitioning amongCoccinella species. For example, to test the hypothesis that reproductive interference prevents a species from occupying its fundamental niche and restricts it to a narrower realized niche, experiments should be performed to examine whether reproductive interference occurs in these species and whether the cost of the interaction is incurred more by C. magnifica than by C. septempunctata.

4.2. Conifer Specialists

It is well known that some species of predatory ladybirds utilize specialized habitats on pine trees and other conifers (Table 3) and that this specific specialization has independently evolved several times in the ladybird lineage [86]. Not all conifer-specialist species coexist in the same region with a congeneric relative (Table 3), which suggests that, assuming that reproductive interference is mainly likely to occur among phylogenetically close species, these specializations might have occurred through mechanisms other than reproductive interference. Some conifer specialists, however, do have sympatrically distributed generalist congeners (Table 3).

In Britain, the habitats of Scymnus nigrinus and Scymnus suturalis are restricted to conifer trees, whereas a congener, Scymnus schmidti, utilizes a wide range of habitat types [35]. A similar pattern has been observed in Japan in specialist and generalist species belonging to genus Oenopia [24]. Therefore, we can hypothesize that, as in Harmonia species in Japan, this specialization to conifers might have been driven by reproductive interference from sympatric congeners. To test this hypothesis, it will be necessary to quantify the cost of interspecific mating interactions as well as the quality of the conifer habitat with respect to larval and adult development. It would be interesting to know if reproductive interference has caused some species to specialize to conifers more than once in the ladybird phylogeny.

Table 3: Conifer-associated specialist ladybird species in the UK and Japan.

Species	Detail	Sympatric congeners
UK		
Exochomus quadripustulatus	Coccids on needleleaf conifers, sallows, and willows	NA[1]
Myrrha octodecimguttata	Aphids on Scots pine	NA
Sospita (Myzia) oblongoguttata	Aphids on Scots pine	NA
Harmonia quadripunctata	Aphids on needleleaf conifers	*H. axyridis*[2]
Anatis ocellata	Aphids on needleleaf conifers, particularly pines	NA
Scymnus nigrinus	Scots pine	*S. schmidti* and other species[3]
Scymnus suturalis	Needleleaf conifers, particularly Scots pine, but occasionally deciduous trees	*S. schmidti* and other species
Japan		
Harmonia yedoensis	Aphids on Japanese red pine and Japanese black pine	*H. axyridis*
Sospita (Myzia) oblongoguttata	Pine trees	NA
Oenopia scalaris	Pine trees	*O. hirayamai*

Information on UK ladybirds was adapted from Roy et al. [35] and Majerus [36]; information on Japanese ladybirds was adapted from Sasaji [24] and Shiyake [37]. [1]NA indicates that no congeneric species is distributed in the same region; [2]*H. axyridis* is an exotic species in the UK; [3]refer to Table 2 for other *Scymnus* species.

5. APHIDOPHAGOUS GUILDS

In this section we consider intraguild predation as an alternative mechanism of the niche partitioning of aphidophagous insects. Multiple species of aphidophagous insect predators, as well as parasitoids that attack aphids, often coexist despite the occurrence of intraguild predation. Osawa [87], for example, studied aphidophagous ladybirds and hoverflies preying on eight species of aphids infecting seven species of deciduous trees in the Botanical Garden of Kyoto University, central Japan. He collected five aphidophagous ladybird species and eight hoverfly species (see also [88]). Notably, both field [89] and laboratory studies [90, 91] have shown that four of the collected ladybird species (except Scymnus posticalis) engage in intraguild predation.Harmonia axyridis, in particular, is an intraguild predator that preys predominantly on other ladybird species [89, 92, 93]. Nevertheless, the abundance of H. axyridis is not negatively correlated with the abundance of the other four ladybird species in this area [88]. Thus, some mechanism must exist that allows these ladybird species to coexist in the same patch.

Similarly, in hoverfly communities generalist predators often use many of the same prey species [94, 95]. Moreover, aphidophagous ladybirds and hoverflies often coexist where their distributions overlap [88, 89, 96], even though predatory ladybirds often attack hoverflies with which they compete for the shared aphid resource (reviewed by [97]). Thus, in aphidophagous communities, predation and resource depletion do not necessarily result in strict niche partitioning.

If parasitoids that attack aphids are assumed to belong to the same guild as aphidophagous insects [13], then aphidophagous communities become even more species-rich and complex. In this case, intraguild predation includes predators such as ladybirds that feed on parasitized aphids (e.g., aphids that have been mummified), as well as different parasitoid species that, when parasitizing the same aphid, attack each other. Despite this complex food web structure, multiple parasitoid species are known to utilize the same aphid colonies simultaneously. For example, Müller et al. [98] reported that, in Silwood Park, southern England, some species of aphids were infected with multiple parasitoid species. Moreover, Osawa [87] identified various primary and secondary parasitoids on aphids on which both predatory ladybirds and hoverflies intensively foraged. These observations suggest that some factor must exist that permits the coexistence of multiple species in aphidophagous communities with multiple and complex prey-predator links.

The conditions that allow the stable coexistence of a predator and its prey, together with a shared prey species, have been extensively analyzed by mathematical models, as well as by some empirical studies. One classic model [12] predicts that stable coexistence should be possible if a trade-off exists between resource consumption and predation ability, but only under a limited range of environmental conditions. In addition, mechanisms such as a temporary refuges for prey [99], optimal predator foraging behavior [100, 101], and intense intraspecific interactions [102] have been proposed to allow species coexistence despite the occurrence of intraguild predation. These mechanisms likely explain the seemingly paradoxical robust persistence of many natural intraguild predation systems.

The local coexistence of phylogenetically closely related species, however, may still be inhibited. In Japan, two sibling aphidophagous ladybirds, Propylea japonica and Propylea quatuordecimpunctata, are distributed parapatrically; the former lives at low elevation and the latter at high elevation, but within a narrow zone they coexist [24]. In

Britain, Coccinella septempunctata utilizes a wide range of prey types and habitats, whereasCoccinella hieroglyphica is mainly found on heather, where it forages on the heather aphid, Aphis callunae, and on larvae of the heather leaf beetle, Lochmaea suturalis [35]. It has often been demonstrated that in both animal and plant communities local assemblages tend to be composed of phylogenetically distant species (e.g., [103–105]), probably owing to exclusive interactions among closely related species. Therefore, to understand the observed spatial distributions of aphidophagous insects, in addition to intraguild predation, some factor that is tightly connected to phylogenetic closeness should be considered.

Interspecific mating behavior is more likely to occur among phylogenetically close species because they often share similar mating signals [56, 106]. The resulting, often costly mating interactions can lead to local species extinction, which in turn can cause diversification of niche use and resource-use traits [49]. In contrast, phylogenetically distant species, which generally have distinctive, easily discriminated mating signals, focus their mating efforts on high-quality conspecifics even when their ecological niches largely overlap. As a result, distantly related species can stably coexist within a patch [49]. These observations suggest that reproductive interference might be a mechanism of niche diversification in closely related species.

6. CONCLUSION

We have shown that the effect of reproductive interference on local population dynamics has probably been underestimated, in contrast to the intraguild predation mechanism, which is specific to the predator community. Therefore, to evaluate the validity and generality of the reproductive interference mechanism, more empirical studies, and extensions of published studies, of various predator species are needed. We have suggested some possible approaches earlier in this paper. Moreover, the relative importance and synergistic effects of predation, exploitative competition, and reproductive interference are largely unknown. Because these mechanisms of species exclusion are not mutually exclusive, the impact of each individually, and of their interactions, on niche partitioning needs to be quantified. A theoretical research priority is the incorporation of reproductive interference into the intraguild predation model to examine how it changes the population dynamics. Together, such empirical and theoretical efforts will enable us to evaluate whether niche partitioning in predators is explicable mainly by reproductive interference, a mechanism that occurs in several functional groups, including herbivores [42, 50] and plants [107], or by a specific mechanism that is unique to carnivores such as intraguild predation.

At present, it cannot be conclusively decided whether the observed niche partitioning between phylogenetically close species has been caused by sympatric speciation or reproductive interference after allopatric speciation, because it is difficult to distinguish between these two evolutionary scenarios by examining current phenotypic traits and fitness. Thus, a fruitful approach might be to identify the locality of the origin of the species by appropriate techniques such as phylogeography and genomics. Sympatric speciation has been frequently assumed to explain niche partitioning among phylogenetically similar relatives such as cryptic species complexes [108], but compelling evidence that two sympatrically distributed species (races) have diverged is still lacking [109]. A theoretical basis now exists for ecological character displacement via reproductive interference [49]. Incorporation of reproductive interference into existing models should make it possible to design more rigorous tests and to further refine the models and thus improve their skills in reconstructing the natural world.

COMPETING INTERESTS

The authors declare that there is no conflict of interests regarding the publication of this paper.

ACKNOWLEDGMENTS

The authors thank Yasuko Kawakami for information on Japanese ladybirds and Jun-Ya Ide for useful comments on the manuscript. This study was funded by support from the Japan Society for the Promotion of Science to Suzuki Noriyuki (no. 24-3112, no. 26840137, and no. 26650154) and to Naoya Osawa (no. 20405047 and no. 24255013).

REFERENCES

1. E. R. Pianka, "The structure of lizard communities," Annual Review of Ecology and Systematics, vol. 4, no. 1, pp. 53–74, 1973.
2. C. J. F. Huckins, "Functional linkages among morphology, feeding performance, diet, and competitive ability in molluscivorous sunfish," Ecology, vol. 78, no. 8, pp. 2401–2414, 1997.

3. P. Schausberger and B. A. Croft, "Nutritional benefits of intraguild predation and cannibalism among generalist and specialist phytoseiid mites," Ecological Entomology, vol. 25, no. 4, pp. 473–480, 2000.
4. J. Terraube, B. Arroyo, M. Madders, and F. Mougeot, "Diet specialisation and foraging efficiency under fluctuating vole abundance: a comparison between generalist and specialist avian predators," Oikos, vol. 120, no. 2, pp. 234–244, 2011.
5. R. Levins, Evolution in Changing Environments, Princeton University Press, Princeton, NJ, USA, 1968.
6. D. Schluter, "Ecological causes of adaptive radiation," American Naturalist, vol. 148, pp. S40–S64, 1996.
7. G. S. Albuquerque, M. J. Tauber, and C. A. Tauber, "Life-history adaptations and reproductive costs associated with specialization in predacious insects," Journal of Animal Ecology, vol. 66, no. 3, pp. 307–317, 1997.
8. S. B. Yamada and G. E. Boulding, "Claw morphology, prey size selection and foraging efficiency in generalist and specialist shell-breaking crabs," Journal of Experimental Marine Biology and Ecology, vol. 220, no. 2, pp. 191–211, 1998.
9. S. Noriyuki, N. Osawa, and T. Nishida, "Prey capture performance in hatchlings of two sibling Harmonia ladybird species in relation to maternal investment through sibling cannibalism," Ecological Entomology, vol. 36, no. 3, pp. 282–289, 2011.
10. S. F. Norton, "Capture success and diet of cottid fishes: the role of predator morphology and attack kinematics," Ecology, vol. 72, no. 5, pp. 1807–1819, 1991.
11. G. A. Polis, C. A. Myers, and R. D. Holt, "The ecology and evolution of intraguild predation: potential competitors that eat each other," Annual Review of Ecology and Systematics, vol. 20, pp. 297–330, 1989.
12. R. D. Holt and G. A. Polis, "A theoretical framework for intraguild predation," The American Naturalist, vol. 149, no. 4, pp. 745–764, 1997.
13. J. A. Rosenheim, H. K. Kaya, L. E. Ehler, J. J. Marois, and B. A. Jaffee, "Intraguild predation among biological-control agents: theory and evidence," Biological Control, vol. 5, no. 3, pp. 303–335, 1995.
14. M. Arim and P. A. Marquet, "Intraguild predation: a widespread interaction related to species biology," Ecology Letters, vol. 7, no. 7, pp. 557–564, 2004.
15. C. Vilà and R. K. Wayne, "Hybridization between wolves and dogs," Conservation Biology, vol. 13, no. 1, pp. 195–198, 1999.
16. E. A. Dame and K. Petren, "Behavioural mechanisms of invasion and displacement in Pacific island geckos (Hemidactylus)," Animal Behaviour, vol. 71, no. 5, pp. 1165–1173, 2006.
17. Y. Okuzaki, Y. Takami, and T. Sota, "Resource partitioning or reproductive isolation: the ecological role of body size differences among closely related species in sympatry," Journal of Animal Ecology, vol. 79, no. 2, pp. 383–392, 2010.
18. H. Van Gossum, K. Beirinckx, M. R. Forbes, and T. N. Sherratt, "Reproductive interference between Nehalennia damselfly species," Ecoscience, vol. 14, no. 1, pp. 1–7, 2007.
19. B. M. Fitzpatrick, J. S. Placyk Jr., M. L. Niemiller, G. S. Casper, and G. M. Burghardt, "Distinctiveness in the face of gene flow: hybridization between specialist and generalist gartersnakes," Molecular Ecology, vol. 17, no. 18, pp. 4107–4117, 2008.
20. J. Gröning and A. Hochkirch, "Reproductive interference between animal species," The Quarterly Review of Biology, vol. 83, no. 3, pp. 257–282, 2008.
21. D. Kyogoku, "Reproductive interference: ecological and evolutionary consequences of interspecific promiscuity," Population Ecology, vol. 57, no. 2, pp. 253–260, 2015.
22. W. E. Snyder and E. W. Evans, "Ecological effects of invasive arthropod generalist predators," Annual Review of Ecology, Evolution, and Systematics, vol. 37, pp. 95–122, 2006.
23. M. Kurisaki, "Studies on the genus Ptychanatis crotch," Kontyu Sekai, vol. 24, pp. 369–372, 1920 (Japanese).
24. H. Sasaji, Natural History of the Ladybirds, University of Tokyo Press, Tokyo, Japan, 1998 (Japanese).

25. P. H. Adler, "Sibling species," in Encyclopedia of Entomology, J. L. Capinera, Ed., pp. 3371–3372, Springer, Dordrecht, The Netherlands, 2008.
26. N. Osawa and K. Ohashi, "Sympatric coexistence of sibling species Harmonia yedoensis and H. axyridis(Coleoptera: Coccinellidae) and the roles of maternal investment through egg and sibling cannibalism," European Journal of Entomology, vol. 105, no. 3, pp. 445–454, 2008.
27. S. Noriyuki and N. Osawa, "Intrinsic prey suitability in specialist and generalist Harmonia ladybirds: a test of the trade-off hypothesis for food specialization," Entomologia Experimentalis et Applicata, vol. 144, no. 3, pp. 279–285, 2012.
28. S. Noriyuki, Y. Kameda, and N. Osawa, "Prevalence of male-killer in a sympatric population of two sibling ladybird species, Harmonia yedoensis and Harmonia axyridis (Coleoptera: Coccinellidae)," European Journal of Entomology, vol. 111, no. 2, pp. 307–311, 2014.
29. S. Noriyuki, N. Osawa, and T. Nishida, "Asymmetric reproductive interference between specialist and generalist predatory ladybirds," Journal of Animal Ecology, vol. 81, no. 5, pp. 1077–1085, 2012.
30. S. Noriyuki, K. Kawatsu, and N. Osawa, "Factors promoting maternal trophic egg provisioning in non-eusocial animals," Population Ecology, vol. 54, no. 3, pp. 455–465, 2012.
31. T. Eisner, K. Hicks, M. Eisner, and D. S. Robson, "'Wolf-in-sheep's-clothing' strategy of a predaceous insect larva," Science, vol. 199, no. 4330, pp. 790–794, 1978.
32. L. R. Milbrath, M. J. Tauber, and C. A. Tauber, "Prey specificity in Chrysopa: an interspecific comparison of larval feeding and defensive behavior," Ecology, vol. 74, no. 5, pp. 1384–1393, 1993.
33. C. A. Tauber, J. R. Ruberson, and M. J. Tauber, "Size and morphological differences among the larvae of two predacious species and their hybrids (Neuroptera: Chrysopidae)," Annals of the Entomological Society of America, vol. 88, no. 4, pp. 502–511, 1995.
34. C. A. Tauber and M. J. Tauber, "Food specificity in predacious insects: a comparative ecophysiological and genetic study," Evolutionary Ecology, vol. 1, no. 2, pp. 175–186, 1987.
35. H. Roy, P. Brown, R. Frost, and R. Poland, Ladybirds (Coccinellidae) of Britain and Ireland, Centre for Ecology and Hydrology, Oxfordshire, UK, 2011.
36. M. E. N. Majerus, Ladybirds, Harper Collins, London, UK, 1994.
37. S. Shiyake, A Guide to Ladybirds of Osaka, Osaka Museum of Natural History, Osaka, Japan, 2005 (Japanese).
38. C. S. Henry, "Sibling species, call differences, and speciation in green lacewings (Neuroptera: Chrysopidae: Chrysoperla)," Evolution, vol. 39, no. 5, pp. 965–984, 1985.
39. C. A. Tauber and M. J. Tauber, "A genetic model for sympatric speciation through habitat diversification and seasonal isolation," Nature, vol. 268, no. 5622, pp. 702–705, 1977.
40. C. A. Tauber and M. J. Tauber, "Inheritance of seasonal cycles in Chrysoperla (Insecta: Neuroptera)," Genetical Research, vol. 49, no. 3, pp. 215–223, 1987.
41. S. Kishi, T. Nishida, and Y. Tsubaki, "Reproductive interference determines persistence and exclusion in species interactions," Journal of Animal Ecology, vol. 78, no. 5, pp. 1043–1049, 2009.
42. M. Friberg, O. Leimar, and C. Wiklund, "Heterospecific courtship, minority effects and niche separation between cryptic butterfly species," Journal of Evolutionary Biology, vol. 26, no. 5, pp. 971–979, 2013.
43. C. A. Tauber and M. J. Tauber, "Sympatric speciation based on allelic changes at three loci: evidence from natural populations in two habitats," Science, vol. 197, no. 4310, pp. 1298–1299, 1977.
44. M. J. Tauber, C. A. Tauber, J. R. Ruberson, L. R. Milbrath, and G. S. Albuquerque, "Evolution of prey specificity via three steps," Experientia, vol. 49, no. 12, pp. 1113–1117, 1993.

45. P. Nosil, Ecological Speciation, Oxford University Press, Oxford, UK, 2012.
46. C. S. Henry and M. L. M. Wells, "Adaptation or random change? The evolutionary response of songs to substrate properties in lacewings (Neuroptera: Chrysopidae: Chrysoperla)," Animal Behaviour, vol. 68, no. 4, pp. 879–895, 2004.
47. S. Noriyuki, "Host selection in insects: reproductive interference shapes behavior of ovipositing females," Population Ecology, vol. 57, no. 2, pp. 293–305, 2015.
48. C. A. Tauber and M. J. Tauber, "Sympatric speciation in insects: perception and perspective," in Speciation and Its Consequences, D. Otte and J. A. Endler, Eds., pp. 307–344, Sinauer Associates, Sunderland, UK, 1989.
49. J. Konuma and S. Chiba, "Ecological character displacement caused by reproductive interference," Journal of Theoretical Biology, vol. 247, no. 2, pp. 354–364, 2007.
50. D. W. Crowder, A. R. Horowitz, H. Breslauer et al., "Niche partitioning and stochastic processes shape community structure following whitefly invasions," Basic and Applied Ecology, vol. 12, no. 8, pp. 685–694, 2011.
51. N. Osawa, "Population field studies on the aphidophagous ladybird beetle Harmonia axyridis(Coleoptera: Coccinellidae): resource tracking and population characteristics," Population Ecology, vol. 42, no. 2, pp. 115–127, 2000.
52. M. L. Niemiller, B. M. Fitzpatrick, and B. T. Miller, "Recent divergence with gene flow in Tennessee cave salamanders (Plethodontidae: Gyrinophilus) inferred from gene genealogies," Molecular Ecology, vol. 17, no. 9, pp. 2258–2275, 2008.
53. J. L. Feder, S. P. Egan, and P. Nosil, "The genomics of speciation-with-gene-flow," Trends in Genetics, vol. 28, no. 7, pp. 342–350, 2012.
54. I. Ohshima, "Genetic mechanisms preventing the fusion of ecotypes even in the face of gene flow," Scientific Reports, vol. 2, article no. 506, 2012.
55. M. M. Wells and C. S. Henry, "The role of courtship songs in reproductive isolation among populations of green lacewings of the genus Chrysoperla (Neuroptera: Chrysopidae)," Evolution, vol. 46, no. 1, pp. 31–42, 1992.
56. K. S. Pfennig, "The evolution of mate choice and the potential for conflict between species and mate-quality recognition," Proceedings of the Royal Society B: Biological Sciences, vol. 265, no. 1407, pp. 1743–1748, 1998.
57. T. Dobzhansky, "Geographical variation in lady-beetles," The American Naturalist, vol. 67, no. 709, pp. 97–126, 1933.
58. T. Komai, M. Chino, and Y. Hosino, "Contributions to the evolutionary genetics of the lady-beetle, Harmonia. I. Geographic and temporal variations in the relative frequencies of the elytral pattern types and in the frequency of elytral ridge," Genetics, vol. 35, no. 5, pp. 589–601, 1950.
59. T. Komai, M. Chino, and Y. Hosino, "Population genetics on the lady-beetle Harmonia axyridis," in Population Genetics, T. Komai and K. Sakai, Eds., pp. 45–60, Baifukan, Tokyo, Japan, 1956 (Japanese).
60. H. Sasaji, "Biosystematics on Harmonia axyridis-complex (Coleoptera: Coccinellidae)," Memoirs of the Faculty of Education, Fukui University, Series II Natural Science, vol. 30, pp. 59–79, 1980.
61. H. Sasaji, "Larval characters of Asian species of genus Harmonia Mulsant (Coleoptera: Coccinellidae)," Memoirs of the Faculty of Education, Fukui University, Series II Natural Science, vol. 27, pp. 1–17, 1977.
62. M. Takizawa, "Some new species of Coccinellidae in Japan. I," Transactions of the Sapporo Natural History Society, vol. 4, pp. 220–224, 1917 (Japanese).
63. H. Sasaji, "On the sibling species of Harmonia," Kontyu to Shizen, vol. 12, no. 11, pp. 4–13, 1977 (Japanese).
64. J. Jaenike, "Resource predictability and niche breadth in the Drosophila quinaria species group," Evolution, vol. 32, no. 3, pp. 676–678, 1978.
65. S. P. Courtney and J. Forsberg, "Host use by two pierid butterflies varies with host density," Functional Ecology, vol. 2, no. 1, pp. 67–75, 1988.
66. C. Wiklund and M. Friberg, "The evolutionary ecology of generalization: among-year variation in host plant use and offspring survival in a butterfly," Ecology, vol. 90, no. 12, pp. 3406–3417, 2009.

67. J. C. Perry and B. D. Roitberg, "Ladybird mothers mitigate offspring starvation risk by laying trophic eggs," Behavioral Ecology and Sociobiology, vol. 58, no. 6, pp. 578–586, 2005.
68. T. Ikeda, F. Matsumura, and D. M. Benjamin, "Chemical basis for feeding adaptation of pine sawflies neodiprion rugifrons and neodiprion swainei," Science, vol. 197, no. 4302, pp. 497–499, 1977.
69. S. G. Codella Jr. and K. F. Raffa, "Host plant influence on chemical defense in conifer sawflies (Hymenoptera: Diprionidae)," Oecologia, vol. 104, no. 1, pp. 1–11, 1995.
70. I. Okada, K. Nijima, and Y. Toriumi, "Comparative studies on sibling species of lady beetles, Harmonia yedoensis and H. axyridis (Coleoptera: Coccinellidae)," Bulletin of the Faculty of Agriculture Tamagawa University, vol. 18, pp. 60–68, 1978 (Japanese).
71. S. Azuma, M. Kinjo, M. Murayama, S. Hori, K. Minato, and K. Uesugi, A Field Guide to Insects in Okinawa, Okinawa Press, Okinawa, Japan, 1996 (Japanese).
72. M. Moriguchi, Ladybirds in Islands, Chijin Shokan, Tokyo, Japan, 2015 (Japanese).
73. Y. Yamada, Larval survival in Harmonia axyridis and Harmonia yedoensis [M.S. dissertation], Yamagata University, Yamagata, Japan, 2010 (Japanese).
74. J. M. Chase, P. A. Abrams, J. P. Grover et al., "The interaction between predation and competition: a review and synthesis," Ecology Letters, vol. 5, no. 2, pp. 302–315, 2002.
75. P. Chesson and J. J. Kuang, "The interaction between predation and competition," Nature, vol. 456, no. 7219, pp. 235–238, 2008.
76. P. M. Brakefield, "Polymorphic Müllerian mimicry and interactions with thermal melanism in ladybirds and a soldier beetle: a hypothesis," Biological Journal of the Linnean Society, vol. 26, no. 3, pp. 243–267, 1985.
77. A. Honěk and I. Hodek, "Distribution in habitats," in Ecology of Coccinellidae, A. Honěk and I. Hodek, Eds., pp. 95–141, Kluwer Academic, Dordrecht, The Netherlands, 1996.
78. J. J. Sloggett and M. E. N. Majerus, "Habitat preferences and diet in the predatory coccinellidae (Coleoptera): an evolutionary perspective," Biological Journal of the Linnean Society, vol. 70, no. 1, pp. 63–88, 2000.
79. N. J. Mills, "Essential and alternative foods for some British Coccinellidae (Coleoptera)," Entomologica Germanica, vol. 32, pp. 197–202, 1981.
80. A. Ferrer, A. F. G. Dixon, and J.-L. Hemptinne, "Habitat or prey specialization in predatory ladybird beetles: a case study of two closely related species," Journal of Applied Entomology, vol. 140, no. 3, pp. 199–208, 2016.
81. M. J. Jeffries and J. H. Lawton, "Enemy free space and the structure of ecological communities," Biological Journal of the Linnean Society, vol. 23, no. 4, pp. 269–286, 1984.
82. H. Ireland, P. W. E. Kearns, and M. E. N. Majerus, "Interspecific hybridisation in the Coccinellidae: some observations on an old controversy," The Entomologist's Record and Journal of Variation, vol. 98, pp. 181–185, 1986.
83. J. J. Sloggett, W. Völkl, W. Schulze, J. H. G. V. D. Schulenburg, and E. N. Majerus, "The ant-associations and diet of the ladybird Coccinella magnifica (Coleoptera: Coccinellidae)," European Journal of Entomology, vol. 99, no. 4, pp. 565–569, 2002.
84. J. J. Sloggett, R. A. Wood, and M. E. N. Majerus, "Adaptations of Coccinella magnifica redtenbacher, a myrmecophilous coccinellid, to aggression by wood ants (Formica rufa Group). I. Adult behavioral adaptation, its ecological context and evolution," Journal of Insect Behavior, vol. 11, no. 6, pp. 889–904, 1998.
85. J. J. Sloggett and M. E. N. Majerus, "Adaptations of Coccinella magnifica, a myrmecophilous coccinellid to aggression by wood ants (Formica rufa group). II. Larval behaviour, and ladybird oviposition location," European Journal of Entomology, vol. 100, no. 3, pp. 337–344, 2003.
86. A. Magro, E. Lecompte, F. Magné, J.-L. Hemptinne, and B. Crouau-Roy, "Phylogeny of ladybirds (Coleoptera: Coccinellidae): are the subfamilies monophyletic?" Molecular Phylogenetics and Evolution, vol. 54, no. 3, pp. 833–848, 2010.

87. N. Osawa, Ecology of Harmonia axyridis in the field population [Ph.D. dissertation], Kyoto University, Kyoto, Japan, 1991 (Japanese).
88. N. Osawa, "Ecology of Harmonia axyridis in natural habitats within its native range," BioControl, vol. 56, no. 4, pp. 613–621, 2011.
89. Y. Hironori and S. Katsuhiro, "Cannibalism and interspecific predation in two predatory ladybirds in relation to prey abundance in the field," Entomophaga, vol. 42, no. 1, pp. 153–163, 1997.
90. B. K. Agarwala and H. Yasuda, "Overlapping oviposition and chemical defense of eggs in two co-occurring species of ladybird predators of aphids," Journal of Ethology, vol. 19, no. 1, pp. 47–53, 2001.
91. S. Sato, A. F. G. Dixon, and H. Yasuda, "Effect of emigration on cannibalism and intraguild predation in aphidophagous ladybirds," Ecological Entomology, vol. 28, no. 5, pp. 628–633, 2003.
92. H. Yasuda, E. W. Evans, Y. Kajita, K. Urakawa, and T. Takizawa, "Asymmetric larval interactions between introduced and indigenous ladybirds in North America," Oecologia, vol. 141, no. 4, pp. 722–731, 2004.
93. R. L. Ware and M. E. N. Majerus, "Intraguild predation of immature stages of British and Japanese coccinellids by the invasive ladybird Harmonia axyridis," BioControl, vol. 53, no. 1, pp. 169–188, 2008.
94. F. Gilbert and J. Owen, "Size, shape, competition, and community structure in hoverflies (Diptera: Syrphidae)," Journal of Animal Ecology, vol. 59, no. 1, pp. 21–39, 1990.
95. M. Mizuno, T. Itioka, Y. Tatematsu, and Y. Itô, "Food utilization of aphidophagous hoverfly larvae (Diptera: Syrphidae, Chamaemyiidae) on herbaceous plants in an urban habitat," Ecological Research, vol. 12, no. 3, pp. 239–248, 1997.
96. Y. Hironori and S. Katsuhiro, "Cannibalism and interspecific predation in two predatory ladybirds in relation to prey abundance in the field," Entomophaga, vol. 42, no. 1-2, pp. 153–163, 1997.
97. É. Lucas, "Intraguild predation among aphidophagous predators," European Journal of Entomology, vol. 102, no. 3, pp. 351–364, 2005.
98. C. B. Müller, I. C. T. Adriaanse, R. Belshaw, and H. C. J. Godfray, "The structure of an aphid-parasitoid community," Journal of Animal Ecology, vol. 68, no. 2, pp. 346–370, 1999.
99. P. Amarasekare, "Trade-offs, temporal variation, and species coexistence in communities with intraguild predation," Ecology, vol. 88, no. 11, pp. 2720–2728, 2007.
100. V. Krivan, "Optimal intraguild foraging and population stability," Theoretical Population Biology, vol. 58, no. 2, pp. 79–94, 2000.
101. R. D. Holt and G. R. Huxel, "Alternative prey and the dynamics of intraguild predation: theoretical perspectives," Ecology, vol. 88, no. 11, pp. 2706–2712, 2007.
102. P. Amarasekare, "Coexistence of intraguild predators and prey in resource-rich environments," Ecology, vol. 89, no. 10, pp. 2786–2797, 2008.
103. N. Cooper, J. Rodríguez, and A. Purvis, "A common tendency for phylogenetic overdispersion in mammalian assemblages," Proceedings of the Royal Society B: Biological Sciences, vol. 275, no. 1646, pp. 2031–2037, 2008.
104. N. J. B. Kraft, R. Valencia, and D. D. Ackerly, "Functional traits and niche-based tree community assembly in an Amazonian forest," Science, vol. 322, no. 5901, pp. 580–582, 2008.
105. C. H. Graham, J. L. Parra, C. Rahbek, and J. A. McGuire, "Phylogenetic structure in tropical hummingbird communities," Proceedings of the National Academy of Sciences of the United States of America, vol. 106, S2, pp. 19673–19678, 2009.
106. T. J. Ord, L. King, and A. R. Young, "Contrasting theory with the empirical data of species recognition," Evolution, vol. 65, no. 9, pp. 2572–2591, 2011.
107. K.-I. Takakura and S. Fujii, "Reproductive interference and salinity tolerance differentiate habitat use between two alien cockleburs: Xanthium occidentale and X. italicum (Compositae)," Plant Ecology, vol. 206, no. 2, pp. 309–319, 2010.

108. K. W. Matsubayashi, I. Ohshima, and P. Nosil, "Ecological speciation in phytophagous insects," Entomologia Experimentalis et Applicata, vol. 134, no. 1, pp. 1–27, 2010.

109. J. A. Coyne and H. A. Orr, Speciation, Sinauer Associates, Sunderland, Mass, USA, 2004.

CHAPTER 2

eButterfly: Leveraging Massive Online Citizen Science for Butterfly Conservation

Kathleen L. Prudic [1,*], Kent P. McFarland [2], Jeffrey C. Oliver [3], Rebecca A. Hutchinson [4,5], Elizabeth C. Long [6], Jeremy T. Kerr [7] and Maxim Larrivée [8,9,*]

[1]*Entomology, University of Arizona, Tucson, AZ 85721, USA*

[2]*Vermont Center for Ecostudies, Norwich, VT 05055, USA*

[3]*Health Sciences Library, University of Arizona, Tucson, AZ 85721, USA*

[4]*Electrical Engineering and Computer Science, Oregon State University, Corvallis, OR 97331, USA*

[5]*Fisheries and Wildlife, Oregon State University, Corvallis, OR 97331, USA*

[6]*Daniel Smiley Research Center, Mohonk Preserve, New Paltz, NY 12561, USA*

[7]*Biology, University of Ottawa, Ottawa, ON K1N 6N5, Canada*

[8]*Insectarium, Montreal Space for Life, Montreal, QC H1X 2B2, Canada*

[9]*Institut de Research en Biologie Végétale, Montreal, QC H1X 2B2, Canada*

ABSTRACT

Data collection, storage, analysis, visualization, and dissemination are changing rapidly due to advances in new technologies driven by computer science and universal access to the internet. These technologies and web connections place human observers front and center in citizen science-driven research and are critical in generating new discoveries and innovation in such fields as astronomy, biodiversity, and meteorology. Research projects utilizing a citizen science approach address scientific problems at regional, continental, and even global scales otherwise impossible for a single lab or even a small collection of academic researchers. Here we describe eButterfly an integrative checklist-based butterfly monitoring and database web-platform that leverages the skills and knowledge of recreational butterfly enthusiasts to create a globally accessible unified database of butterfly observations across North America. Citizen scientists, conservationists, policy makers, and scientists are using eButterfly data to better understand the biological patterns of butterfly species diversity and how environmental conditions shape these patterns in space and time. eButterfly in collaboration with thousands of butterfly enthusiasts has created a near real-time butterfly data resource producing tens of thousands of observations per year open to all to share and explore.

Keywords: biodiversity; butterfly; Citizen Science; crowd-sourced; *Danaus plexippus*; data management; pollinator conservation; Lepidoptera; Papilionoidea; *Vanessa atalanta*

1. INTRODUCTION

Identifying species is a challenging skill that can require specialized training and years of experience. Human observers still outperform computer algorithms in species identification because of the need to process impressions of shape, size, color, and behavior under variable conditions all while comparing this with another list of possible species based on time and location [1]. This difficult task of identification is central to many scientific fields and critical to conservation and preservation efforts; and humans, not current computer vision algorithms, do this task accurately and efficiently. Without accurate knowledge of which species occur in a given area at a particular time or season, developing strategies to manage biodiversity will remain a formidable challenge [2].

Most data on species-level occurrence are gathered by direct observation in the field. This necessitates innovative approaches for wide-scale data collection and analysis, such as massive online citizen science initiatives [3,4]. For example, the National Phenology Network [5] and Plant Watch Canada [6] focus on phenology of a subset of plants and animals in North America; iNaturalist [7] focuses on single observations of all taxa across the globe and is one of the Global Biodiversity Information Facility's (GBIF) [8] contributors; eBird [9] focuses on bird checklists around the globe and has the most extensive database of avian distribution and abundance.

The checklist approach utilized in the field of ornithology since the early 1950s [10] is a powerful way to monitor a specific taxon. In its essence, a checklist in the field of biodiversity monitoring constitutes a list of species and their abundance recorded at a particular location. While checklists can take several forms and collect various pieces of information, they will generally collect spatiotemporal information about the observations reported, such as the date, time, and effort measured in duration, number of people participating, and distance travelled. As such, the data arising from checklists can be standardized in terms of effort and thus compared in space and time, dramatically increasing the scientific value of monitoring data when compared to single species "dot on the map" monitoring efforts. Butterflies are well suited to checklisting; they are well documented, and identification tools such as field guides are readily available for all regions of North America.

For more than 300 years, the public has contributed to the understanding of butterfly taxonomy, ecology, biogeography, evolution, and conservation [11]. In North America, The Lepidopterists' Society was originally founded as an inclusive scientific society of both amateurs and professionals and still has almost equal membership between people who earn their living as lepidopterists and those who pursue their passion for butterflies out of private interest [12]. eButterfly (www.e-butterfly.org) builds on this tradition and provides a web-platform for citizen scientists to submit their checklists and photographs of butterfly observations across North America via the Internet. eButterfly engages a vast network of human observers (citizen scientists), now over 5000, to report butterfly checklists and photographic observations using standardized protocols analogous to eBird and compatible with iNaturalist and GBIF. eButterfly's mission is to harness the power of butterfly enthusiasts in an effort to better understand butterfly distribution and abundance across broad spatiotemporal scales and identify factors that influence these patterns.

Butterflies are ideal insects for citizen science research [13]. There are nearly 10,000 species linked to biotic processes at multiple trophic levels. As larvae, butterflies are herbivores and are an important food source for other animals in the community. As adults, butterflies are pollinators and are easily trackable on the wing. Much is known about butterfly larval host plants and adult nectar plant affiliations across North America thanks to the combined efforts of professionals and amateurs. Butterflies are diurnal, behaviorally and morphologically conspicuous, and abundant. All of these characteristics mean butterflies are easily observed, counted, and are among the most studied of free-living insect species. Butterflies such as the monarch (*Danaus plexippus*) even engage in long-distance migrations and in doing so, demonstrate the biological integration of seemingly disparate ecosystems. Most importantly, butterflies are sensitive environmental indicators, signaling changes in environmental processes and ecosystem health [14,15,16].

2. EBUTTERFLY IS AN ONLINE CITIZEN SCIENCE COMMUNITY AT A CONTINENTAL SCALE

eButterfly (www.e-butterfly.org), launched by Max Larrivee and Jeremy Kerr at the University of Ottawa in 2012, uses standardized web-based protocols. Initially, eButterfly was limited to observations from Canada but quickly expanded to the United States with the addition of Kent McFarland at the Vermont Center for Ecostudies and Katy Prudic and Jeff Oliver, both from the University of Arizona. eButterfly is currently available to citizen scientists in English and French, with plans underway for adding Spanish.

Butterflies are perhaps the most popular group of insects and are frequently observed and admired by almost everyone. Approximately 68.6 million people watch wildlife around their homes, with 24.1 million enjoying wildlife photography and 21.3 million indicating that they specifically watch insects [17]. Wildlife observation is rapidly growing, and becoming more affordable as higher performing digital photography technology is making the documentation of wildlife easier. However, the data butterfly enthusiasts collect are often stored on personal computers, and in file cabinets, notebooks, or specimen collections, thus being unavailable for science, conservation, and education. eButterfly provides a way for butterfly observational data to be organized, verified, and shared.

eButterfly serves multiple stakeholders (e.g., butterfly enthusiasts, conservation managers, policy makers, scientists, educators) as an open data source to gather, organize, verify, and disseminate checklists of butterflies across North America. These data provide information on species occurrence, migration timing, community structure, and relative abundance at a variety of spatial and temporal scales. Through the process of informal science education, eButterfly participants become more familiar with scientific approaches, data collection and analyses by using standardized data-gathering techniques, exploring data through the website's data visualization tools, and interacting with scientists. eButterfly provides a permanent repository for checklist observations and a method for keeping track of each participant's butterfly observations, photographs, and collection effort.

eButterfly data are proving vital for scientific use in broad scale ecological and conservation research. eButterfly data are available in real-time, creating new opportunities for rapid integration of butterfly data with other kinds of information such as climate, land cover, plant phenology, or even the timing of butterfly species' seasonal activities, such as migration. For example, spatial variation in land cover and climatic conditions strongly structure butterfly species assemblages [18], but new butterfly observations have enabled detection of how warming temperatures are shifting butterfly species composition across broad areas of North America [19]. As a conservation tool, eButterfly is being used to monitor butterflies locally and regionally by providing new data on butterfly distribution, phenology, and relative abundance. eButterfly data can also contribute to improved understanding and models of species distribution, which helps prioritize and manage areas for conservation [20]. eButterfly's broad spatial and temporal coverage complements more intensive monitoring programs, such as the Ontario Butterfly Atlas [21], allowing scientists to generate new hypotheses and direct future research efforts based on large amounts of data, beyond the scope of a single principal investigator or whole institute. Importantly, all eButterfly data contributes to scientific knowledge by joining networked data available through larger global biodiversity initiatives such as Bison [22], Canadensis [23], and GBIF [8].

While it is continental in scope, the majority of eButterfly data still come from northeastern North America. Since its release in 2012, over 30,000 people have visited eButterfly. Over 5500 unique participants have entered data into eButterfly, from more than 28,000 locations across North America. The project has gathered over 230,000 observations for more than 39,000 checklists (Figure 1). eButterfly will also upload over 300,000 additional historical observations from museum collections in Canada and the United-States in 2017.

Funding to develop and operate eButterfly cyberinfrastructure was $60,000/year, making the cost per observation incurred by the science community very low. For example, eButterfly gathered over 60,000 observations in 2016, so each observation costs ~$1 (USA). With increasing citizen scientist contributions, eButterfly's per-observation costs are likely to continue to decline. eButterfly relies on the work of its contributors, and measuring their contributions is essential; we estimate these efforts to have required (conservatively) 56,000 h. This in-kind donation of time and expertise by our citizen scientists is valued at ~$1,030,000 (USA) [24].

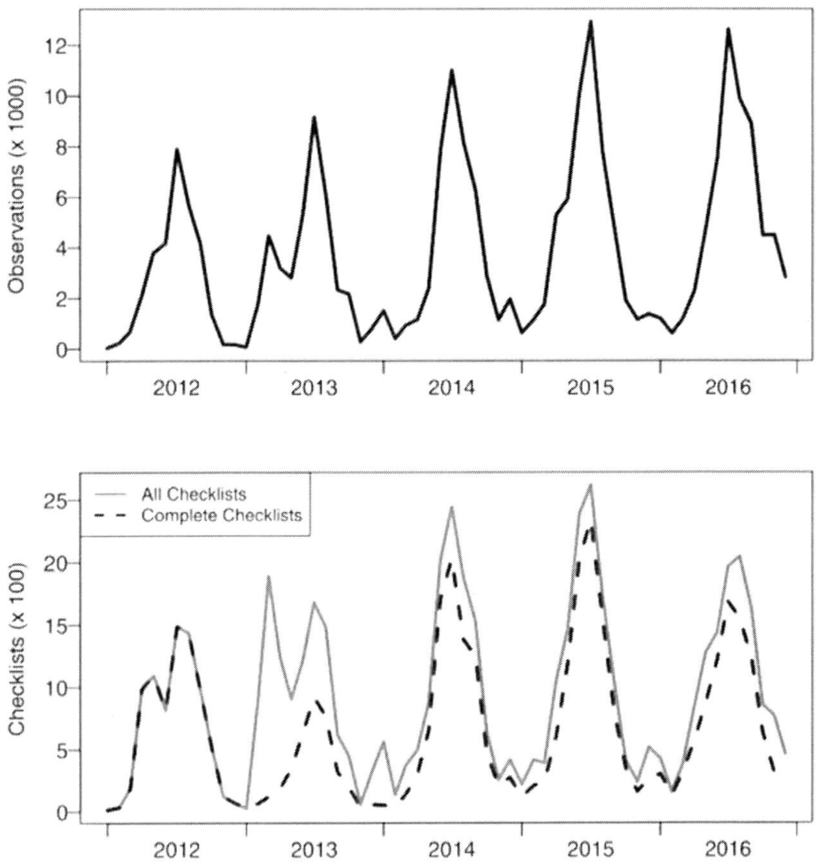

Figure 1. Monthly observations and checklists submitted to eButterfly since 2012. Over 39,000 checklists by 5500 registered participants have been gathered. Within those checklists there are 230,000 observations and 682 species. Note the heavy usage during summer followed by much lower reports in winter months due to the phenology of butterfly species in the northern hemisphere.

3. EBUTTERFLY'S PROJECT DESIGN LEVERAGES CITIZEN SCIENTISTS' OBSERVATIONS

eButterfly was created based upon two simple ideas. First, many people are passionate about observing butterflies and, second, every butterfly observation has potential research value for fundamental and conservation research. Even observations of common species from well-sampled areas have value for monitoring population change, phenology, and for other spatiotemporal studies. Using informal science education, eButterfly steers participants into standardized data collection and provides extensive resources to improve observers' capacities for butterfly detection and identification. As with eBird [25] and other citizen science programs [26], eButterfly documents the presence or presumed absence of species as well as abundance through checklist data. To report butterfly observations, a web interface engages participants to submit observations through three interactive steps.

After logging into their eButterfly account to enter observations, participants are asked where they observed butterflies using an online mapping tool to select an existing or new location. Each citizen scientist's existing location is stored in the database so that participants can more easily record new observations from established collection sites, which are also accessible using a drop-down menu or "favorites" list.

Step two asks participants to indicate which of four different protocols they used to count butterflies. Traveling and area counts are effort-based sampling protocols, which require submitting both the amount of time spent butterfly watching and the distance traveled or area searched. Timed counts lack the spatial component and only record the total amount of time spent observing butterflies. The fourth protocol is a less rigorous option, called "casual observation", which requires only date, location, and species observed to describe the sampling event. Participants are encouraged to use the most rigorous protocols when possible.

The final step presents a checklist of the butterflies known from the state or province where the outing occurred. The participant fills in the number of individuals seen for each species and can upload photographs they captured of those species during the outing. Although "sight only" observations are accepted, eButterfly encourages participants to include photos with their observations for species verification. Observations are tagged as photograph, specimen, or sight observations. Other ancillary information such as counts of each life stage, host plants, sex, behaviors, and other notes, can also be added. At the end of the checklist, participants are presented with a final question that asks if they are submitting a checklist of all butterflies observed during the count. This is an important question that allows data consumers to determine if this is a complete presence/presumed absence checklist, which is more informative than presence-only data for many emerging machine learning and other analytical approaches in biogeography and conservation research [27].

4. DATA QUALITY AND QUANTITY CAN BE IMPROVED AND MANAGED IN CONCERT

Data collection design is critical to any successful scientific endeavor. Citizen science project designs vary in their intent and ability to collect high quality data. Many citizen science programs are focused on informal learning outcomes and participant experience rather than producing data that can be used for scientific inquiry. For eButterfly, project design is based on the formation of scientific questions, followed by the recruitment of participants to collect the data necessary to answer those questions. eButterfly focuses on scientific inquiry in changes in biodiversity, effects of climate change, and butterfly conservation from regional to continental scales [14,28,29]. Specifically, we are interested in how climate and land use change alters butterfly distributions, abundances, phenologies, and community structure in space and time. Our protocols and participant incentives encourage this data collection priority over others, such as new or rare species discovery. Open data and community vetting improves data quality. eButterfly is building a network of regional experts to verify each observation submitted. These experts are often "super-users" of eButterfly, with years of expertise on local butterfly species and their phonologies.

eButterfly encourages participants to submit photos of their observations as vouchers for species verification. Photographs are processed in a way that is comparable to how a natural history museum traditionally works with a specimen. The eButterfly participant is analogous to the collector, the regional data expert is analogous to the determiner, and eButterfly acts as the curator. Regional data experts validate or suggest corrections to observations with and without photographs based on species' known ranges and phenology. Regional data experts can add a new determination for the observation base, which is held in the database as the correct identification. Our regional experts also vet submissions of collection specimens databased through the eButterfly interface, sometimes improving the data quality of the museum collection through our verification process. Once verification is made, the participant is alerted and has the opportunity to respond or correct their field identification. Additionally, eButterfly participants often report and comment on others' observations, especially if they are incorrect, and this provides another mechanism for increased data quality.

To date, our vetters have reviewed over 190,000 butterfly observations. As eButterfly grows, we will build filters to automatically flag unusual records outside reported ranges and flight times to help streamline the data verification process as the data quantity exceeds the capacity for our regional experts. This combination of human computation and artificial learning algorithms has worked well in keeping data quality high in eBird as they expanded to millions of observations per year [30]. In the future, we hope to partner with experts in pattern recognition and machine learning to further automate butterfly species identifications from images [31].

Engaging citizen scientists is critical for increasing data quality and quantity. The data collection protocols and participation incentives of eButterfly engage large numbers of participants with a range of expertise from highly trained observers to young students, similar to other citizen science programs [1,25]. eButterfly particularly targets people with high levels of butterfly identification expertise, who in their spare time, collect, photograph, and watch butterflies for recreation. Many citizen science projects struggle to engage participants, and it can be especially hard to sustain participation when projects incorporate little or no rewards. In these instances, the number of participants generally plateaus once the threshold of like-minded individuals is reached [4,25]. eButterfly uses a variety of participant incentives such as life lists, year lists, species distribution maps and phenology graphs, and other tools to continue building and maintaining participation. We also try to communicate frequently with our super-users, who contribute the majority of data, and feature them in news stories and social media.

While crowd-sourced programs like eButterfly collect large amounts of data with broad spatiotemporal coverage, the data are influenced by more noise and bias than data collected through rigorous protocols conducted exclusively by expert-level observers. Imperfect detection of organisms is a pervasive source of noise in biodiversity surveys [32,33], affecting both citizen science and more specifically targeted data collection. In citizen science data, the effects of imperfect detection may be more variable due to differences in observers' skill and expertise, but extensions of common methods to correct for detectability can account for this variability [34]. Citizen science data are more affected by spatiotemporal sampling bias than studies with careful sampling designs

[35], and analyses of these data benefit from modeling methods that account for this bias [36,37].

Sampling effort by participants is an important source of bias in eButterfly data. eButterfly data are more robust near urban centers in the Northeast than in areas with sparse spatial coverage, such as in the Midwest (Figure 2). Similar to other crowd-sourced biodiversity projects (e.g., iNaturalist and eBird), observations tend to be clustered around major urban areas. Like eBird [4], we attempt to capitalize on butterfly watchers' strong desire to find and identify as many species as possible by providing tools that gamify butterfly watching. For example, we provide a system that automatically displays participants' life lists at multiple spatiotemporal scales and a 'Top 100' list that displays each citizen scientist's scores for the total number of species, photographs and checklists. Additionally, we have started collaborating with Adventure Scientists [38] to specifically target remote areas that are unreachable by most citizen scientists. Adventure Scientists recruits and trains backcountry specialists to collect data for a variety of scientific ventures. We have also targeted specific under-collected and under-engaged regions, such as the Arctic, in 2011 and 2012 (Larrivée and Kerr, unpublished data). Finally, we control for several aspects of this data collection bias by including appropriate covariates in analytic models. Variation in detection rates is modeled as a function of observation effort, both the total time and the distance or area of the count, underscoring the importance of steering citizen scientists to these more rigorous protocols.

eButterfly has found a balance between data quantity and quality that has allowed us to gather a volume of useful butterfly data at spatial and temporal scales previously unattainable. Our experience has shown us that a simplified protocol and incentives for volunteer participation are critical for the recruitment and retention of large numbers of volunteers. We found crowd-sourcing verification is critical to both our data quality and participant engagement. Previous methods of data analysis may not always be suitable for use even after verification and vetting, so new analytical approaches are being created in cooperation with other ecologists, statisticians, and computer scientists.

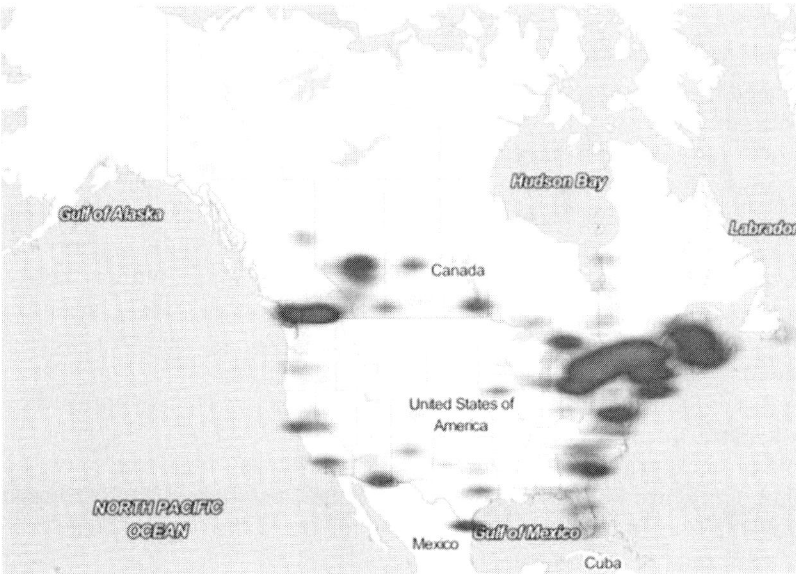

Figure 2. Spatial density map of all checklists reported to eButterfly. High density of checklists is shown in red, with areas of low density shown in blue. Note the spatial bias in relation to areas of high human population density such as urban centers and corridors.

5. EBUTTERFLY PROVIDES PRESENCE/PRESUMED ABSENCE DATA FOR BUTTERFLY CONSERVATION

eButterfly is a source of presence/presumed absence data, not just presence-only data. Presence-only data, which includes much of the data from natural history specimen collections, can be less informative in complex modeling such as species distribution models (SDMs) [27,39]. These models play a critical role in conservation decision-making and ecological or biogeographical inference, but can lead to suboptimal conservation outcomes and misguided theory if the underlying data do not suit the intended application. Building models with unsuitable data can waste valuable resources and deliver outputs that do not inform conservation and management decisions. eButterfly data collection protocols offer more and better modeling outcomes for conservation and ecology.

With these data, we are able to address previously difficult ecological and conservation questions. A long-standing challenge in insect conservation has been the difficulty of studying annual cycles for species that undergo broad-scale and complex geographic movements. Massive online citizen science web-platforms have started to make these movements trackable. The migration of the red admiral (*Vanessa atalanta* Nymphalidae) in Spring 2012 provides an example. Benefiting from an historically warm winter and spring in their southern United States overwintering grounds [40,41], record numbers of red admiral butterflies migrated northward following warm southern wind currents in early spring. Observations of red admirals were submitted to eButterfly so rapidly and over such a broad area that we could track the leading edge of the migration in real-time across southern Ontario, Canada. Using observations captured by participants from 15–17 April 2012, we estimated that the migrant population approached ~300 million individuals migrating over an 800,000 km^2 area (Figure 3). The estimated migration speed of red admirals closely matched wind speeds recorded in the region (mean = 26 km/h ± 7 km/h) (Larrivée et al., unpublished data).

Another example of a large spatial movement captured in near real-time was recorded with the monarch butterfly (*Danaus plexippus* Nymphalidae) across Canada. This species appeared unexpectedly and in high numbers during the summer of 2012 in Alberta, Canada (Figure 4). Monarch butterflies normally range across the eastern half of Canada, but beginning in early summer, participants began reporting large numbers of monarchs farther west, in the province of Alberta, documenting what may be a rare east to west dispersal. These rare dispersal events may be a mechanism for gene flow between eastern and western monarch populations, which show little genetic subdivision across North America [42]. However, since eButterfly was only collecting data in Canada in 2012, we cannot rule out a south to north dispersal coming from the United States to Alberta. This uncertainty illustrates the importance of a continental approach to understanding species dispersal and movement in space and time. The Great Basin, in the western United States, is undersampled in eButterfly and other monarch citizen science programs such as Journey North [43], so the movement of monarchs in the summer of 2012 remains shrouded in uncertainty. Despite an abundant monarch population in the northern latitudes (southern Canada) over the summer of 2012, milkweed (*Asclepias* spp.) abundance is low or absent where monarchs were recorded in 2012, and fewer monarchs were detected returning to southern latitudes (central Mexico) relative to historical expectations [44].

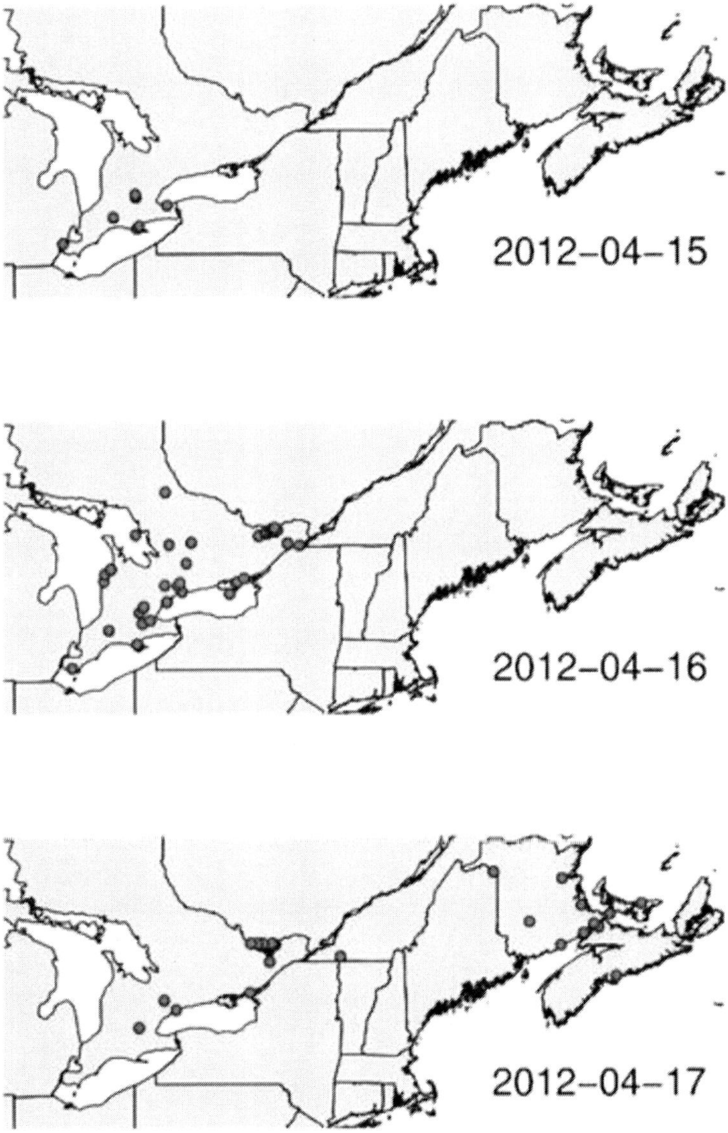

Figure 3. Observations of the red admiral (*Vanessa atalanta*) reported to eButterfly over three days in April 2012 in eastern Canada.

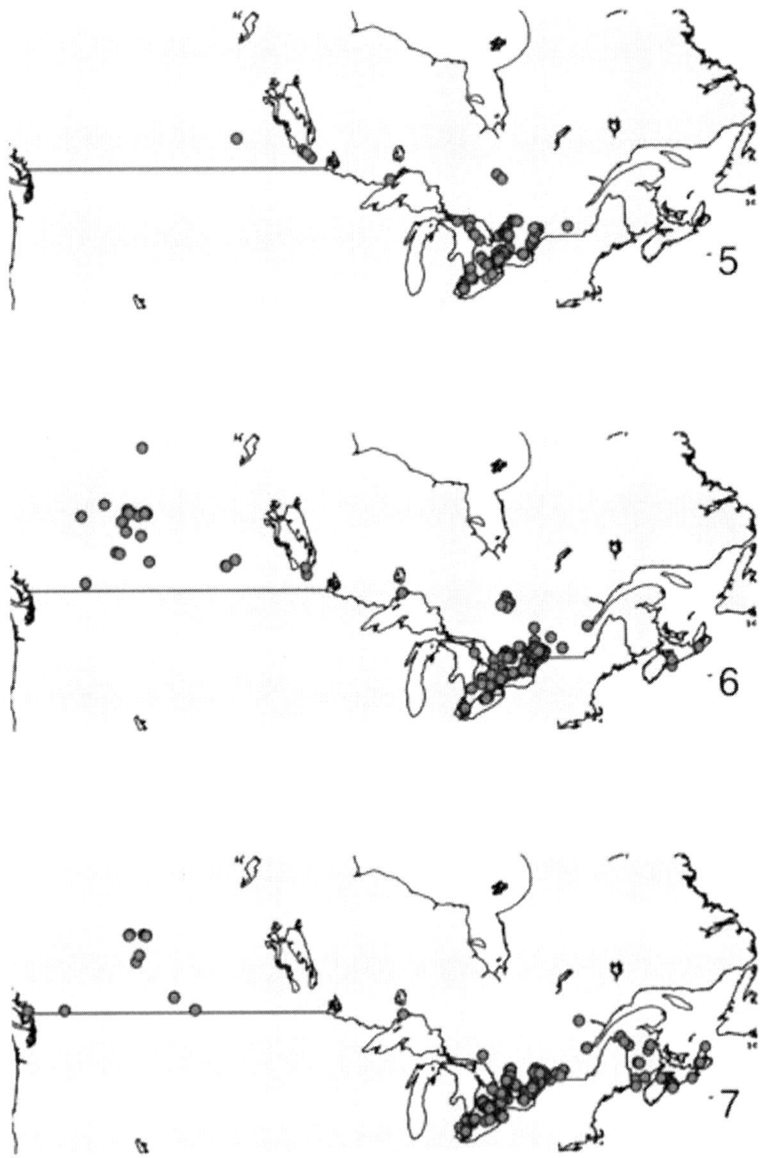

Figure 4. Observations of the monarch (*Danaus plexippus*) reported to eButterfly across Canada over three months in 2012. Note the appearance of monarchs in western Canada in June and July.

Environmental change has potential to shift phenologies and impact butterfly survival and reproduction, so establishing multi-species migration and distribution monitoring capability is increasingly important and complements remote sensing-based measurements for ecology [45,46]. Mass migrations and range expansions of butterflies often involve important pollinator or pest species. Real-time measurement of the timing and magnitude of migration and range expansion, such as those provided through eButterfly, allows scientists to study the impacts of global change on butterflies and relate these findings to conservation priorities.

6. NEW COLLABORATIONS AND OPPORTUNITIES ARE ON THE HORIZON

Despite the growing volume of species occurrence data collected by eButterfly, species observations are often sparsely distributed in space and through time, similar to museum collection data. This has motivated us to begin developing models of species occurrence by relating environmental features that may help elucidate a species' preferred niche (e.g., habitat, climate, elevation) to observational data. These models can then be used to make statistical predictions of occurrence at unsampled locations and times. To facilitate this process, eButterfly observations have been linked through observation location and time to a large number of local environmental variables, such as remotely sensed habitat information from the National Land Cover Database, and vegetation phenology from NASA's Moderate Resolution Imaging Spectroradiometer. We use these data to estimate species distributions designed to utilize both the broad extent and fine resolution information in the eButterfly database. Moving forward, we are exploring ways to estimate year-round distributions for all North American butterfly species with a wide variety of distinct migration pathways and local habitat associations. With this information, scientists and land managers will be better able to identify, prioritize, and coordinate conservation actions across broad landscapes.

Butterfly conservation is becoming more important as the general public becomes more interested in their success and survival, and as butterflies are increasingly a model system for understanding threats to insect pollinators. A significant strength of eButterfly as a conservation tool lies in its near real-time collection and dissemination of data, the reliability of which is often traceable to associated photographic vouchers. We are working on how to leverage the increasing quantity of photographic data on eButterfly to improve data verification, collection, and modeling. Up-to-the-minute butterfly observations could be coupled with other disparate data sources to provide powerful on-the-ground conservation tools. For example, a visualization tool which alerts interested stakeholders which species could be vulnerable to pesticide usage could help growers and landowners to weigh the risks of particular management options. Other conservation tools may alert stakeholders about nectar and host plant requirements for local butterfly species assemblages at different times of year, providing on-demand updates for land management decisions.

The eButterfly database structure and code has been instrumental in the development of other entomological citizen science monitoring programs, namely Bumblebee Watch [47] and Mission Monarch [48]. eButterfly has a similar data collection protocol to eBird, so efforts are underway to leverage the data of both initiatives to produce conservation data products with both birds and butterflies in mind. eButterfly has been actively collaborating with Nature's Notebook sponsored by the National Phenology Network with the Nectar Connector campaign [49] to monitor the phenology of nectar plants across the United States and to relate plant phenology with butterfly phenology at various locations.

7. CONCLUSIONS

The power of eButterfly and other massive online citizen science programs lies in the strength and diversity of its participants and stakeholders. Anyone with an interest in butterflies can participate—from the new enthusiast, to the backyard gardener, to the seasoned expert. As more participants submit data, an environment of sharing and free data exchange will become the norm between butterfly enthusiasts, scientists, and conservationists. There is an urgent need for butterfly enthusiasts to begin to explore new and uncharted territory. Visiting under-sampled habitats will improve understanding of species-habitat relationships and help prioritize management interventions.

Global climate change, land conversion, and human population growth are presenting significant conservation challenges that require understanding processes at continental and global scales. Monitoring systems across full annual cycles in near real-time to identify current and impending threats and to evaluate the outcomes of conservation actions can only be done with massive online, crowd-sourced efforts at multiple spatial and temporal scales. For terrestrial insect pollinators and herbivores, eButterfly is well poised to provide this information.

The ability to create, manage, and manipulate vast, real-time data resources is changing how biological research is conducted and has the potential to revolutionize conservation practices. Critical to these new tools is the cyberinfrastructure of eButterfly, which depends on database, networking, and computational technologies. eButterfly is an excellent example of how these technologies can be used to grow, manage, and sustain a community of citizen scientists, whose cooperation may be the only way to gather large amounts of data needed to detect key biological impacts of broad-scale environmental change. eButterfly synthesizes massive volumes of information in real-time, providing the critical data that is necessary to answer questions about pollinator diversity and distributions and to address potential issues on how anthropogenic and environmental factors affect said diversity. With a continental perspective on insect dynamics, we can better engage policymakers and stakeholders and propose reasonable, understandable, and accountable management solutions. We believe the result of such efforts will be a better understanding of the critical conservation issues facing insect communities across the planet.

ACKNOWLEDGMENTS

We would like to thank all eButterfly participants for whose passion, knowledge, and expertise make this project successful. We appreciate funding from NatureServe Canada and Agriculture Canada which helped eButterfly take flight. The founding of eButterfly, were supported by generous research support from National Science and Engineering Research Council of Canada Discovery Grants, the Canadian Foundation for Innovation, Ontario Innovation Trust, an Early Researcher Award, and University of Ottawa Research Chair in Macroecology and Conservation to JT Kerr. We are grateful to the butterfly experts and "super-users" who provide their regional expertise to our data quality process through reviewing records and critiquing eButterfly itself. Without such ongoing support and assistance, eButterfly would not exist.

AUTHOR CONTRIBUTIONS

For All authors have been involved in conceiving and designing the eButterfly data collection protocol; Maxim Larrivée and Kent P. McFarland downloaded and processed the eButterfly data presented here; Jeffrey C. Oliver and Maxim Larrivée summarized the eButterfly data presented here; Jeffrey C. Oliver wrote the code for the figures;

Kathleen L. Prudic conceived and wrote the paper with all authors editing and commenting extensively on several versions of the manuscript and figures.

CONFLICTS OF INTEREST

The authors declare no conflict of interest. The founding sponsors had no role in the design of the study; in the collection, analyses, or interpretation of data; in the writing of the manuscript; and in the decision to publish the results.

REFERENCES

1. Wood, C.; Sullivan, B.; Iliff, M.; Fink, D.; Kelling, S. eBird: Engaging birders in science and conservation. PLoS Biol.2011, 9, e1001220.
2. Valdecasas, A.G.; Camacho, A.I. Conservation to the rescue of taxonomy. Biodivers. Conser. 2003, 12, 1113–1117.
3. Hochachka, W.M.; Fink, D.; Hutchinson, R.A.; Sheldon, D.; Wong, W.K.; Kelling, S. Data-intensive science applied to broad-scale citizen science. TREE 2012, 27, 130–137.
4. Sullivan, B.L.; Aycrigg, J.L.; Barry, J.H.; Bonney, R.E.; Bruns, N.; Cooper, C.B.; Damoulas, T.; Dhondt, A.A.; Dietterich, T.; Farnsworth, A.; et al. The eBird enterprise: An integrated approach to development and application of citizen science. Biol. Conserv. 2014, 169, 31–40.
5. National Phenology Network. Nature's Notebook. Available online: https://www.usanpn.org/natures_notebook (accessed on 1 May 2017).
6. Plant Watch Canada. What is plantwatch? Available online: www.naturewatch.ca/plantwatch (accessed on 1 May 2017).
7. iNatualist. Available online: http://www.inaturalist.org (accessed on 2 May 2017).
8. Global Biodiversity Information Facility. Available online: http://www.gbif.org (accessed on 3 May 2017).
9. eBird. Lab of Ornithology. Cornell University. Available online: http://www.ebird.org (accessed on 3 May 2017).
10. Cyr, A.; Larivee, J. A Checklist Approach for Monitoring Neotropical Migrant Birds: Twenty-Year Trends in Birds of Québec Using ÉPOQ; General Technical Report; U.S. Department of Agriculture, Forest Service: Washington, DC, USA, 1993; pp. 229–236.
11. Pollard, E.; Yates, T.J. Monitoring Butterflies for Ecology and Conservation: The British Butterfly Monitoring Scheme; Springer: Berlin, Germany, 1994.
12. Munroe, E. The President's Address. J. Lepidopterists Soc. 1960, 14, 1–4.
13. Cohn, J.P. Citizen Science: Can volunteers do real research? Bioscience 2008, 58, 192–197.
14. Collinge, S.K.; Prudic, K.L.; Oliver, J.C. Effects of local habitat characteristics and landscape context on grassland butterfly diversity. Cons. Biol. 2003, 17, 178–187.
15. Prudic, K.L.; Oliver, J.C.; Bowers, M.D. Soil nutrient effects on oviposition preference, larval performance, and chemical defense of a specialist insect herbivore. Oecologia 2005, 143, 578–587.
16. Forister, M.L.; Cousens, B.; Harrison, J.G.; Anderson, K.; Thorne, J.H.; Waetjen, D.; Nice, C.C.; De Parsia, M.; Hladik, M.L.; Meese, R.; et al. Increasing neonicotinoid use

and the declining butterfly fauna of lowland California. Biol. Lett.2016, 12, 20160475.
17. United States Fish and Wildlife Service. National Survey of Fishing, Hunting, and Wildlife-Associated Recreation. 2011. Available online: https://www.census.gov/prod/2012pubs/fhw11-nat.pdf (accessed on 1 May 2017).
18. Kerr, J.T.; Southwood, T.R.E.; Cihlar, J. Remotely sensed habitat diversity predicts butterfly species richness and community similarity in Canada. Proc. Nat. Acad. Sci. USA 2001, 98, 11365–11370.
19. Lewthwaite, J.M.; Debinski, D.M.; Kerr, J.T. High community turnover and dispersal limitation relative to rapid climate change. Glob. Ecol. Biogeogr. 2017, 26, 459–471.
20. Coristine, L.E.; Soroye, P.; Soares, R.N.; Robillard, C.; Kerr, J.T. Dispersal Limitation, Climate Change, and Practical Tools for Butterfly Conservation in Intensively Used Landscapes. Nat. Areas J. 2016, 36, 440–452.
21. Macnaughton, A.; Layberry, R.; Jones, C.; Edwards, B. Ontario Butterfly Atlas Online. Available online: http://www.ontarioinsects.org/atlas_online.htm (accessed on 24 March 2017).
22. BISON. United State Geological Survey. Available online: https://usgs.bison.gov (accessed on 3 May 2017).
23. Canadensys. Available online: http://data.candensys.net (accessed on 3 May 2017).
24. Independent Sector. The Value of Volunteer Time. Available online: http://www.independentsector.org/resource/the-value-of-volunteer-time (accessed on 24 March 2017).
25. Sullivan, B.L.; Wood, C.L.; Iliff, M.J.; Bonney, R.E.; Fink, D.; Kelling, S. eBird: A citizen-based bird observation network in the biological sciences. Biol. Conserv. 2009, 142, 2282–2292.
26. United Kingdom Butterfly Monitoring Scheme. Available online: http://www.ceh.ac.uk (accessed on 1 May 2017).
27. Guillera-Arroita, G.; Lahoz-Monfort, J.J.; Elith, J.; Gordon, A.; Kujala, H.; Lentini, P.E.; McCarthy, M.A.; Tingley, R.; Wiintle, B.A. Is my species distribution model fit for purpose? Matching data and models to applications. Glob. Ecol. Biogeogr. 2015, 24, 276–292.
28. Kharouba, H.M.; Paquette, S.; Kerr, J.T.; Vellend, M. Predicting the sensitivity of butterfly phenology over the past century. Glob. Change Biol. 2014, 20, 504–514.
29. Robillard, C.M.; Coristine, L.E.; Soares, R.N.; Kerr, J.T. Facilitating climate-change-induced range shifts across continental land-use barriers. Conserv. Biol. 2015, 29, 1586–1595.
30. Kelling, S.; Gerbracht, J.; Fink, D.; Lagoze, C.; Wong, W.-K.; Yu, J.; Gomes, C. eBird: A Human/Computer Learning Network for Biodiversity Conservation and Research. In Proceedings of the 24th Innovative Applications of Artificial Intelligence Conference, Toronto, ON, Canada, 22–26 July 2012.
31. Kaya, Y.; Kayci, L.; Tekin, R.; Faruk Ertuğrul, Ö. Evaluation of texture features for automatic detecting butterfly species using extreme learning machine. J. Exp. Theor. Art. Intell. 2014, 26, 267–281.
32. MacKenzie, D.I.; Nichols, J.D.; Lachman, G.B.; Droege, S.; Royle, J.A.; Langtimm, C.A. Estimating Site Occupancy Rates When Detection Probabilities Are Less Than One. Ecology 2002, 83, 2248–2255.
33. Bailey, L.L.; MacKenzie, D.I.; Nichols, J.D. Advances and applications of occupancy models. Methods Ecol. Evol. 2014, 5, 1269–1279.
34. Yu, J.; Wong, W.-K.; Hutchinson, R. Modeling experts and novices in citizen science data for species distribution modeling. Proc. 2010 IEEE Int. Conf. Data Min. 2010, 1157–1162.
35. Boakes, E.H.; McGowan, P.J.K.; Fuller, R.A.; Chang-qing, D.; Clark, N.E.; O'Connor, K.; Mace, G.M. Distorted Views of Biodiversity: Spatial and Temporal Bias in Species Occurrence Data. PLoS Biol. 2010, 8, e1000385.

36. Fink, D.; Hochachka, W.M.; Zuckerberg, B.; Winkler, D.W.; Shaby, B.; Munson, M.A.; Kelling, S. Spatiotemporal exploratory models for broad-scale survey data. Ecol. Appl. 2010, 20, 2131–2147.
37. Phillips, S.J.; Dudik, M.; Elith, J.; Graham, C.H.; Lehmann, A.; Leathwick, J.R.; Ferrier, S. Sample selection bias and presence-only distribution models: Implications for background and pseudo-absence data. Ecol. Appl. 2009, 19, 181–197.
38. Adventure Scientists. Available online: http://www.adventurescientist.org (accessed on 1 May 2017).
39. Fithian, W.; Elith, J.; Hastie, T.; Keith, D.A. Bias Correction in Species Distribution Models: Pooling Survey and Collection Data for Multiple Species. Methods Ecol. Evol. 2015, 6, 424–438.
40. NASA Earth Observatory. Historic Heat in North America Turns Winter to Summer March 21, 2012. Available online:http://earthobservatory.nasa.gov/IOTD/view.php?id=77465 (accessed on 9 May 2012).
41. National Climatic Data Center. State of the Climate Drought. Available online: http://www.ncdc.noaa.gov/sotc/drought (accessed on 22 April 2012).
42. Zhan, S.; Zhang, W.; Niitepold, K.; Hsu, J.; Haeger, J.F.; Zalucki, M.P.; Altizer, S.; De Roode, J.C.; Reppert, S.M.; Kronforst, M.R. The genetics of monarch butterfly migration and warning colouration. Nature 2014, 514, 317–321.
43. Spring monarch migration visualization. Journey North. Available online: http://www.learner.org/jnorth/maps/galleries/2012/monarch_an_spring2012.html (accessed on 1 May 2017).
44. Vidal, O.; Rendón-Salinas, E. Dynamics and trends of overwintering colonies of the monarch butterfly in Mexico. Biol. Conserv. 2014, 180, 165–175.
45. Kerr, J.T.; Ostrovsky, M. From space to species: Ecological applications for remote sensing. TREE 2003, 18, 299–305.
46. Chapman, J.W.; Nesbit, R.L.; Burgin, L.E.; Reynolds, D.R.; Smith, A.D.; Middleton, D.R.; Hill, J.K. Flight orientation behaviors promote optimal migration trajectories in high-flying insects. Science 2010, 327, 682–685.
47. Bumble Bee Watch. Xerces Society. Available online: http://www/bumblebeewatch.org (accessed on 2 May 2017).
48. Mission Monarch. Montreal Space for Life. Available online: http://www.mission-monarch.org (accessed on 1 May 2017).
49. Nectar Connector Campaign. National Phenology Network. United States Geological Survey. Available online: http://www.usanpn.org/nn/NectarConnectors (accessed on 1 May 2017).
50. R Core Team. R: A Language and Environment for Statistical Computing; R Foundation for Statistical Computing: Vienna, Austria, 2015.
51. Wickham, H. ggplot2: Elegant Graphics for Data Analysis; Springer: New York, NY, USA, 2009.
52. Kahle, D.; Wickham, H. ggmap: Spatial Visualization with ggplot2. R J. 2013, 5, 144–161.

CHAPTER 3

Insecticidal Activity of the Essential Oils from Different Plants Against Three Stored-Product Insects

Abdurrahman Ayvaz, Osman Sagdic,Salih Karaborklu, Ismet Ozturk

[1]*Erciyes University, Faculty of Arts and Sciences, Department of Biology, 38039, Kayseri, Turkey*

[2] *Erciyes University, Collage of Engineering, Food Engineering Department, 38039, Kayseri, Turkey*

[3] *Osmaniye Korkut Ata University, Faculty of Arts and Sciences, Department of Biology, 80020,Osmaniye, Turkey*

ABSTRACT

This study was conducted to determine the insecticidal activity of essential oils from oregano, *Origanum onites* L. (Lamiales: Lamiaceae), savory, *Satureja thymbra* L. (Lamiales: Lamiaceae), and myrtle, *Myrtus communis* L. (Rosales: Myrtaceae) against three stored-product insects. Essential oils from three species of plants were obtained by Clevenger-type water distillation. The major compounds in these essential oils were identified using gas chromatography-mass spectrometry and their insecticidal activity was tested against adults of the Mediterranean flour moth *Ephestia kuehniella* Zeller (Lepidoptera: Pyralidae), the Indian meal moth *Plodia interpunctella* Hübner (Lepidoptera: Pyralidae) and the bean weevil *Acanthoscelides obtectus* Say (Coleoptera: Bruchidae). While the major compound found in oregano and savory was carvacrol, the main constituent of the myrtle was linalool. Among the tested insects, *A. obtectus* was the most tolerant species against the essential oils. However, the insecticidal activity of the myrtle oil was more pronounced than other oils tested against *A. obtectus* adults. The essential oils of oregano and savory were highly effective against *P. interpunctella* and *E. kuehniella*, with 100% mortality obtained after 24 h at 9 and 25 μl/l air for *P. interpunctella* and *E. kuehniella*, respectively. LC_{50} and LC_{99} values of each essential oil were estimated for each insect species.

Keyword: Botanical based fumigants, stored product insect management, savory, myrtle, oregano

INTRODUCTION

The bean weevil, *Acanthoscelides obtectus* (Say) (Coleoptera: Bruchidae) is one of the most damaging pests of the kidney bean, Phaseolus vulgaris L. (Fabales: Fabaceae) in the Mediterranean area. It causes losses of up to 30% of stored beans (Pemonge et al. 1997). Its oviposition and growth are continuous, and the larvae feed on the seeds. After emergence from the seeds, the adults reproduce either in the field or in the stored seeds in a continuous cycle (Labeyrie 1962).

The Mediterranean flour moth, *Ephestia kuehniella* Zeller (Lepidoptera: Pyralidae), is one of the major pests in industrial flour mills in temperate climates (Jacob and Cox 1977). Larvae reduce product quality by their presence and by the production of frass and webbing, and they also cause direct damage by feeding (Johnson et al. 1997). The Indian meal moth, *Plodia interpunctella* (Hübner) (Lepidoptera: Pyralidae), is a cosmopolitan pest that infests a wide range of stored products including nuts, beans, processed foods and dried fruits (Simmons and Nelson 1975). The control of these pests in storage systems mainly depends on fumigants such as methyl bromide or phosphine. However, methyl bromide was banned in many countries starting in 2004 because of its ozone depleting properties (Hansen and Jensen 2002). Many alternatives have been tested to replace methyl bromide fumigation for stored-product and quarantine uses. There is an urgent need to develop safe alternatives that have the potential to replace the toxic fumigants, yet are effective, economical and convenient to use (Ayvaz et al. 2008). Many spices and herbs, and their extracts are known to possess insecticidal properties that are frequently present in the essential oil fraction (Brattsten 1983; Schmidt et al. 1991; Shaaya et al. 1991). Most of the essential oil constituents are monoterpenoids, which are secondary plant chemicals and considered to be of little metabolic importance. The toxicity of a large number of essential oils and their constituents have been evaluated against a number of stored-product insects.

Over the past 15 years, interest in botanical insecticides has increased as a result of environmental concerns and insect populations becoming resistant to conventional chemicals. Botanical insecticides are naturally occurring insecticides that are derived from plants (Isman 2000). The insecticidal activity of essential oils and plant extracts against different stored-product pests has been evaluated (Shaaya et al. 1991; Sarac and Tunc 1995; Tunc et al. 2000; Kim et al. 2003; Lee et al. 2003; Aslan et al. 2005; Cetin and Yanikoglu 2006; Negahban et al. 2007, Ayvaz et al. 2009). In spite of the wide-spread recognition that many plants possess insecticidal properties, only a handful of pest control products directly obtained from plants are in use because the commercialization of new botanicals can be hindered by a number of issues (Isman 1997). Botanicals used as insecticides presently constitute 1% of the world insecticide market (Rozman et al. 2007). Essential oils from different plant species possess ovicidal, larvicidal, and repellent properties against various insect species and are regarded as environmentally compatible pesticides (Isman 2000; Cetin et al. 2004).

In the present study, the chemical constituents of essential oils from *Satureja thymbra* L. (Lamiales: Lamiaceae), *Origanum onites* L. (Lamiales: Lamiaceae), and *Myrtus communis* L. (Rosales: Myrtaceae) were determined, and the insecticidal activity of these essential oils was tested against the adult stages of the stored-products pests: *E. kuehniella*, *P. interpunctella* and *A. obtectus.* No study has been reported previously concerning the activity of these compounds as fumigants against these stored product insects. The essential oils were applied primarily on adults to prevent mass egg production and further damages from larvae.

MATERIALS AND METHODS

Insect cultures

The founding insect culture of *A. obtectus* was collected in infested kidney beans (*P. vulgaris*) that were stored in 5-liter plastic containers at a storehouse from Kayseri Province of Turkey. Adults oviposited on *P. vulgaris* beans, and the larvae developed inside the beans until adult emergence in l-liter glass jars. To allow air passage, a hole 2 cm in diameter was opened in the center of each jar lid, and a sterile cloth was glued to the underside of each lid.

The Mediterranean flour moth, *E. kuehniella,* culture was obtained from the Department of Plant Protection, Faculty of Agriculture of Ankara University. *E. kuehniella* larvae were reared using a mixture consisting of one kg wheat flour, 55 g yeast and 30 g germs of wheat (Marec et al. 1999).

The Indian meal moths, *P. interpunctella*, used in this experiment were taken as larvae from naturally infested dried apricot collected from Kayseri province. The larvae of *P. interpunctella* were maintained continuously on a diet containing 10% glycerol, 50% dried apricot and 40% wheat flour with wheat bran mixture. Throughout the experiments insect cultures were maintained at constant temperature (27 ± 1°C), photoperiod (14L:10D) and relative humidity (60% ± 5) (Ayvez et al. 2007, 2009).

Plant materials

The plants savory (*S. thymbra*) and myrtle (*M. communis*) were collected in the middle of July 2007 from the fields of Mersin (Southern Turkey), and oregano (*O. onites*) was collected in the middle of August 2007 from the fields of Canakkale (Western Turkey). Leaves were randomly collected from plant parts and shade dried.

Extraction of essential oil

The plant leaves were air dried at room temperature and chopped into small pieces using a mill with rotary knives. The essential oil was extracted from the plants using a Clevenger-type water steam distillation apparatus (Papachristos and Stamopoulos 2004). The distilled essential oils were stored in a refrigerator at 4°C until being used in the treatments.

Analysis of volatile compounds

The composition of the volatile constituents was established by gas chromatography-mass spectrometry/Quadropole detector analyses using a Shimadzu QP 5050 system fitted with a Free Fatty Acid Phase (50 m × 0.32 mm (i.d.) film thickness: 0.25 μm) capillary column. Detector and injector temperature were set at 230°C. The temperature program for the column was from 120°C (1 min) to 230°C at a rate of 6°C/min and than held at 200°C for 35 min. Helium was used as a carrier gas at a flow of 14 psi (Split 1:10), and the injection volume of each sample was one μl. The identification of the components was based on comparison of their mass spectra with those of Wiley and Nist Tutore Libraries (Adams 2001). The ionization energy was set at 70 eV.

Insecticidal activity

In order to test the toxicity of essential oils on the adults (< 24 h age) of *E. kuehniella* and *P. interpunctella,* ten adults were put into the 1000 ml glass jars. Essential oils were applied on a filter-paper strip measuring 3 x 3 cm that was attached to the lower side of the jar's lid. Doses were calculated based on nominal concentrations and assumed 100% volatilization of the oils in the exposure vessels (Papachristos and Stamopoulos 2004). The same procedure was applied for the adults (< 48 h age) of *A. obtectus,* but this species was provided with kidney beans as food. All the insecticidal activity experiments were conducted at constant temperature (27 ± 1°C), photoperiod (14L: 10D) and relative humidity (60% ± 5). The adults of *E. kuehniella* and *P. interpunctella* were exposed to essential oil vapors (1.5; 3; 6; 9; and 25 µl/l air) for 24 h. However, because of high tolerance of *A. obtectus* adults were treated with higher doses (65; 130; and 195 µl/l air) and longer exposure time (24, 72 and 144 h) than that of *E. kuehniella* and *P. interpunctella.* A dose-mortality line depending on the exposure time(s) was developed, and the lethal concentration of essential oil needed to kill 50 or 99% of the pest population (LC_{50} and LC_{99}, respectively) was determined. Three replicates were set up for each dose and exposure time. A complete set of controls was maintained and replicated three times for each treatment. All replicates ran simultaneously during the experiments.

Statistical analysis

The data were corrected using Abbott's formula (Abbot 1925) for the mortalities in the controls, and the data were subjected to probit analyses using SPSS (2001) for Windows to estimate LC_{50} and LC_{99} values of the essential oils against each stored-product insect species. Percentage mortality values for different exposure times were subjected to analysis of variance (one-way ANOVA) using the same statistical program (SPSS 2001) for probit analysis. Data were transformed using arcsine $\sqrt{x}$ transformation to meet normality, which is recommended for analysis of variance (ANOVA) (Steel and Torrie 1980). Means were separated at the 5% significance level by the least significant difference (LSD) test.

RESULTS

Chemical composition of essential oils

The chemical constituents of three different essential oils are given in Table 1. The results of analysis showed that the chemical contents of oregano and savory were similar, but myrtle had a different chemical composition. The main compounds found in the savory were characterized as carvacrol (53.7%), γ -terpinen (17.6%), thymol (13%) and *p*-cymene (10.1%). Of all the compounds found in the oregano, the percentage of the carvacrol was the highest (70.3%) as in the savory. The main other compounds were linalool (11.9%) and thymol (9.3%). The chemical constituents of the myrtle were different from other oils tested, and the main components were linalool (31.3%), linalyl acetate (17.8%), and 1.8-cineole (14.7%). The other constituents of the savory, oregano and myrtle are given in Table 1.

Insecticidal activity of essential oils against the *E. kuehniella* adults

The increasing doses of essential oils caused a significant increase in the mortality when the *E. kuehniella* adults were exposed to these oils for 24 h (for myrtle: $F = 55.98$; d.f. = 5, 12; $p < 0.0001$; for savory: $F = 23.90$; d.f. = 5, 12; $p < 0.0001$; and for oregano: $F = 80.08$; d.f. = 5, 12; $p < 0.0001$). Percent mortality of *E. kuehniella* was 90% at the 25 μl/l air dose when exposed to myrtle, and the same dose caused complete mortality (100%) when exposed to savory and to oregano (Figure 1).

Probit analysis showed that LC_{50} values of these essential oils against *E. kuehniella* were 7.52, 10.34 and 12.74 μ l/l air for oregano, savory and myrtle, respectively. LC_{99} values revealed that *E. kuehniella* was more susceptible to oregano (12.72 μl/l air) than savory and myrtle (21.27 and 29.43 μl/l air for savory and myrtle, respectively) (Table 2).

Insecticidal activity of essential oils against the *P. interpunctella* adults

The mortality values significantly increased depending on the increasing essential oil concentration when the *P. interpunctella* adults were exposed to myrtle, savory and oregano (for myrtle: $F = 135.32$; d.f. = 5, 12; $p < 0.0001$; for savory: $F = 267.73$; d.f. = 5, 12; $p < 0.0001$; and for oregano: $F = 100.48$; d.f. = 5, 12; $p < 0.0001$). The mortality values reached 96.7 and 100% when the adults were exposed to 6 μ l/l air concentrations of savory and oregano, respectively, and all the adults were killed by a 9 μ l/l air or higher concentration (Figure 2). The mortality effect of myrtle was lower than those of the other oils, and the rate of adult mortality was 66.7% at the 25 μ l/l air myrtle concentration. LC_{50} and LC_{99} values of the essential oils tested are summarized in Table 2, and these values revealed that the most effective essential oil against *P. interpunctella* adults was oregano, the same as for *E. kuehniella*. LC_{50} and LC_{99} values of the oregano against *P. interpunctella* were 4.06 and 5.77 μ l/l air, respectively (Table 2).

Table 1. Chemical constituent of the essential oils obtained from three plants

Compound	% Composition		
	Savory	Turkish oregano	Myrtle
α-Terpinen	1.78	0.61	-
Carvacrol	53.68	70.27	-
Caryophyllene	1.52	0.96	-
γ-terpinen	17.55	2.61	-
Geranyl acetate	-	-	9.1
Linalool	-	11.85	31.3
Linalyl acetate	-	-	17.8
Methyl eugenol	-	-	2.3
Myrcene	2.32	0.47	0.9
Myrtenol	-	-	1.4
p-cymene	10.05	2.93	-
Terpinen-4-ol		0.92	-
Thymol	12.96	9.34	-
α-Terpineol	-	-	6.5
α-Terpinyl acetate	-	-	1.7
α–Pinene	-	-	8.4
1.8- Cineole	-	-	14.7

%: Calculated from TIC (Total ion cromotograms) data.

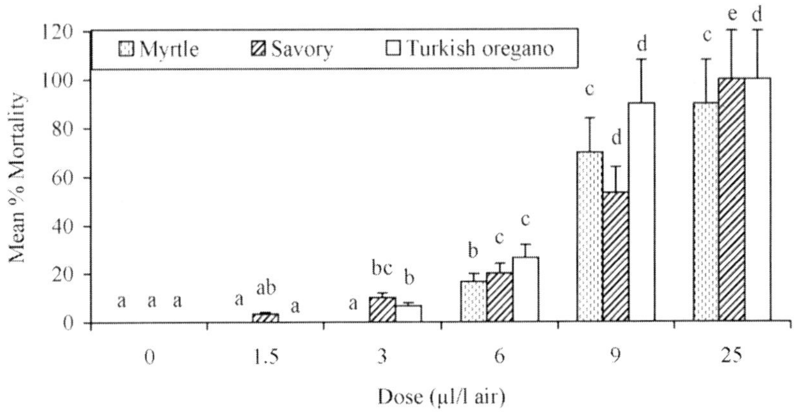

Figure 1. Percent mortality of the /Ephistia kuehniella/ adults after exposure to three different essential oils. Letters above bars indicate significant differences between doses. Bars with the same letter are not significantly different. Error bars indicate SD of means.High quality figures are available online.

Table 2. LC50 and LC99 values of essential oils from different plants against the adults of *Ephisitia kuehniella*and *Plodia interpunctella*

Time (24 h)	(N)	LC_{50} (µl/l air)	LC_{99} (µl/l air)	df	c²
		E. kuehniella			
Myrtle	30	12.74	29.43	16	38.22[a]
Turkish oregano	30	7.52	12.72	16	10.44
Savory	30	10.34	21.27	16	8.53
		P. interpunctella			
Myrtle	30	22.61	41.74	16	7.3
Turkish oregano	30	4.06	5.77	16	5.48
Savory	30	3.43	7.72	16	6.83

N: Number of the tested insects
[a]Since goodness-of-fit test is significant (P<0.150), a heterogeneity factor is used in the calculation of fiducial limits (f.l.).

Insecticidal activity of essential oils against the *A. obtectus*adults

Among the tested insects, *A. obtectus* was the most tolerant species against the essential oils. To kill the adults of this insect by using these three essential oils, higher concentrations and exposure times are required than are required for the other two species. The concentration (25 µ l/l air) that caused complete mortality for *E. kuehniella* and *P. interpunctella* did not affect *A. obtectus* adults. The myrtle was more effective than the other essential oils against *A. obtectus* adults. Increasing mortality was observed for the *A. obtectus* adults when the concentrations and exposure times increased. Although 100% mortality was obtained after 144 h at 195 µl/l air when *A. obtectus* was exposed to oregano and savory, the myrtle showed the same effect at 65 µl/l air and 72 h of exposure (Figure 3).

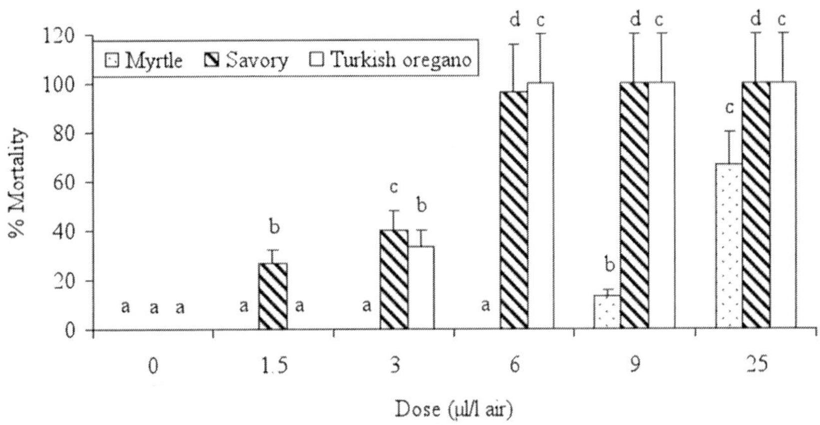

Figure 2. Percent mortality of the *Plodia interpunctella* adults after exposure to three different essential oils. Letters above bars indicate significant differences between doses. Bars with the same letter are not significantly different. Error bars indicate SD of means. High quality figures are available online

LC_{50} and LC_{99} values of the essential oils tested for the length of the time are summarized in Table 3, and the LC_{50} and LC_{99} values of the myrtle against *A. obtectus* were 33.56 and 50.97 µ l/l air, respectively, for the highest exposure time. The doses required for 99% mortality (LC_{99}) for using oregano and savory against *A. obtectus* were 127.36 and 76.05 µ l/l for the 144 h exposure, respectively (Table 3).

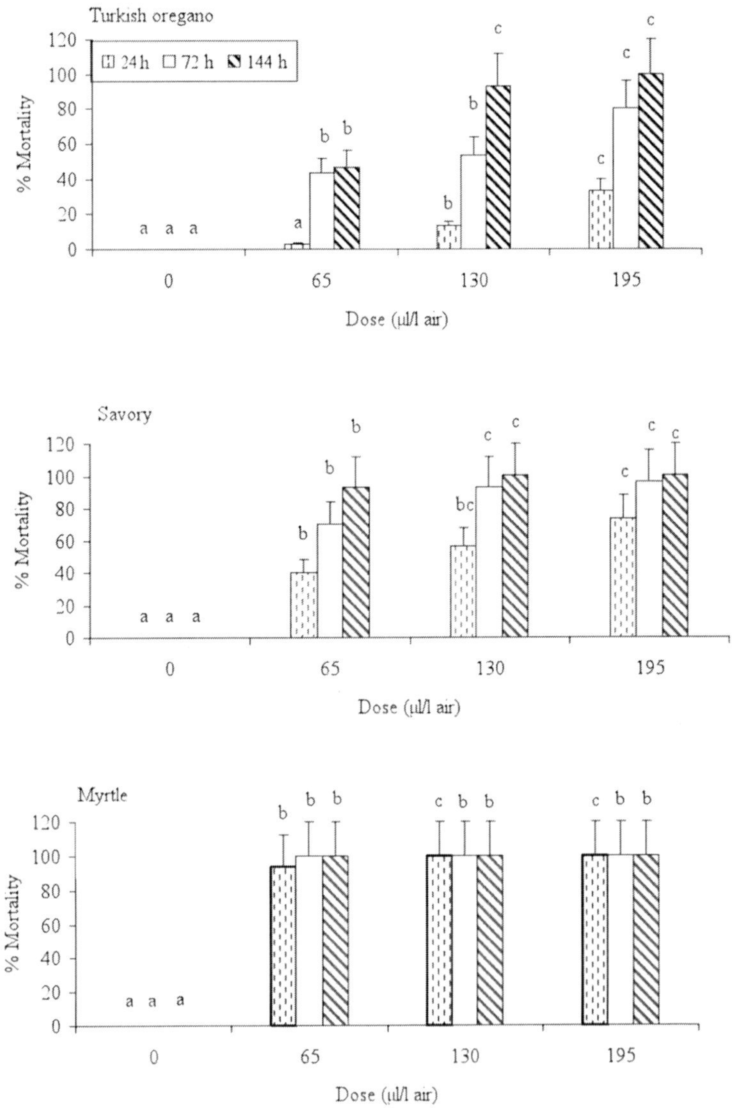

Figure 3. Percent mortality of the *Acanthoscelides obtectus* adults after exposure to three different essential oils. Letters above bars indicate significant differences between doses. Bars with the same letter are not significantly different. Error bars indicate SD of means. High quality figures are available online.

Table 3. LC_{50} and LC_{99} values of essential oils from different plants against the adults of *Acanthoscelides obtectu*

Time (h)	(N)	LC_{50} (µl/l air)	LC_{99} (µl/l air)	df	c^2
Myrtle					
24	30	49.58	76.07	10	1.06
72	30	33.56	50.97	10	0
144	30	33.56	50.97	10	0
Turkish oregano					
24	30	236.42	444.68	10	3.45
72	30	119.42	341.13	10	17,84[a]
144	30	55.94	127.36	10	19,63[a]
Savory					
24	30	124.75	336.64	10	11.51
72	30	63.14	174.85	10	54.76
144	30	49.59	76.05	10	1.06

N: Number of the tested insects

[a]Since goodness-of-fit test is significant ($P<0.150$), a heterogeneity factor is used in the calculation of fiducial limits (f.l.).

DISCUSSION

The insecticidal constituents of many plant extracts and essential oils are monoterpenoids. Due to their high volatility they have fumigant activity that might be of importance for controlling stored-product insects (Konstantopoulou et al. 1992; Regnault-Roger and Hamraoui 1995; Ahn et al. 1998). In the current study, the essential oils obtained from savory, oregano and myrtle showed insecticidal activity against the adults of *E. kuehniella*, *P. interpunctella* and *A. obtectus*. *A. obtectus* was the most tolerant species against these oils as the doses required to kill the adults of this species were much higher than those required for the other two pest species for all treatments. The adults of *A. obtectus* were more susceptible to myrtle oil than oregano or savory. The toxic effects of the myrtle could be attributed to major constituents such as linalool (31.3%), linalyl acetate (17.8%) and 1.8-cineole (14.7%). The high toxicity of linalool, linalyl acetate and 1.8-cineole was reported against the rice weevil *Sitophilus oryzae* and *Rhyzopertha dominica*(Rozman et al. 2007). Due to the linalool, linalyl acetate and 1.8-cineole constituents, the myrtle could also be used effectively against *S. oryzae* and *R. dominica*. These results, and those reported earlier, indicate that the insecticidal activity of the essential oils varies depending on the stage of the insect, the species and the plant origin of the essential oil (Tunc et al. 2000; Chiasson et al. 2001; Choi et al. 2003; Sedy and Koschier 2003; Negahban et al. 2007).

The essential oils of oregano and savory were highly effective against *P. interpunctella* and *E. kuehniella*, and 100% mortality was obtained after 24 h at 9 and 25 μl/l air for these species, respectively. The major components of these two essential oils are monoterpenes, primarily carvacrol and thymol. Reported biological activities of plant terpenoids include repellency and deterrence, reduced palatability, growth inhibition through altered protein availability, enzyme inhibition, and direct toxicity (Harborne 1993). The plants oregano and savory belong to the family Lamiaceae. Jacobson (1989) pointed out that the most promising botanical insect-control agents are in the families of Annonaceae, Asteraceae, Canellaceae, Lamiaceae, Meliaceae and Rutaceae.

The insecticidal activity against *A. obtectus* of the essential oils obtained from oregano and savory was higher at the lowest dose and longest exposure period than at the highest dose and the lowest exposure period (Figure 3). El-Nahal et al. (1989) stated that the period of exposure appears to be more important than dosage in affecting the efficiency of the vapors of *Acorus calamus*essential oil to adults of five stored-product insect species. Similar results have been reported for the toxicity of methanol extract of the rhizome from *Acorus gramineus* to adults of *S. oryzae* and *Lasioderma serricorne* (Park 2000).

Similarly, the essential oils of oregano and savory have been found to be lethal to *P. interpunctella* and *E. kuehniella* but less toxic to *A. obtectus*. Carvacrol and thymol are usually extracted from *Satureja, Coridothymus, Thymbra*, and *Origanum* species and have been found to be lethal to turnip aphids (Chiasson et al. 2001). There are numerous reports on the insecticidal activity of the essential oils from *Origanum* species, and the major components of this species, such as carvacrol, thymol, γ-terpinen and terpinen-4-ol, are based on fumigant and repellent activity rather than contact toxicity (Tunc et al. 2000; Erler and Tunc 2005; Erler 2005).

Carvacrol has broad insecticidal and acaricidal activity against agricultural, stored-product and medical pests, and it acts as a fumigant. It is highly toxic to nymphs of the termite *Reticulitermes speratus*, adults of the rice weevil *S. oryzae*, the pulse beetle *Callosobruchus chinensis*, the cigarette beetle, *L. serricorne* (Ahn et al. 1998), and the mite, *Tetranychus urticae* (Isman 2000).

The insecticidal activity of the myrtle was lower than that of the oregano and savory against *E. kuehniella* and *P. interpunctella*. The percent mortality of both moth adults was 100% at the 25 μl/l air and 24 h exposure when treated with oregano and savory. However, percent mortality of *E. kuehniella* and *P. interpunctella* was 90% and 67%, respectively, when exposed to myrtle.

If the cost-effective commercial problems can be solved, the essential oils obtained from these plants can be effectively used as part of integrated pest management strategies (Rozman et al. 2007). The essential oil content of aromatic plants is about 1-3% (Cakir 1992). Therefore large quantities of plant material have to be processed in order to obtain the essential oils in quantities sufficient for commercial-scale tests (Tunc et al. 2000). It would be useful to breed the plants containing desired essential oils in elevated quantity.

The high activity of these compounds could make it a potential substitute for methyl bromide in various uses in stored-product control programs and can be used in coordination with microbial insecticides, attractants and traps, and beneficial insects and mites (natural enemies of pests) as a component of the integrated pest management. The observed fumigant activity shows that essential oils are sources of biologically active vapors that are potentially efficient insecticides. Consequently, the possibility of employing these natural fumigants to control insects in stored products may be worthy of further investigation.

The authors thank Dr. Dogan Bulut for language improvement, and to the editors for critical review of the manuscript.

REFERENCES

1. Abbott W, 1925. A method for computing the effectiveness of an insecticide. Journal of Economic Entomology 18:265–267.
2. Adams RP. 2001. Identification of essential oil components by gas chromatography/ quadrupole mass spectroscopy. Allured Publishing Corporation. Ahn YJ, Lee SB, Lee HS, Kim GH. 1998. Insecticidal and acaricidal activity of caravacrol and β-thujaplicine derived from Thujopsis dolabrata var. hondai sawdust. Journal of Chemical Ecology 24: 81-90.
3. Aslan I, Calmasur O, Sahin F, Caglar O. 2005. Insecticidal effects of essential plant oils against Ephestia kuehnielia (Zell.), Lasioderma serricome (F.) and Sitophilus granarius (L.). Journal of Plant Diseases and Protection 112: 257-267.
4. Ayvaz A, Albayrak S, Tuncbilek AS. 2007. Inherited sterility in Mediterranean flour moth Ephestia kuehniella Zeller (Lepidoptera: Pyralidae): effect of gamma radiation doses on insect fecundity fertility and developmental period. Journal of Stored Product Research 43:234–239.
5. Ayvaz A, Albayrak S, Karaborklu S. 2008. Gamma radiation sensitivity of the eggs, larvae and pupae of Indian meal moth Plodia interpunctella (Hübner) (Lepidoptera: Pyralidae). Pest Management Science 64:505–512.
6. Ayvaz A, Karaborklu S, Sagdic O. 2009. Fumigant Toxicity of Five Essential Oils Against the Eggs of Ephestia kuehniella Zeller and Plodia interpunctella (Hübner) (Lepidoptera: Pyralidae). Asian Journal of Chemistry 21: 596-604.

7. Brattsten LB. 1983. Cytochrome P-450 involvement in the interactions between plant terpenes and insect herbivores. In: Hedin PA, editor. Plant Resistance to Insects. pp. 173-195.
8. ACS Symposium Series No. 208, American Chemical Society. Cetin H, Erler F, Yanikoglu A. 2004. Larvicidal activity of a botanical natural product, AkseBio2, against Culex pipiens. Fitoterapia 75: 724-728.
9. Cetin H, Yanikoglu A. 2006. A study of the larvicidal activity of Origanum (Labiatae) species from Southwest Turkey. Journal of Vector Ecology 31: 118-122.
10. Chiasson H, Belanger A, Bostanian N, Vincent C, Poliquin A. 2001. Acaricidal properties of Artemisia absinthium and Tanacetum vulgare (Asteraceae) essential oils obtained by three methods of extraction. Journal of Economic Entomology 94: 167- 171.
11. Choi WI, Lee EH, Choi BR, Park HM, Ahn YJ. 2003. Toxicity of plant essential oils to Trialeurodes vaporariorum (Homoptera: Aleyrodidae). Journal of Economic Entomology 96: 1479-1484.
12. Cakir C. 1992. Investigations on the fungitoxic potentials of some plants occurring in Antalya. M.S. Thesis, Akdeniz University, Antalya (in Turkish). El-Nahal AKM, Schmidt GH, Risha EM. 1989. Vapours of Acorus calamus oil-a space treatment for stored-product insects. Journal of Stored Products Research 25: 211-216.
13. Erler F, Tunc I. 2005. Monoterpenoids as fumigants against greenhouse pests: toxic, development and reproduction inhibiting effects. Zeitschrift für Pflanzenkrankheiten und Pflanzenschutz 112: 181-192.
14. Erler F. 2005. Fumigant activity of six monoterpenoids from aromatic plants in Turkey against the two stored-product insects confused flour beetle, Tribolium confusum, and Mediterranean flour moth, Ephestia kuehniella. Zeitschrift für Pflanzenkrankheiten und Pflanzenschutz 112: 602–611.
15. Hansen LS, Jensen KMV. 2002. Effect of temperature on parasitism and host-feeding of Trichogramma turkestanica (Hymenoptera: Trichogrammatidae) on Ephestia kuehniella (Lepidoptera: Pyralidae). Journal of Economic Entomology 95: 50-56.
16. Harborne JB. 1993. Introduction to Ecological Biochemistry, 4th ed. Academic Press. Isman MB. 1997. Neem and other botanical insecticides commercialization. Phytoparasitica 25: 339-344.
17. Isman MB. 2000. Plant essential oils for pest and disease management. Crop Protection 19: 603-608.
18. Jacob TA, Cox PD. 1977. The influence of temperature and humidity on the life-cycle of Ephestia kuehniella Zeller (Lepidoptera: Pyralidae). Journal of Stored Products Research 13: 107-118.
19. Jacobson M. 1989. Botanical pesticide: past, present, and future. In: Arnason JT, Philogene BJR, Morand P, editors. Insecticides of Plant Origin. pp. 1-10. ACS Symposium Series No. 387. American Chemical Society.
20. Johnson JA, Valero KA, Hannel MM. 1997. Effect of low temperature storage on survival and reproduction of Indian meal moth (Lepidoptera: Pyralidae). Crop Protection 16: 519-523.
21. Kim SI, Park C, Ohh MH, Cho HC, Ahn YJ. 2003. Contact and fumigant activities of aromatic plants extracts and essential oils against Lasioderma serricorne (Coleoptera: Anobiidae). Journal of Stored Products Research 39: 11-19.
22. Konstantopoulou LL, Vassilopoulou L, Mavragani-Tsipidou P, Scouras ZG. 1992. Insecticidal effects of essential oils. A study of the effects of essential oils extracted from eleven Greek aromatic plants on Drosophila auraria. Experientia 48: 616-619.
23. Labeyrie V. 1962. Les Acanthoscelides. In: Balachowsky AS, editor. Entomologie Appliqub a l'dgriculture. Vol. 1, pp. 469- 490.
24. Masson. Lee SE, Peterson CJ, Coats JR. 2003. Fumigation toxicity of monoterpenoids to several stored-product insects. Journal of Stored Products Research 39: 77-85.
25. Marec F, Lollarova I, Pevelka J. 1999. Radiation induced inherited sterility combined with a genetic sexing system in Ephestia kuehniella (Lepidoptera: Pyralidae). Annals of Entomological Society of America 92: 250-259.

26. Negahban M, Moharramipour S, Sefidkon F. 2007. Insecticidal activity of essential oil from Artemisia sieberi Beser against three stored-product insects. Journal of Stored Products Research 43: 123-128.
27. Papachristos DP, Stamopoulos DC. 2004. Fumigant toxicity of three essential oils on the eggs of Acanthoscelides obtectus (Say) (Coleoptera: Bruchidae). Journal of Stored Products Research 40: 517-525.
28. Park C. 2000. Insecticidal activity of basarone derived from Acorus gramineus root against insect pests. MS Thesis, Seoul National University.
29. Pemonge J, Pascual-Villalobos MJ, Regnault-Roger C. 1997. Effects of material and extracts of Trigonella foenum-graecum L. against the stored product pests Tribolium castaneum (Herbst) (Coleoptera: Tenebrionidae) and Acanthoscelides obtectus (Say) (Coleoptera: Bruchidae). Journal of Stored Products Research 33: 209-217.
30. Regnault-Roger C, Hamrouni A. 1995. Fumigant toxic activity reproductive inhibition induced by monoterpenes on Acanthoscelides obtectus (Say), a bruchid of kidney bean. Journal of Stored Products Research 31: 291-299.
31. Rozman V, Kalinovic I, Korunic Z. 2007. Toxicity of naturally occurring compounds of Lamiaceae and Lauraceae to three storedproduct insects. Journal of Stored Products Research 43:349-355.
32. Sarac A, Tunc I. 1995. Toxicity of essential oil vapours to stored-product insects. Zeitschrift für Pflanzenkrankheiten und Pflanzenschutz 102: 429-434.
33. Schmidt GH, Risha EM, Nahal AKM. 1991. Reduction of progeny of some storedproducts Coleoptera by vapours of Acorus calamus oil. Journal of Stored Products Research 21: 121-127.
34. Sedy KA, Koschier EH. 2003. Bioactivity of carvacrol and thymol, against Frankliniella occidentalis and Thrips tabaci. Journal of Applied Entomology 127: 313-316.
35. Shaaya E, Ravid U, Paster N, Juven B, Zisman U, Pissarev V. 1991. Insecticidal activity of essential oils against four major stored product insects. Journal of Chemical Ecology 17: 499-504.
36. Shaaya E, Kostjukovski M, Eilberg J, Sukprakarn C. 1997. Plant oils as fumigants and contact insecticides for the control of stored-product insects. Journal of Stored Products Research 33: 7-15.
37. Simmons P, Nelson HD. 1975. Insects on dried fruits. USDA Agricultural Handbook, vol. 464.
38. SPSS. 2001. SPSS Version 10.0. SPSS Inc, 233 S. Wacker Drive, Chicago, Illinois.
39. Steel RGD, Torrie JH. 1980. Principles and Procedures of Statistics: A Biometrical Approach. 2nd Edition. McGraw-Hill.
40. Tunc I, Berger BM, Erler F, Dagli F. 2000. Ovicidal activity of essential oils from five plants against two stored-products insects. Journal of Stored Products Research 36: 161-168.

CHAPTER 4

Insect Artifacts Are More than Just Altered Bloodstains

David Rivers * and Theresa Geiman

Department of Biology, Loyola University Maryland, Baltimore, MD 21210, USA

ABSTRACT

The bases for forensic entomology are that insects and their arthropod relatives can serve as evidence in criminal, medical and civil legal matters. However, some of the very same species that provide utility to legal investigations can also complicate crime scenes by distorting existing body fluid evidence (e.g., bloodstains, semen, saliva) and/or depositing artifacts derived from the insect alimentary canal at primary or secondary crime scenes. The insect contaminants are referred to as insect stains, artifacts, specks or spots, and are most commonly associated with human bloodstains. This review will discuss the different types of insect artifacts that have been described from crime scenes and laboratory experiments, as well as examine insect contaminates (non-blood based artifacts, transfer patterns, meconium, and larval fluids) that have received little research or case attention. Methods currently used for distinguishing insect stains from human body fluids will also be discussed and compared to presumptive tests used for identification of human body fluids. Since all available methods have severe limitations, areas of new research will be identified for the purpose of development of diagnostic techniques for detection of insect artifacts.

Keywords: forensic entomology; insect artifacts; fly spots; blow flies; calliphorids; sarcophagids; bloodstain evidence; crime scene investigation

1. INTRODUCTION

Insects do not commit crimes, but they can be instrumental in solving them. This is especially true with necrophagous Diptera that are attracted to human remains. Foraging adults from several families of flies seek out a corpse as a source of nutrients for egg provisioning and oviposition/larviposition [1]. For these same species, larval nutriment is largely derived from the remains, establishing a linkage between immature development, the corpse, and ambient conditions. With such information in hand, an estimate of the minimum post mortem interval (PMI_{min}) can be made [2]. This is considered the primary focus of forensic entomology within the subdiscipline of medicocriminal entomology [3]. The utility of insects to criminal investigations is not limited, however, to time estimates and may include determination of whether a corpse has been moved from another location inferred from faunal, developmental or seasonal information [4]; detection of illicit or prescribed medications bioaccumulated within insects that fed on the corpse [5]; person's identification based upon DNA profiling of tissues consumed by necrophagous insects [6]; and discovery of gunshot or bomb residues in larval tissues when otherwise undetectable in human remains [7]. Though more obscure, adult flies can also be a source of DNA from imbibed body fluids, which are then deposited as artifacts in other locations, the latter being especially important if the primary crime scene has already been cleaned following body removal [6].

In contrast to their roles as physical and trace evidence, necrophagous insects are not always helpful at crime scenes. The foraging activity of necrophagous Diptera has the potential to be counterproductive to criminal investigations, largely as a result of the way they feed. For flies in the families Calliphoridae and Sarcophagidae, adults land on or beside a corpse, walking across the surfaces or through wet body fluids. Gustatory receptors located at the tips of tarsi (pulvilli) and on the sponging mouthparts are used to assess the nutritional value of the fluids and tissues. Applying Locard's Exchange Principle to the interaction between necrophagous flies and a corpse, evidence of this association will be left behind at the crime scene [8]. Foraging activity is known to cause mechanical disruption of pooled blood and body fluid stains that have not dried [9,10]. It can also lead to transfer patterns, created by tarsi or the abdomen leaving impressions after passing through wet fluids, either at the primary scene or at other sites [11,12]. As adult flies consume body fluids, they regurgitate and defecate some of the ingested food onto surfaces at or near the crime scene, creating unique stains and/or an intermixing of fly artifacts with bloodstains and other human body fluids [11,13]. Fly contaminates are not restricted to the primary crime scene, as adults display positive phototaxis, and thus are attracted to windows and lights, locations in which wet blood may be transferred or artifacts deposited. In essence, false secondary crime scenes are established as a direct consequence of foraging activity on a corpse.

The problems with fly artifacts are magnified by the fact that regurgitate and defecate are virtually indistinguishable from human bloodstains. Fly stains are morphologically very similar to impact (i.e., forward, back, and mist-like spatter), projected, sneezed, and expirated bloodstains [10], and cannot be reliably distinguished using presumptive or confirmatory tests available for identification of human blood [13,14,15]. The use of molecular methods, namely DNA typing, for person's identification does not overcome these limitations since complete DNA profiles can be obtained of an individual from blood consumed by flies [6,16]. A few methods have been reported to be useful in differentiating fly artifacts from human bloodstains [9,10,13,17], but all have limitations that prevent each from being consistently reliable for use in crime scene investigations. The reality is that artifacts from very few fly species have been examined to come to any consensus on the typical classification of regurgitate and fecal stains, or accurate methods of detection.

In this review, a discussion of the different types of insect artifacts that have been described from crime scenes and laboratory experiments will be presented, along with an examination of insect contaminates (non-blood based artifacts, transfer patterns, meconium, and larval fluids) that have received little research or case attention. Methods currently used for distinguishing insect stains from human body fluids will also be discussed and compared to presumptive tests used for identification of human body fluids. All available methods have severe limitations, and thus, new areas of research will be identified for the purpose of development of diagnostic techniques for detection of insect artifacts.

2. INSECT ARTIFACTS

In theory any insect that interacts with a corpse or associated exuded body fluids can potentially create artifacts that confuse reconstruction efforts at a crime scene. However, the reality is that several species of necrophagous Diptera are the chief culprits in producing insect artifacts [11]. As a consequence, this review will focus on stains that result from the activity of flies in the families Calliphoridae and Sarcophagidae, which coincidentally also correspond to the vast majority of the literature focused on insect artifacts. By definition, only one type of insect artifact is officially recognized by bloodstain pattern analysts: insect stains. The Scientific Working Group for Bloodstain Pattern Analysis (SWGSTAIN) has defined insect stains as those bloodstains produced as a result of insect activity [18]. This definition leaves open the possibility of producing insect stains by two methods: insect modification of existing bloodstains or creation of new stains. It is the latter that is most frequently cited by forensic entomologists, since both regurgitation and fecal elimination can yield insect stains containing human blood. The reality is that necrophagous flies can produce stains or artifacts as a result of feeding on several types of fluids (e.g., blood, saliva, semen, vaginal fluids, decomposition fluids), and which yield artifacts that vary widely in terms of shape, color, and size [12,14,19,20]. Deposition of artifacts is also not restricted to just foraging adults, as post-feeding larvae and newly emerged adults have the potential to contaminate crime scenes with unique artifacts. The small literature base that exists for fly artifacts is predominantly focused on regurgitate and defecatory stains, which arguably are the most frequently encountered at crime scenes and most likely to compromise bloodstain pattern analysis. What follows is a brief description of five types of fly artifacts (regurgitate, defecatory stains, transfer patterns, meconium, and larval stains) that have been identified or could potentially be found at crime scenes, including the mechanisms used to produce each type and predicted chemical composition of the stains (Table 1).

Table 1. Morphological characteristics of fly artifacts.

Stain Type	Shape	Color	Dimensions *
Regurgitate			Area in mm^3
	round to asymmetrically round, occasionally with small tails to form tear drop shape	highly variable and dependent on food source. May be clear, red, green, gray tan/light, dark brown/black	3.0–16.2 (human blood)
			1.9–19.2 (bovine blood)
Defecatory			Tail length (mm)
	round to asymmetrically round, some possessing long tails that form tad pole, tear-drop, and sperm-like shapes	highly variable and dependent on food source. May appear creamy, dark brown/black, gray, tan/light #	3.0–16.2 (human blood)
			1.9–19.2 (bovine blood)
			4.8–9.2 (chicken blood) +
Translocation			Stain length (mm)
	asymmetrically linear	same as food source	0.98–10.4 (human blood)
			0.74–9.7 (bovine blood)
Tarsal tracks	small and round, impressions of tarsi or pulvilli	same as food source	>0.2 mm in diameter

* Data from [12] using *Sarcophaga bullata* (Sarcophagidae), *Calliphora vicina* (Calliphoridae), *Chrysomya rufifacies*, *Ch. megacephala*, and *Phormia regina*. + Data from [20] using *Ch. megacephala* feed chicken blood. # Data from [19] using *Lucilia cuprina* fed human blood.

2.1. Regurgitate

Regurgitation is considered to be the expulsion of food from any location within the foregut out the oral opening [21]. It is a component of bubbling behavior that leads to a food droplet or bubble forming on the distal tip of the pseudotrachea. Regurgitation is in contrast to vomiting, in which food from the midgut is forced into the foregut, and then passed out of the mouth. Distinguishing between the two would seem to have little forensic value, but in fact the importance lies in composition differences, which can be the bases for development of confirmatory tools used to detect fly artifacts. Additionally, at times the terms have been used interchangeably in the forensic entomology literature [20,22], but clearly the two are separate physiological processes. Regurgitate stains should be considered far more significant as potential contaminants at crime scenes, since bubbling behavior is quite common among necrophagous Diptera, especially following consumption of a meal. After imbibing a liquid diet, the ingested food is pushed through the anterior foregut to the crop. The crop serves as the initial site of mixing of food with salivary enzymes [21]. If the crop was already full when the meal was ingested, regurgitation of the crop contents occurs [23]. This is manifested as bubble formation at the tips of the mouthparts. The bubble itself is highly variable in color, presumably a reflection of the food consumed. That said food bubbles might appear clear despite the recent consumption of carrion, blood, or feces [11] (Figure 1).

The fate of the regurgitated droplet appears to be dependent on the composition of the meal consumed, fly species, and whether the adult is disturbed during bubbling. Bubbling behavior is thought to serve two primary functions; decrease water content of the food prior to enzymatic digestion or flight, and to permit extra-oral digestion [13,21]. It is the latter function that causes regurgitation stains to achieve status as trace evidence at crime scenes. Adult flies will simply drop the food bubbles from the mouthparts onto the substrate they are resting. This can occur around a corpse following feeding on tissues, decomposition fluids, blood, or any other type of exposed body fluid, with the exception of urine [19]. Regurgitate stains are typically round or asymmetrically round (i.e., elliptical or oval) owing to the food bubbles dropping essentially perpendicular to the substrate [12], although occasional tear-drop shaped artifacts have been reported for *Calliphora vicina*, *Sarcophaga bullata*, *Lucilia sericata*, and *Chrysomya megacephala* [6,10,12,17,20] and the tails appear to result from flies beginning to move before release of the droplet [12]. A third morphological pattern reported is dome shaped craters that supposedly result from the sucking process of fly mouthparts [24]. This would seem

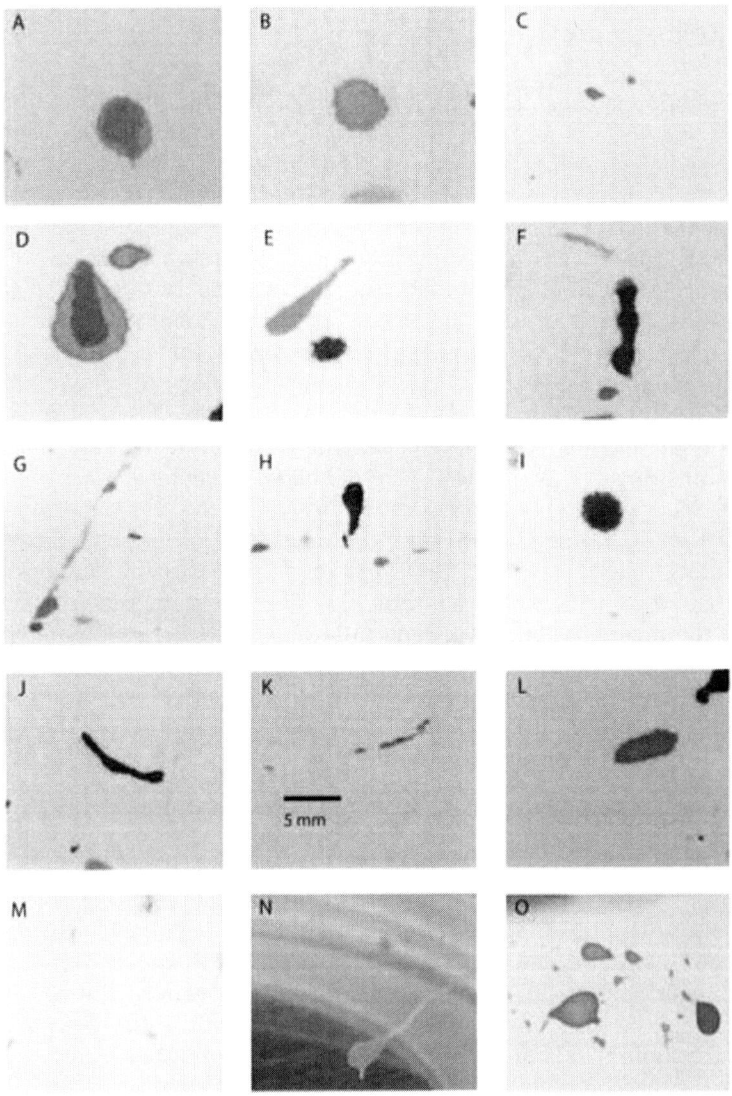

Figure 1. Types of fly artifacts produced by adult calliphorids and sarcophagids. All artifacts were produced following feeding or exposure to human blood, unless otherwise noted. Regurgitate stains deposited by (**A**,**D**,**E**) *Sarcophaga bullata* (Sarcophagdae), (**B**) *Calliphora vicina* (Calliphoridae), (**C**) *Lucilia sericata* (Calliphoridae); defecatory stains deposited by (**F**,**G**) *S. bullata*, (**H**,**I**) *C. vicina*; translocation stains deposited by (**J**,**K**) *S. bullata*, (**L**) *C. vicina*; tarsal tracks produced by *S. bullata*; and meconium deposited by (**N**) *S. bullata*, (**O**) *C. vicina*. Images were captured using a ChemiDoc Imaging System (BioRad).

to implicate regurgitate stains, since adult flies are not known to consume their own feces. However, the idea that the hydrostatic action of the cibarial pump and associated muscles of the foregut can create sufficient negative pressure to suck a stain from a surface or modify dry fly artifacts has not been demonstrated experimentally. In fact, cratered fly stains have only been reported for two species of calliphorids (*Ch. megacephala* and *L. cuprina*), both under laboratory conditions [19,20], but with *L. cuprina*, the stain morphology was attributed to drying on a smooth, non-porous surface, and not due to feeding activity of adult flies. This type of stain is not commonly encountered [25].

Regurgitate stains are composed predominantly of the meal consumed. However, the food is modified by the addition of salivary enzymes that are mixed into the gut fluids within the crop. In adult *Protophormia terraenovae*, enzymes consistent with trypsin, chymotrypsin and pepsin have been detected in the crop and exogenously deposited regurgitate [26]. Presumably a similar enzyme profile would be expected in other carrion-inhabiting calliphorids, although these same proteases have not been observed in *Phormia regina* [21]. Interesting, *P. regina* apparently does not typically drop bubbles and is more apt to reabsorb the regurgitated food [23]. When regurgitate is released by this species, it is twice as likely to occur following a protein meal than after sugar consumption [23]. By contrast, most other species of calliphorids observed bubbling do release regurgitate [24], and the event coincides with feeding on tissues and fluids from human remains or carrion. In other words, diets high in protein [1]. It is also quite likely that antimicrobial compounds produced by the labellar glands are released into the ingested food since uptake of bacteria from decomposing food sources is inevitable [21,27]. Endogenous bacteria and possibly other microorganisms conceivably are present in the crop of necrophagous Diptera, or are introduced to ingested food if vomiting occurs to permit intermixing of enzymes from the midgut. Regardless of the precise composition, stains resulting from regurgitated food are expected to be chemically similar yet distinct from the original food source.

The discussion of regurgitate (or any other insect artifact) composition is based on the assumption that adult flies consume fluids and tissues found at the crime scene. In reality, many flies may be introduced to the corpse during the crime scene investigation; that is, the adults gained access as a result of doors, containers, etc. being opened by the first responders to the scene [11]. In such scenarios, deposition of regurgitate and/or feces chemically distinct from fluids or tissues of the deceased may be introduced at the crime scene. This includes the possibility of fly artifacts containing DNA from an individual not associated with the current crime scene.

2.2. Defecatory Stains

Defecatory stains result from adult flies eliminating liquid feces. These stains are varied in terms of shape and color based on species. Eliminated feces on smooth surfaces commonly appear round or asymmetrically round, sometimes with tails, and on occasion may appear elongate (linear or sausage shaped) [12,19,20]. In most cases, if a fly artifact possesses a tail, the stain is derived from feces. The tail originates from either the fly beginning to walk prior to completion of defecation, producing a tail tapered in the direction of fly movement, or results from forcible expulsion of liquid feces from a protruded anus. The latter also yields tails pointed in the direction of fecal droplet movement. In theory, artifacts with tails permit distinction between defecate and regurgitate, and between bloodstains and defecatory stains. The idea is based on the premise that defecatory stains are often formed into shapes—sperm-shape, tadpole, or tear-drop—unique from other types of stains found at crime scenes [11,19]. However, as will be discussed later in this review, the uniqueness between these types of stains is not quite as distinct as originally thought.

Stains resulting from fecal release are the most commonly produced artifacts by *L. cuprina* and *C. vicina* (collected in Frankfurt, Germany) following feeding on human blood [6,14]. In contrast, regurgitate stains are the most abundant artifacts released by *S. bullata*, *L. sericata*, *Ch. rufifacies*, *Ch. megacephala*, and *P. regina* following consumption of a range of foods, including human blood [10,12,17,20]. Interestingly, Rivers and McGregor [12] demonstrated that *C. vicina* collected in Baltimore, Maryland predominantly deposits regurgitate stains, and that nearly all types of artifacts produced by this fly are round and lack tails. The latter differs from the observations of Striman et al. [17], in which *C. vicina* obtained in Nebraska frequently deposited defecatory stains that were tear-drop shaped with long tails. Durdle et al. [19] reported that defecatory stains of *L. cuprina* are more varied in shapes, colors, textures, viscosity, and translucence than observed with any other fly species. Varied fecal stain morphologies have also been reported for *Ch. megacephala* collected in Malaysia, in which tear-drop, sperm-like, snake-like, and irregular tadpole-like defecate with long tails (4.8 to 9.2 mm) were produced on porous surfaces following *ad libitum* consumption of chicken blood [20]. Such differences in regurgitate and defecatory artifact morphology appears to reflect species and diet dependency, as well as within species variation based on geographic location [12].

Fly feces is a unique mixture of partially and undigested food, and various metabolic wastes deposited from the Malpighian tubules and hindgut [28]. Defecate contains relatively high concentrations of ammonia, allantoin/allantoic acid, and uric acid [29], all of which are absent from regurgitate. Several proteolytic enzymes (e.g., pepsin, trypsin, α-glucosidase, β-galactosidase, β-glucosidase, amylase, fructofuranosidase) released by the midgut epithelia can also be found in feces [30,31]. Enzyme type, quantity, and level of activity are dependent on food composition and most likely fly species. Like with the adult crop, endogenous bacteria and other microorganisms are present in the hindgut, and possibly exogenously acquired microorganisms, presuming that they can survive the highly acid environment of the midgut [29]. Defecatory stains are expected to be chemically similar yet distinct from regurgitate and the original food source. That said Weiss [32] states that it is possible in certain instances for ingested food to pass through the alimentary canal unmodified. In which case, feces, regurgitate, and the original food source may be virtually indistinguishable from each other based on chemical composition. Variable degrees of food processing by an adult fly may also account in part for the wide variation in physical features of defecatory stains.

2.3. Transfer Patterns

Transfer patterns are defined as bloodstain patterns resulting from contact between a wet bloody object (surface) with that of another object or surface [18]. This definition can be extended to that of insects that interact with a corpse or exuded body fluids. For example, adult flies can generate transfer patterns by dragging the abdomen through or across food and then transferring a wet impression to another location. This form of transfer pattern is referred to as translocation [12], and is known to occur with adult flies, cockroaches and ants that frequent crime scenes [11]. Translocation stains typically are asymmetrical linear stains and appear the same color as the food source that was transferred by the insect's body. A second type of transfer artifact is called tarsal tracks. These transfer patterns are formed by adult flies or other insects walking through wet food and leaving impressions of tarsi or pulvilli on another surface. Tarsal tracks produced by adult flies are typically small (<0.2 mm in diameter), round stains that appear the same color as the food source walked through by the insect [12]. These fly footprints are often randomly distributed across the surface of an object but have also been shown to occur in clusters with some species [12]. Since both translocation stains and tarsal tracks are produced from simply walking through or across a food source, the composition of the artifacts will be identical to the original source. It should also be noted that the behaviors of flies and other insects that lead to transfer patterns also alters the morphology of existing stains (wipe patterns) and insect artifacts [19].

2.4. Meconium

Meconium is a creamy, viscous fluid released via the anus of newly emerged adult flies [22]. It is produced during intrapuparial development and represents metabolic waste products that accumulate in the partially formed adult gut. Meconium is thus a form of storage excretion since waste removal is not possible until after the adult emerges from the puparium [28]. As a consequence, storage of nitrogenous end products occurs in non-toxic forms, predominantly as uric acid and allantoin [33], which in turn are the major components of excreted meconium. The presence of uric acid is also largely responsible for the creamy white to yellow coloration of the deposited fluid.

The occurrence of meconium at a crime scene is indicative of a long association between the fly and the remains, since all phases of fly development outside of adult maturation have occurred at that location or close by. Meconium is deposited within minutes of emergence from the puparia but prior to expansion [22]. What this means is that adult flies have yet to gulp air into the crop, so the wings appear shriveled and the abdomen narrow. Consequently, meconium is usually deposited on the substrate near the site of extrication and would be less commonly found on walls or furniture. The size and shape of meconium stains vary based on species but are typically large (>2 mm in diameter), round to asymmetrically round (ellipse or oval) stains that often have long tails. Like with defecatory stains, the tails result from the fly beginning to walk prior to completion of elimination, and thus point in the direction of fly movement.

2.5. Larval Stains

Necrophagous flies deposit unique forms of artifacts at crime scenes that are readily distinguishable from all other forms: eggs and larvae. Juvenile forms of Diptera are never referred to as contaminates or artifacts but they do potentially confound crime scene investigations in at least two ways: through consumption of physical evidence, including the corpse, and by modification of existing stains via transference or deposition of larval fluids. Larval stains generally would not be expected to be confused for human bloodstains or other body fluids, but the activity of maggots potentially can alter the physical appearance and/or chemical composition of existing stains. The latter obviously has the potential to compromise the utility of presumptive or confirmatory tests performed during an investigation. Here, the focus is on fluids derived from larvae since the chemical composition of each type is unique from human body fluids. Two types of larval-derived fluids will be considered, secretions from the oral opening and excretions released from the hindgut via the anus.

2.5.1. Secretions

Larvae from the families Calliphoridae and Sarcophagidae, and to a lesser extent Muscidae, feed on human remains and carrion in larval feeding aggregations or maggot masses [34]. For most species, the aggregations form during the late second to early third stage of larval development. Within the feeding masses, larvae release oral secretions on to the food substrate to promote extra-oral digestion [34]. The food substrate obviously can be human remains, body fluids or stains, and feces. Thus the potential exists for modification of trace evidence by larval secretions. The composition of the secretions is the key to this potential for alteration. Larval salivary glands produce amylase, invertase, proteases, and hydrolases [31], with chymotrypsin functioning as the major orally secreted protease in larvae of *L. cuprina* [35]. Several other species, including *L. sericata*, *C. erythrocephala*, and *Sarconesiopsis magellanica*, release a cocktail of proteases (e.g., trypsin-like, chymotrypsin-like, leucine aminopeptidase, aspartyl proteinase, metalloproteinase, and carboxypeptidase A and B) in secretions that promote digestion of necrotic tissues and other food sources [36,37,38]. Larvae of *L. sericata* also secrete collagenase, although the source of the enzyme may be the midgut rather than salivary glands [39]. Larval secretions also contain a range of antimicrobial compounds [40,41,42], however, only the peptides lucifensin and lucifensin II have been identified thus far [43,44].

Like with meconium, the presence of larval secretory stains suggests a long association with the corpse as only older (late 2nd or 3rd instars) larvae can survive any significant length of time off the remains before returning to feed [45,46]. Such stains are also most likely to occur on non-porous surfaces like tile, wood, or vinyl floors since carpets and fabrics are potentially desiccating conditions to any age larva.

2.5.2. Excretions

Larval feces are broadly comparable to the composition of adult defecate. In practical terms, this means larval feces contains partially and undigested food, various metabolic wastes, digestive enzymes, endogenous bacteria, and the remains of lysed or digested exogenously acquired bacteria [28,47]. Larvae excrete high concentrations of ammonia, nonionic ammonia and allantoin throughout the feeding stages [29,48] and especially when purging the gut prior to pupariation [49,50].

Other nitrogenous products (i.e., uric acid and allantoic acid) are present in feces but in lower concentrations [28,33]. An array of digestive enzymes is released into the lumen of the midgut, including trypsin, chymotrypsin, pepsin, collagenase, lysozyme, β-galactosidase, α-glucosidase, fructofuranosidase, maltase, amylase, and lipases [31,51,52,53], with the potential to pass out of the gut in feces. Indeed, excretions of *L. cuprina* contain several proteases, with chymotrypsin being the most abundant [35]. However, in larvae of *Musca domestica*, only 20% of the enzymes are released in excreta, as an endo-ectoperitrophic circulation in the midgut facilitates enzyme recovery before entering the hindgut [54]. Trypsin and amylase are the most prevalent hydrolases in excretions of *M. domestica* [54], but neither are distinctive from human fluids based on enzymatic activity alone. Larval defecate is also comprised of a number of antimicrobial agents including small peptides (lucifensin and lucifensin II), urea, phenylacetic acid, phenylacetaldehyde, and calcium carbonate [42,43,44].

It is during the postfeeding stage of larval development in which most species will wander from the food to seek refuge to initiate pupariation [50]. Larvae will disperse from the food source and initiate random crawling, during which time the immatures will migrate through existing fluids or stains in their path. This behavior not only distorts stain morphology (wipe pattern) but also alters the chemical composition as transfer of fluids adhering to their body occurs, as does release of excreta from the anal opening. Larval trails are evident from human remains across non-porous surfaces, revealing the initial path of travel of the maggots.

3. METHODS OF DETECTION

A variety of methods have been purported to permit some differentiation of fly artifacts from human bloodstains. While there is limited success with each technique, at present, no empirical methods exist for reliable distinction between insect-derived artifacts and body fluids. Langer and Illes [27] reviewed the techniques available for identifying insect artifacts at crime scenes, as well as delineating the limitations of each method. Since no new developments in methodology have occurred since their review, we will not attempt to duplicate their efforts here. Instead, a brief description of each method will be presented to provide context for areas in need of new research. The methodology is grouped according to the categories proposed by Langer and Illes [27]: visual, contextual, and chemical methods of detection.

3.1. Visual Methods

Visual methods rely on comparative morphology and alternate light detection. The morphological features of fly artifacts, especially because they may be construed as irregular shaped by comparison to human bloodstains, are believed by some to be distinctive, and thus would not be confused by a trained expert in bloodstain pattern analysis with true bloodstains [19]. While this is undoubtedly true in some scenarios, many crime scene analysts do not have a background in forensic entomology to readily recognize fly artifacts [6]. Importantly, subjective analysis is not a satisfactory means for distinguishing fly artifacts from other forms of trace evidence for a number of reasons. For one, the morphology of regurgitate can be distinct from that of defecatory stains for some species [13,19,20], but not others [10,12,17]. A semi-quantifiable method has been proposed as one means of visually identifying fecal stains [13]. The method depends on the ratio of the length of stain tail to the length of stain body, which if greater than one, supposedly excludes bloodstains. Thus, the method relies on a process of elimination of stain suspects to identity defecatory stains. One problem with this approach is that the ratios generated to evaluate this technique are from limited sample sizes at crime scenes and from laboratory tests that did not utilize blood as a food source for the flies, and only one species of fly was tested for validation [13]. However, the authors have reported using the technique many times since and remain convinced of its validity in recognizing potential defecatory stains [25]. A second concern is that for several species, tails are commonly absent from defecatory spots, yielding fecal and regurgitate stains that are indistinguishable from each other [10,12,17]. A third issue is that the underlying premise that a tail length to body ratio exceeding one excludes all forms of human bloodstains is not correct [55]. In reality, fly artifacts are highly variable in size, color, and morphology due to unique species behaviors, size of blood meal, and time taken to consume the meal, as well as being dependent on the physical surfaces on which they have been deposited [19].

A second method of visual identification of fly artifacts involves the use of alternate light detection. The method has been used with only two species (*C. vicina* and *L. sericata*), in which expelled defecate could be visualized by alternate lighting at 465 nm with an orange contrast filter [9,10]. The authors indicate that neither regurgitate or human blood fluoresced under the same conditions. However, this method is not entirely satisfactory since in the absence of tails and depending on fly species, regurgitate and defecatory stains can be very difficult to distinguish from each other based on morphology [6,17]. The alternate light technique also does not make a distinction between other forms of body fluids that may be present at a crime scene, and there have been no reports as to whether the method is effective at detection of translocation stains or tarsal tracks. Also lacking is an understanding of the mechanisms that lead to fluorescence of defecatory but not regurgitate stains and human blood. Fujikawa et al. [10] speculate that the presence of urea in defecate accounts for detection at 465 nm. This seems unlikely in that only trace amounts of urea are present in adult excreta [28,33], human blood contains small quantities of urea yet did not fluoresce [10], and other body fluids that do not contain nitrogenous end products can be detected by the same technique [8].

3.2. Contextual Methods

Contextual methods of fly artifact detection rely on visual analysis coupled with where the insect stains occur in relation to one another and with respect to human body fluid stains [27]. There are two ways to consider contextual analysis of suspect stains: one is to examine the stains based on physical location with respect to the crime scene or other stains, and the other is to compare the directionality of multiple stains located in close proximity. The former is relevant to stains being located in seemingly unusual or atypical circumstances by comparison to other trace evidence. For example, adult flies display positive phototaxis, so they are attracted to artificial and natural light sources. As a consequence, fly artifacts may be deposited on lampshades, light fixtures, windows, curtains, window shades, windows casings, and other objects in close proximity to the light sources [11]. The net effect can be that stains consistent (e.g., morphology and/or chemistry) with human blood are present in locations not consistent with other physical or trace evidence associated with a crime scene or pathology reports [27]. Necrophagous flies will also forage for other food sources, most prominently located in kitchen and food storage areas. Again, fly artifacts may be deposited in these locations, which would be especially suspicious if no other evidence is found in such areas. Despite the irregularity of locations, trained bloodstain analysts only can rely on subjective determination to include or exclude these stains in their analysis.

When multiple stains are present with some showing directionality not consistent with other stains, the possibility exists that insect artifacts are intermixed with human bloodstains [27]. An alternative explanation is that cast off patterns are present from multiple impacts or trauma events. In the latter scenario, multiple patterned bloodstains are recognizable to trained bloodstain analysts who could group the stains based on consistent directional patterns and angles of impact [56]. However, when insect artifacts are intermixed with castoff or impact bloodstains, the insect-derived stains can only be identified based on discretionary interpretation by individual analysts. Fly activity may be suspected based on the presence of stains with random directionality, round stains intermixed with bloodstains with tails that can be grouped based on consistent directionality, or by the occurrence of irregular shaped stains typical of translocation, defecatory stains with distinctive tails (i.e., sperm-shaped, tadpole shaped, tear-drop shaped) or dome shaped craters. Thus, visual analysis is used to detect irregular or atypical stains in relation to other stains or evidence found at the scene. At present, there is no means to confirm that suspected fly artifacts are in fact derived from necrophagous flies or other insects.

3.3. Chemical Methods

Chemical analyses in the form of presumptive and confirmatory tests provide the most definitive evidence that fly artifacts cannot be reliably distinguished from human bloodstains or other body fluids. This assertion is based on the idea that fly artifacts are the direct result of adult flies' interaction with a corpse and/or associated body fluids. The most obvious examples are regurgitate and defecatory stains produced after consumption of human blood. Presumptive chemical tests designed to detect animal blood will test positive for both fly artifacts and true human bloodstains [9,10,13]. This is true regardless of the type of presumptive blood test employed. That said Durdle et al. [15] demonstrated that some differentiation could be made with presumptive tests used on artifacts (regurgitate and defecate) produced by adults of *L. cuprina* following feeding on human blood or semen if the stains were relatively fresh (3-days-old) but not if 2-weeks-old nor if flies consumed saliva. Confirmatory tests for semen and saliva demonstrated similar trends in that fresh but not older artifacts could be distinguished from the original food sources. Confirmatory tests designed for human blood do not permit distinction between fly artifacts and bloodstains, and DNA typing does not overcome this limitation since complete DNA profiles of the victim or offender have been obtained from regurgitate and defecatory stains [6,16]. Despite some limited differentiation using chemical analysis, Durdle et al. [15] concluded that presumptive and confirmatory testing could not be used reliably to detect fly artifacts. However, in combination with visual analysis, the utility of chemical analysis may potentially improve.

Translocation and tarsal tracks have not been tested by chemical analysis, but based on the mechanism of production, should not be distinctive from the original food source. Necrophagous flies generally do not consume saliva and semen found at crime scenes, and even when they are fed upon, the resulting artifacts are often difficult to visualize [6]. The latter poses a potential serious problem since fly artifacts derived from such fluids contain human DNA and thus are a source of extraneous DNA whose origin would be unknown since detection of the stain is limited.

4. NEW RESEARCH

The inability to consistently and reliably distinguish insect artifacts from human bloodstains and other body fluids represents the biggest issue with respect to entomological contaminants at crime scenes. There has been modest success with a few methods designed for visual, contextual, or chemical analysis of fly artifacts, but none are satisfactory based on several limitations. The deficiencies include a lack of reliability, no single technique is suitable for all fly species, none make a distinction from other forms of body fluids that may also be present at crime scenes, all are presumptive not confirmatory tests, assessment of artifact morphology is dependent on a very small pool of forensic experts, and very few forensically important species known worldwide have been examined by the reported methods for discerning fly artifacts from human bloodstains and other bodily fluids. The latter makes it very difficult to come to any consensus on the typical classification of fly artifacts or accurate methods of detection. Durdle et al. [19] suggested that the use of two methods in conjunction with one another (i.e., presumptive blood testing coupled with visual analysis) may improve the precision in distinguishing fly artifacts from human bloodstains. Nonetheless, the techniques alone or in combination still should be viewed as inconsistent and non-quantifiable, especially in terms of the visual analysis component. A lack of diagnostic tests for the identification of insect artifacts at a crime scene means that only subjective interpretation is currently used to distinguish fly evidence from bloodstains. The take home message is that at present, insufficient data and methodology are available to make consistently reliable and quantifiable distinctions between insect artifacts and stains from human body fluids. This obviously means that new research is needed to overcome the stated deficiencies and develop new methods for detection of fly artifacts.

4.1. Chemical Analysis Based on Composition of Fly Fluids

The majority of presumptive and chemical tests available for forensic serological analysis at crime scenes relies on chemical testing. In the context of human body fluid stains, discernment is based on unique chemical properties of each type of fluid that minimally leads to broad classification. Confirmatory tests, especially in the cases of human blood, semen, and saliva, depend on immunoassays employing polyclonal or monoclonal antibodies that recognize unique human-specific antigens [57]. In this respect, very little research has been done to analyze the chemical composition of fly artifacts, aside from DNA analysis for the purpose of identifying the source of blood, or inferences made based on the results of presumptive blood tests [6,9,10,13]. Rivers et al. [26] demonstrated that regurgitate stains deposited by *P. terraenovae* possess at least three (trypsin-like, chymotrypsin-like, and pepsin-like) digestive enzymes that were also found in the crop of the adult fly, independent of the food source. These observations point to further research into the potential of fly digestive enzymes as the bases for new chemical methods of detection of insect artifacts. Most likely presumptive tests in the form of enzymatic assays would not be sufficient to distinguish artifacts from human body fluids, as many fly enzymes overlap in substrate specificity with vertebrate enzymes [31]. However, in several instances, though enzyme functionality is similar to vertebrate counterparts, enzyme structure is substantially different [58,59,60]. This is especially true for the various forms of trypsin (i.e., earlier, late) that are produced in the midgut of several Diptera [61]. Consequently, fly digestive enzymes, potentially from all feeding stages that interact with human remains and body fluids, could be used as antigens for the development of immunological tools used in confirmatory tests. Similarly, the small antimicrobial peptides lucifensin and lucifensin II are potential candidates for antibody development to use in immunoassays that recognize larval stains or distortions of existing human body fluid stains caused by maggot activity. Further research is needed to determine if adult labellar glands synthesize these peptides as well, and if so, this would greatly increase the utility of the peptides in development of immunological diagnostic tools.

Two potential limitations of any of the described potential antigens are that (1) they likely would not be able to distinguish regurgitate from defecatory stains; and (2) would also not recognize translocation and tarsal tracks as being separate stains from the original source, meaning human body fluids. In terms of the former, this should be viewed as a minor weakness since the primary need is to differentiate fly artifacts from human bloodstains in a reliable and quantifiable manner. Much less information is derived from knowing precisely what type of fly artifact is present.

The composition of defecatory stain offers insight into the possible development of new presumptive and confirmatory tests to detect fly fecal stains. Uric acid and allantoin are present in the excreta of adult and larval flies, although the concentration varies based on developmental stage [28,33]. In the absence of hyperuricemia condition, human blood contains only trace amounts of uric acid, and humans are incapable of naturally synthesizing allantoin [62,63]. Thus, like occurs with acid phosphatase in presumptive testing of human semen [64], concentration (high levels) differences of uric acid in irregular shaped 'bloodstains' may be indicative of defecatory stains. Similarly, allantoin detection through enzymatic assays [65] could potentially be developed into a confirmatory test for fly defecate.

4.2. Molecular Methods

The development of molecular methods for identifying fly artifacts would allow for more conclusive determination of artifacts than most current methods. Design of these methods will likely depend on identification of DNA sequences present in the artifacts, rather than RNA or protein as these macromolecules are more difficult to detect in extremely small quantities and require additional manipulation before analysis. Since human blood stains, as well as fly regurgitate and defecatory stains could all have human DNA present, distinguishing between these types of artifacts at a crime scene would benefit from being able to identify insect DNA within artifacts. Fly gut epithelial cells are routinely shed over the lifespan of the organism leading to frequent intestinal regeneration [66,67]. While much of this research comes from the study of *Drosophila melanogaster* [68,69,70], there is a high likelihood that this developmental process occurs with many species of necrophagous Diptera. Research into the existence of epithelial cells shed from the gut in either regurgitate or excrement, could determine the presence of insect DNA in these artifacts. Identification of this insect DNA could be performed by looking for high copy number DNA sequences within the fly genome following genomic sequencing. The use of high copy number sequences, such as ribosomal genes and transposons, to make an organism determination would result in detection that is more feasible given the low number of cells likely present in an individual fly artifact. Additionally, quantitative polymerase chain reaction (qPCR) could be used to amplify insect DNA from artifacts as a means for organism detection. The use of degenerative primers for qPCR would allow for an assay in which the specific fly species does not need to be known prior to assay initiation, making the assay easy to use and less costly than trying primers for each specific fly species in an attempt to determine which flies were present.

4.3. Fly Microbiome

The endogenous microbiome of necrophagous flies is an attractive area of study to differentiate fly artifacts from human bloodstains. Investigation into the microbial flora of insects commonly found at crime scenes may lead to a catalog of microbes that could be identified within fly artifacts, and presumably absent from human bloodstains and other bodily fluids, at least at the time of death. Detection could be accomplished using DNA extraction followed by bacterial 16S rDNA tag encoded FLX-titanium amplicon pyrosequencing (bTEFAP), generating short DNA sequence reads [71,72]. These DNA sequences could then be classified by bacteria taxonomy based on sequence. Analysis of the normal microbiome of two calliphorids, *L. sericata* and *L. cuprina*, has been undertaken for various life stages [72]. The feasibility of obtaining enough microbial DNA from fly artifacts for analysis will need to be studied, but identification of even a low number of microbial sequences from adult flies common to a crime scene location may be enough to distinguish the spot from human bloodstains. With this type of analysis, the microbiome likely will change with timing of colonization and developmental stage for both for the fly and microbes. The anticipated result is deposition of endogenous fly microbes as well as microbes newly ingested from a decaying corpse in fly artifacts [73].

5. CONCLUSIONS

Necrophagous Diptera that colonize a corpse, carrion, or an animal or person with open wounds may confound crime scenes by altering body fluid stains. Both adults and larvae have the potential to distort existing body fluid stains or depositing contaminants in the form of artifacts that are morphologically similar in size, color and shape to human body fluids. Adult flies deposit regurgitate and defecatory stains released from opposing ends of the digestive tract. Meconium is released almost immediately upon emergence from puparia, but is typically restricted to locations near the site of extrication. Transfer patterns are produced as either translocation (from dragging body parts) or tarsal tracks, which are literally fly footprints. By contrast, crawling larvae can distort existing stains as well as yield transfer patterns. Additional larval artifacts can be derived from the alimentary canal in the form of secretions or excretions. At present, fly artifact are detected based on morphological and contextual criteria, which are not consistently reliable in distinguishing insect stains from human body fluids. New research should focus on the development of diagnostic tools for distinguishing fly artifacts from human body stains based on the chemical composition of fly contaminants. This idea is based on the premise that each type of fly-derived fluid contains chemical constituents unique to the condition that produced them.

ACKNOWLEDGMENTS

A portion of the research discussed in the review was supported in part by a Lucas Grant from the Forensic Sciences Foundation, Inc. and a senior sabbatical award from Loyola University Maryland (David Rivers).

CONFLICTS OF INTEREST

The authors declare no conflict of interest.

REFERENCES

1. Rivers, D.B.; Dahlem, G.A. The Science of Forensic Entomology; Wiley-Blackwell: West Sussex, UK, 2014.
2. Villet, M.H.; Richards, C.S.; Midgley, J.M. Contemporary precision, bias and accuracy of minimum post-mortem intervals estimated using development of carrion-feeding insects. In Current Concepts in Forensic Entomology; Amendt, J., Campobasso, C.P., Goff, M.L., Grassberger, M., Eds.; Springer: Dordrecht, The Netherlands, 2010; pp. 109–137.
3. Byrd, J.H.; Castner, J.L. (Eds.) Forensic Entomology: The Utility of Arthropods in Legal Investigations, 2nd ed.; CRC Press: Boca Raton, FL, USA, 2010.
4. Anderson, G.S. Factors that influence insect succession on carrion. In Forensic Entomology: The Utility of Using Arthropods in Legal Investigations, 2nd ed.; Byrd, J.H., Castner, J.L., Eds.; CRC Press: Boca Raton, FL, USA, 2010; pp. 201–250.
5. Goff, M.L.; Lord, W.D. Insect as toxicological indicator and the impact of drugs and toxin on insect development. In Forensic Entomology: The Utility of Using Arthropods in Legal Investigations, 2nd ed.; Byrd, J.H., Castner, J.L., Eds.; CRC Press: Boca Raton, FL, USA, 2010; pp. 427–434.
6. Kulstein, G.; Amendt, J.; Zehner, R. Blow fly artifacts from blood and putrefaction fluid on various surfaces: A source for forensic STR typing. Entomol. Exp. Appl. 2015, 157, 255–262.
7. LaGoo, L.; Schaeffer, L.S.; Szymanski, D.W.; Smith, R.W. Detection of gunshot residue in blowfly larvae and decomposing porcine tissue using inductively coupled plasma mass spectrometry (ICP-MS). J. Forensic Sci. 2010, 55, 624–632.
8. James, S.H.; Nordby, J.J.; Bell, S. (Eds.) Forensic Science: An Introduction to Scientific and Investigative Techniques, 4th ed.; CRC Press: Boca Raton, FL, USA, 2014.
9. Fujikawa, A.; Barksdale, L.; Carter, D.O. Calliphora vicina (Diptera: Calliphoridae) and their ability to alter the morphology and presumptive chemistry of bloodstain patterns. J. Forensic Ident. 2009, 59, 502–512.

10. Fujikawa, A.; Barskdale, L.; Higley, L.G.; Carter, D.O. Changes in the morphology and presumptive chemistry of impact and pooled bloodstain patterns by Lucilia sericata (Meigen) (Diptera: Calliphoridae). J. Forensic Sci. 2011, 56, 1315–1318.
11. Parker, M.A.; Benecke, M.; Byrd, J.H.; Hawkes, R.; Brown, R. Entomological alteration of bloodstain evidence. In Forensic Entomology: The Utility of Using Arthropods in Legal Investigations, 2nd ed.; Byrd, J.H., Castner, J.L., Eds.; CRC Press: Boca Raton, FL, USA, 2010; pp. 539–580.
12. Rivers, D.B.; McGregor, A. Morphological features of regurgitate and defecatory stains deposited by five species of necrophagous flies are influenced by adult diets and body size. J. Forensic Sci. 2017, in press.
13. Benecke, M.; Barksdale, L. Distinction of bloodstain patterns from fly artifacts. Forensic Sci. Int. 2003, 137, 152–159.
14. Durdle, A.; Mitchell, R.J.; van Oorschot, R.A.H. The human DNA content in artifacts deposited by the blowfly Lucilia cuprina fed human blood, semen, and saliva. Forensic Sci. Int. 2013, 233, 212–219.
15. Durdle, A.; Mitchell, R.J.; van Oorschot, R.A.H. The use of forensic tests to distinguish blowfly artifacts from human blood, semen, and saliva. J. Forensic Sci. 2015, 60, 468–470.
16. Durdle, A.; Mitchell, R.J.; van Oorschot, R.A.H. The change in human DNA content over time in the artefacts of the blowfly Lucilia cuprina (Meigen) (Diptera: Calliphoridae). Forensic Sci. Int. 2011, 3, e289–e290.
17. Striman, B.; Fujikawa, A.; Barksdale, L.; Carter, D.O. Alteration of expirated bloodstain patterns by Calliphora vicina and Lucilia sericata (Diptera: Calliphoridae) through ingestion and deposition of artifacts. J. Forensic Sci. 2011, 53, S123–S127.
18. Scientific Working Group on Bloodstain Pattern Analysis (SWGSTAIN). Recommended Terminology; I.A.B.P.A. Newsletter: 2008. Available online: http://iabpa.org/uploads/files/iabpa%20publications/June%202008%20News.pdf (accessed on 10 February 2017).
19. Durdle, A.; van Oorschot, R.A.H.; Mitchell, R.J. The morphology of fecal and regurgitation artifacts deposited by the blow fly Lucilia cuprina fed a diet of human blood. J. Forensic Sci. 2013, 58, 897–903.
20. Zuhu, R.M.; Supriyani, M.; Omar, B. Fly artifact documentation of Chrysomya megacephala (Fabricius) (Diptera: Calliphoridae)—A forensically important blowfly species in Malaysia. Trop. Biomed. 2008, 25, 17–22.
21. Stoffolano, J.G.; Haselton, A.T. The adult dipteran crop: A unique and overlooked organ. Annu. Rev. Entomol. 2013, 58, 205–225.
22. Greenberg, B.; Zunich, J.C. Entomology and the Law; Cambridge University Press: Cambridge, UK, 2002.
23. Stoffolano, J.G.; Acaron, A.; Conway, M. "Bubbling" or droplet regurgitation in both sexes of adult Phormia regina (Diptera: Calliphoridae) fed various concentrations of sugar and protein solutions. Ann. Entomol. Soc. Am. 2008, 101, 964–970.
24. James, S.H.; Sutton, T.P. Medium- and high-velocity impact blood spatter. In Interpretation of Bloodstain Evidence at Crime Scenes, 2nd ed.; James, S.H., Eckert, W.G., Eds.; CRC Press: Boca Raton, FL, USA, 1998; pp. 59–83.
25. Benecke, M.; Barksdale, L. In response to: "Commentary on: Mark Benecke and Larry Barksdale, Distinction of bloodstain patterns from fly artifacts: Forensic Science International 137 (2003) 152–159" [Forensic Sci. Int. 149 (2/3) (2005) 293–294]. Forensic Sci. Int. 2007.
26. Rivers, D.B.; Acca, G.; Fink, M.; Brogan, R.; Schoeffield, A. Spatial characterization of proteolytic enzyme activity in the foregut region of the adult necrophagous fly, Protophormia terraenovae. J. Insect Physiol. 2014, 67, 45–55.
27. Langer, S.V.; Illes, M. Confounding factors of fly artefacts in bloodstain pattern analysis. Can. Soc. Forensic Sci. J. 2015.
28. Chapman, R.F. The Insects: Structure and Function, 4th ed.; Cambridge University Press: Cambridge, UK, 1998.

29. Wigglesworth, V.B. The Principles of Insect Physiology, 7th ed.; Chapman and Hall: London, UK, 1982.
30. Evans, W.A.L. Studies on the digestive enzymes of the blowfly Calliphora erythrocephala: I. The carbohydrates. Exp. Parasitol. 1956, 5, 191–206.
31. Terra, W.R.; Ferreira, C. Insect digestive enzymes: Properties, compartmentalization and function. Comp. Biochem. Physiol. B 1994, 109, 1–62.
32. Weiss, M.R. Defecation behavior and ecology of insects. Annu. Rev. Entomol. 2006, 51, 635–661.
33. O'Donnell, M.J.; Maddrell, S.H.P.; Gardiner, B.O.C. Transport of uric acid by the Malpighian tubules of Rhodnius prolixus and other insects. J. Exp. Biol. 1983, 103, 169–184.
34. Rivers, D.B.; Thompson, C.; Brogan, R. Physiological trade-offs of forming maggot masses by necrophagous flies on vertebrate carrion. Bull. Entomol. Res. 2011, 101, 599–611.
35. Casu, R.E.; Eisemann, C.H.; Vuocolo, T.; Tellam, R.L. The major excretory/secretory protease from Lucilia cuprina larvae is also a gut digestive protease. Int. Parasitol. 1996, 26, 623–628.
36. Vistnes, L.M.; Lee, R.; Ksander, G.A. Proteolytic activity of blowfly larvae secretions in experimental burns. Surgery1981, 90, 835–841.
37. Chambers, L.; Woodrow, S.; Brown, A.P.; Harris, P.D.; Philips, D.; Hall, M.; Church, J.C.T.; Pritichard, D.I. Degradation of extracellular matrix components by defined proteinases from the greenbottle larvae Lucilia sericata used for the clinical debridement of non-healing wounds. Br. J. Dermatol. 2003, 148, 14–23.
38. Pinilla, Y.T.; Moreno-Perez, D.A.; Patarroyo, M.A.; Bello, F.J. Proteolytic activity regarding Sarconesiopsis magellanica (Diptera: Calliphoridae) larval excretions and secretions. Acta Trop. 2013, 128, 686–691.
39. Ziffren, S.E.; Heist, H.E.; May, S.C.; Womack, N.A. The secretion of collagenase by maggots and its implication. Ann. Surg. 1953, 138, 932–934.
40. Kerridge, A.; Lappin-Scott, H.; Stevens, J.R. Antibacterial properties of larval secretions of the blowfly, Lucilia sericata. Med. Vet. Entomol. 2005, 19, 333–337.
41. Bexfield, A.; Nigam, Y.; Thomas, S.; Ratcliffe, N.A. Detection and partial characterisation of two antibacterial factors from the excretions/secretions of the medicinal maggot Lucilia sericata and their activity against methicillin-resistant Staphylococcus aureus (MRSA). Microbes Infect. 2004, 6, 1297–1304.
42. Poppel, A.-K.; Vogel, H.; Wiesner, J.; Vilcinskas, A. Antimicrobial peptides expressed in medicinal maggots of the blow fly Lucilia sericata show combinatorial activity against bacteria. Antimicrob. Agents Chemother. 2015, 59, 2508–2514.
43. Čeřovský, V.; Žďárek, J.; Fučík, V.; Monincová, L.; Voburka, Z.; Bém, R. Lucifensin, the long-sought antimicrobial factor of medicinal maggots of the blowfly Lucilia sericata. Cell. Mole. Life Sci. 2010, 67, 455–466.
44. El Shazely, B.; Veverka, V.; Fučík, V.; Voburka, Z.; Žďárek, J.; Čeřovský, V. Lucifensin II, a defensin of medicinal maggots of the blowfly Lucilia cuprina (Diptera: Calliphoridae). J. Med. Entomol. 2013, 50, 571–578.
45. Greenberg, B. Flies as forensic indicators. J. Med. Entomol. 1991, 28, 565–577.
46. Christopherson, C.; Gibo, D.L. Foraging by food deprived larvae of Neobellieria bullata (Diptera: Sarcophagidae). J. Forensic Sci. 1997, 42, 71–73.
47. Greenberg, B. Model for destruction of bacteria in the midgut of blow fly maggots. J. Med. Entomol. 1968, 5, 31–38.
48. Guerrini, V.H. Excretion of ammonia by Lucilia cuprina larvae suppresses immunity in sheep. Vet. Immunol. Immunopathol. 1997, 56, 311–317.
49. Fraenkel, G.; Bhaskaran, G. Pupariation and pupation in cyclorrhaphous flies (Diptera): Terminology and interpretation. Ann. Entomol. Soc. Am. 1973, 66, 418–422.
50. Greenberg, B. Behavior of postfeeding larvae of some Calliphoridae and a muscid (Diptera). Ann. Entomol. Soc. Am.1990, 83, 1210–1214.

51. Hobson, R.P. On an enzyme from blow-fly larvae (Lucilia sericata) which digests collagen in alkaline solution. Biochem. J. 1931, 25, 1458–1463.
52. Waterhouse, D.F.; Irzkiewicz, H. An examination of proteolytic enzymes from several insects for collagenase activity. J. Insect Physiol. 1957, 1, 18–22.
53. Pendola, S.; Greenberg, B. Substrate-specific analysis of proteolytic enzymes in the larval midgut of Calliphora vicina. Ann. Entomol. Soc. Am. 1975, 68, 341–345.
54. Espinoza-Fuentes, F.P.; Terra, W.R. Physiological adaptations for digesting bacteria. Water fluxes and distribution of digestive enzymes in Musca domestica larval midgut. Insect Biochem. 1987, 17, 809–817.
55. Ristenblatt, R.R., III; Pizzola, P.A.; Shaler, R.C.; Sorkin, L.N. Commentary on: Mark Benecke and Larry Barksdale, Distinction of bloodstain patterns from fly artifacts. Forensic Sci. International. 137 (2003) 152–159. Forensic Sci. Int.2005, 14, 293–294.
56. Bevel, T.; Gardner, R. Bloodstain Pattern Analysis: With an Introduction to Crime Scene Reconstruction, 3rd ed.; CRC Press: Boca Raton, FL, USA, 2008.
57. Li, R. Forensic Biology; CRC Press: Boca Raton, FL, USA, 2015.
58. Dametto, M.; David, A.P.; Azzolini, S.S.; Campos, I.T.N.; Tanaka, A.M.; Gomes, A.; Andreotti, R.; Tanaka, A.S. Purification and characterization of a trypsin-like enzyme with fibrinolytic activity present in the abdomen of horn fly, Haematobia irritans irritans (Diptera: Muscidae). J. Protein Chem. 2000, 19, 515–521.
59. Terra, W.R. Physiology and biochemistry of insect digestion: An evolutionary perspective. Braz. J. Med. Biol. Res. 1988, 21, 675–734.
60. Padilha, M.H.P.; Pimentel, A.C.; Ribeiro, A.F.; Terra, W.R. Sequence and function of lysosomal and digestive cathepsin D-like proteinases of Musca domestica midgut. Insect Biochem. Mol. Biol. 2009, 39, 782–791.
61. Barrilas-Mury, C.V.; Noriega, F.G.; Wells, M.A. Early trypsin activity is part of the signal transduction system that activates transcription of the late trypsin gene in the midgut of the mosquito, Aedes aegypti. Insect Biochem. Mol. Biol.1995, 35, 241–246.
62. Young, E.G.; Wentworth, H.P.; Hawkins, W.W. The absorption and excretion of allantoin in mammals. J. Pharmacol. Exp. Ther. 1944, 81, 1–9.
63. Fujiwara, S.; Noguchi, T. Degradation of purines: Only ureidoglycollate lyase out of four allantoin-degrading enzymes is present in mammals. Biochem. J. 1995, 312 Pt 1, 315–318.
64. Elkins, K.M. Forensic DNA Biology: A Laboratory Manual; Academic Press: New York, NY, USA, 2013.
65. Young, E.G.; Conway, C.F. On the estimation of allantoin by the Rimini-Schruver reaction. J. Biol. Chem. 1942, 142, 839–853.
66. Li, H.; Jasper, H. Gastrointestinal stem cells in health and disease: From flies to humans. Dis. Model Mech. 2016, 9, 487–499.
67. Buchon, N.; Broderick, N.A.; Lemaitre, B. Gut homeostasis in a microbial world: Insights from Drosophila melanogaster. Nat. Rev. Microbiol. 2013, 11, 615–626.
68. Tauc, H.M.; Tasdogan, A.; Pandur, P. Isolating intestinal stem cells from adult Drosophila midguts by FACS to study stem cell behavior during aging. J. Vis. Exp. 2014.
69. Jiang, H.; Patel, P.H.; Kohlmaier, A.; Grenley, M.O.; McEwen, D.G.; Edgar, B.A. Cytokine/Jak/Stat signaling mediates regeneration and homeostasis in the Drosophila midgut. Cell 2009, 137, 1343–1355.
70. Jiang, H.; Edgar, B.A. Intestinal stem cells in the adult Drosophila midgut. Exp. Cell Res. 2011, 317, 2780–2788.
71. Dowd, S.E.; Callaway, T.R.; Wolcott, R.D.; Sun, Y.; McKeehan, T.; Hagevoort, R.G.; Edrington, T.S. Evaluation of the bacterial diversity in the feces of cattle using 16S rDNA bacterial tag-encoded FLX amplicon pyrosequencing (bTEFAP). BMC Microbiol. 2008.
72. Singh, B.; Crippen, T.L.; Zheng, L.; Fields, A.T.; Yu, Z.; Ma, Q.; Wood, T.K.; Dowd, S.E.; Flores, M.; Tomberlin, J.K.; et al. A metagenomic assessment of the bacteria

associated with Lucilia sericata and Lucilia cuprina (Diptera: Calliphoridae). Appl. Microbiol. Biotechnol. 2015, 99, 869–883.

73. Javan, G.T.; Finley, S.J.; Can, I.; Wilkinson, J.E.; Hanson, J.D.; Tarone, A.M. Human thanatomicrobiome succession and the time since death. Sci. Rep. 2016.

CHAPTER 5

Ant-Mimicking Spiders: Strategies for Living with Social Insects

Fadia Sara Ceccarelli

Department of Environmental Sciences, University of Basel, 4056 Basel, Switzerland

ABSTRACT

Mimicry is a fascinating topic, in particular when viewed in terms of selective forces and evolutionary strategies. Mimicry is a system involving a signaller, a signal receiver, and a model and has evolved independently many times in plants and animals. There are several ways of classifying mimicry based on the interactions and cost-benefit scenarios of the parties involved. In this review, I briefly outline the dynamics of the most common types of mimicry to then apply it to some of the spider-ant associative systems known to date. In addition, this review expands on the strategies that ant-associating (in particular ant-mimicking) spiders have developed to minimise the costs of living close to colonies of potentially dangerous models. The main strategy that has been noted to date is either chemical mimicry or actively avoiding contact with ants. If these strategies warrant protection for the spider (living close to potentially dangerous models), then the benefits of ant associations would outweigh the costs, and the association will prevail.

1. INTRODUCTION

The phenomenon of mimicry has intrigued numerous biologists, prompting studies from natural history to behaviour, ecology, evolution, and most recently genomics, to name but a few [1]. Perhaps mimicry so readily attracts attention because it is an evident example of natural selection in action. Mimicry—or the resemblance of one organism (or certain aspects of) to another, taxonomically unrelated one—almost always involves three parties: the signaller (mimic), the signal receiver (or operator), and the model. The mimics in these cases must have a selective advantage over nonmimics, and therefore the particular phenotype is fixed in these populations. The classification of mimicry largely depends on the functions of the parties involved and has, based on this scheme, been subdivided down to 40 theoretical classes, or types of mimicry [2], though the focus is generally on the most common types: Batesian, Müllerian, and aggressive mimicry.

Batesian mimicry, named after H. W. Bates, pioneer in the study of mimicry in Amazonian butterflies [3], is defined by a palatable mimic gaining protection from predators (the signal receiver in this case), by resembling a noxious or unpalatable model organism. In Müllerian mimicry, the line of "palatability" between mimic and model is less clear, with emphasis being placed on a certain phenotype of various organisms being reinforced and acting as a deterrent for predators. A third type of mimicry commonly encountered in nature is aggressive mimicry, so-called because the mimic, rather than gaining protection from potential predators, more easily gains access to resources or prey (sometimes the model itself) through its resemblance to another organism. Although many cases of mimicry can easily be categorised, sometimes an organism displays different strategies, either at the same time or at different stages of its life, such as the cuckoo which was found to be a Batesian mimic as an adult, and an aggressive mimic in other birds' nests [4].

In Batesian mimicry, the mimic is under predator-mediated selection thus resembling a noxious or unpalatable model, whereas traits of aggressive mimics are under pressure to deceive their prey. This means that the sensory channel of the receiver (be it visual, chemosensory, or other) greatly influences the evolution of the mimic [5]. In cases where learning by the signal receiver is involved, it is also important that the mimics do not outnumber the models and that both models and mimics live in sympatry [6]. Be it for protection from predators or access to resources, mimicry has arisen numerous times throughout animals and plants as a recurrent evolutionary stable strategy [6]. This is evidence for strong selection for traits associated with mimicry, where the fitness of the mimic is expected to increase with a closer resemblance to the model [7, 8]. Studies based on theoretical population genetics have modelled Batesian mimicry traits and polymorphism within populations [9, 10]. The fact that Batesian mimicry may be a costly trait must also be considered together with an increased number of parameters such as the cognitive constraints of the signal receivers [11, 12]. Selection pressure on mimics to resemble a model very much depends on the visual system of the receiver [5]. In the particular case of Batesian mimicry, where mathematical models predict greater protection from predators with increasing resemblance to the model organism, the main question that arises is why are there still "imperfect" mimics or those that bear only a slight resemblance to any one model? One explanation given is that the term "imperfect" is subjective, dependent on the signal receiver; what may appear imperfect to a human observer may in fact be seen otherwise by potential predators [13]. Alternatively, an imperfect mimic may be an intermediate phenotype or one of polymorphism [14]. Certain conditions may relax the selection pressure towards a "perfect" phenotype, for example, if the model is very noxious [15] or if behavioural traits reenforce morphology [16]. The selection force towards one "perfect" phenotype is countered by polymorphism, which may arise due to kin selection [17], in some cases the potential cost of being too conspicuous [18] or through selection from receivers with opposing predatory preferences [19].

Mimicry occurs in all forms of terrestrial and aquatic plant and animal life [6]. For example, among vertebrates, marine fishes count with at least 98 cases of mimicry, including Batesian, Müllerian, aggressive, and social (or cases where the mimic aggregates with the school of models) [20]. Perhaps the most diverse and varied forms of mimicry can be found in arthropods, due to their impressive diversity resulting from relatively short generation times, which increase the recombination events, which in turn allow for more genetic diversity. Among terrestrial arthropods, ants are a common model system [21, 22]. Here, I intend to focus on an exceptional group of arthropods, namely, the spiders, and their varied forms of ant mimicry. Even though the majority of spiders are web builders [23], the most striking examples of ant associations can be found in cursorial spiders. Thorough and up-to-date reviews of ant-mimicry in spiders already exist [24–27], so my aim here is not to replicate the information found in these papers, but rather to focus on the various strategies that can be found in these spiders minimising the costs and maximising the benefits of living with or close to ants. I will do this by first talking briefly about ant association and then introducing various examples of benefits and costs to the spiders. Throughout, ant-mimicry will refer to cases of morphological and/or chemical mimicry and "ant associations" include mimics as well as spiders that do not mimic ants but nevertheless gain some advantage living close to ant colonies.

2. ANT ASSOCIATIONS IN ARTHROPODS

Being social insects, ants form large colonies with numerous individuals, thus satisfying the condition of mimicry where any mimic should be at lower densities than the model [6]. For the purpose of Batesian mimicry, ants are also good model organisms because they are unpalatable for many other animals due to characteristics, or combinations thereof, such as formic acid, stings, strong mandibles that bite, and in general an aggressive nature [21, 22, 28]. So acquiring morphological and/or behavioural resemblance to ants confers a certain degree of protection from predation to otherwise palatable arthropods.

Morphological and/or behavioural resemblance to ants, also known as myrmecomorphy, has evolved at least 70 times in more than 2000 described species belonging to 54 arthropod families in groups such as spiders, plant bugs, and staphylinid beetles [21]. In spiders alone, myrmecomorphy can be found in numerous species belonging to 13 different families [24, 25]. Myrmecomorphic spiders have morphological and/or behavioural modifications that increase their resemblance to ants. These include a generally narrower body and longer legs compared to other spiders: at times a constricted carapace or abdomen giving the impression of a three- instead of two-segmented body. The cuticular surface of myrmecomorphic spiders is often strikingly similar to that of their model ant species as well, including hairs and coloration and fake eye spots. As spiders have four pairs of legs while ants have three and one pair of antennae, myrmecomorphic spiders often raise their first pair of legs and wave them as an "antennal illusion" [29, 30] and also carry out an up-and-down movement of the gaster, akin to some ants when they are recruiting nestmates [30–32].

The family of spiders with perhaps the most striking examples of myrmecomorphy is the jumping spiders (Araneae: Salticidae). Here again, myrmecomorphy has evolved independently various times [33], and the most speciose genus of myrmecomorphic salticids is Myrmarachne, which has more than 200 described species and many more undescribed [34].

Arthropods that are not morphological mimics of ants can nevertheless form close associations with colonies. These arthropods are generally referred to as myrmecophiles, and their association to ant colonies can vary in extent [24, 25]. The ecological advantage for myrmecophiles is that the nests of many ant species are relatively stable microhabitats where resources can be readily available, and a certain degree of protection is conferred as well [24, 25]. Some examples of this will be given in the following section.

3. Benefits of Ant Associations for Spiders

The fact that ant mimicry exists in such varied forms across many invertebrate taxa implies that the benefits must outweigh the costs. As social insects, ants form colonies, often containing thousands and in some cases millions of individuals [22], and in many cases their nests are sophisticated structures and spaces in the environment. This has advantages for invertebrates that associate so closely with ants that they actually live inside the ants' nests. The nest provides a stable environment, often with plenty of resources to feed on, be it other inquilines, materials the ants gathered or bred, or the ants/larvae themselves [35]. For example, the linyphiid spider Masoncus pogonophilus feeds on collembolans that also live inside its host ant nests [36], while the salticid spider Cosmophasis bitaeniata enters ants' nests to feed on their larvae [37].

In the case of myrmecomorphic spiders, the main benefit is that they gain protection from ant-averse predators that would otherwise feed on them. Several experiments have been carried out to show that myrmecomorphic spiders are Batesian mimics because they gain protection from potential predators such as wasps [38], mantises [39], and other spiders [40–42] and that ant-aversion is even innate in some predators [39, 43]. Salticids as predators alone were suggested to be a driving force for myrmecomorphy in jumping spiders [44]. To date, there is little evidence that myrmecomorphy serves in protecting the spider directly from the ant, as the ants' primary sensory channel seems to be chemical [45]. On the other hand, most myrmecomorphs do not routinely prey on ants, although there have been cases reported where the myrmecomorphs do prey on ants [46–48].

Within Batesian ant-mimicking spiders, several alternative or supplementary strategies have been described that confer protection from potential predators. One of these strategies is transformational mimicry, meaning that the model mimics different species as it grows [49]. Several Myrmarachne species are transformational mimics, thus always being approximately the same size as their model ants [50]. Another strategy involves the common occurrence among males of several Myrmarachne species that have enlarged chelicerae (thought to be a sexually selected character [51]), a phenotype that could be seen as reducing their resemblance to ants. However, these males were found to be "compound mimics" resembling ants carrying a "parcel" in their mandibles [52]. Additionally, Myrmarachne melanotarsa, a spider unusual in that it lives in aggregations, resembles, as a group, a whole ant colony [53]. Selection has acted on these varied strategies found among myrmecomorphs, increasing their resemblance to ants, yet forces countering the selection of "perfect" resemblance to ants also exist, as polymorphism has been recorded in various Myrmarachne species [54, 55].

So the benefits for spiders of associating with ants come mainly in the form of increased chances of survival for the individuals. These increased survival chances are either due to an easier access to readily available resources or heightened protection from predators. If these benefits did not exist, selection would not have favoured the traits allowing these spiders to associate with ants. However, for the spiders there are not only benefits to these associations, but also costs. For the associations to persist in evolutionary terms, the benefits must still outweigh the costs, meaning that the costs are kept minimal. The next section deals with the spiders' strategies that minimise the various costs.

4. MINIMISING COSTS OF ANT MIMICRY

The costs of ant mimicry for spiders come in varied forms. First of all, for myrmecomorphs there is the fact that morphological modifications, such as a restriction of the abdomen, mean that females can lay fewer eggs than nonmyrmecomorphic spiders ([25] and references therein). A major problem that myrmecomorphic spiders face is that while their resemblance to ants confers protection from ant-averse predators, they are more prone to fall victims to predators that specialise on eating ants [19, 56]. To counter this problem, jumping spiders of the genus Myrmarachne have developed signals using their first pair of legs, aimed at deterring ant-eating salticids from attacking [57]. This "display posture" of holding the first pair of legs almost fully extended, elevated 45°, and held out to the side 45° [57] was also noted in other studies on Myrmarachne when the spiders were in the presence of ants [58], and it resembles the aggressive display posture of worker ants from certain species such as Oecophylla smaragdina (see Figure 1). This display posture, while being efficient in deterring salticid predators, seems to be adopted by Myrmarachne as a general measure when threatened, before fleeing, and may also affect ants—such as O. smaragdina—that have a more sophisticated visual system [59, 60].

Perhaps the biggest challenge for ant-associating spiders comes from living close to ant species, most of which would react aggressively towards inquilines or mimics themselves. In fact, spiders may easily be killed or injured by their own model [61]. The negative effects of ants on spiders are not only restricted to the individuals' survival, but also the spiders' reproductive success in some cases, in that they are less likely to mate if the ants are close by [62]. Certain spiders that have developed a close association with ants deploy chemical mimicry to be able to live among and at times exploit the ants [63]. Cosmophasis bitaeniata even acquires the hosts' cuticular hydrocarbons specific to the ant colony with which it lives [64], as the cuticular hydrocarbons are transferred to the spider while feeding on the ant larvae [65]. In the case of this spider, the host ant species, Oecophylla smaragdina, is particularly aggressive [59, 60, 66] (see Figure 1), and chemical mimicry is a form of protection. Through chemical mimicry, many nonant nestmates are able to enter ant colonies and take advantage of the ants and/or their mutualistic relationships [67]. Some myrmecophiles are small enough to live among the ants undetected without chemical mimicry [36], while others, such as Gamasomorpha maschwitzi, have alternative strategies to chemical mimicry which are to date poorly known but could consist of acoustical, behavioural, and/or morphological adaptations [68].

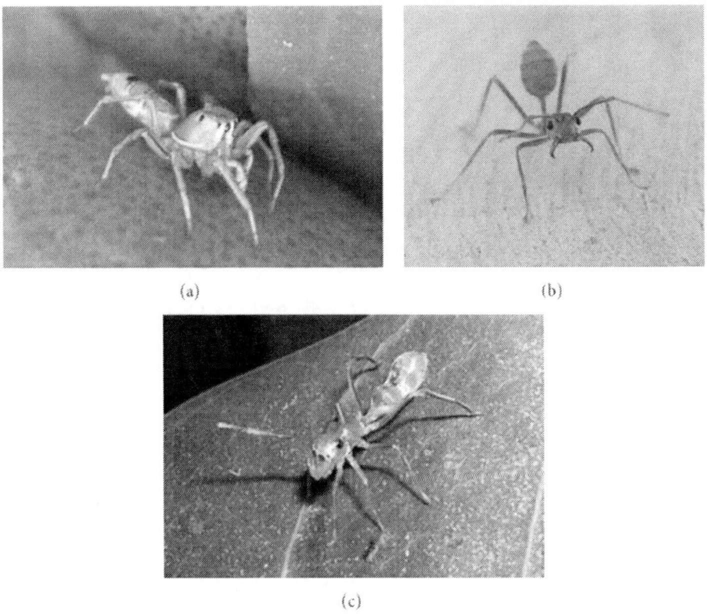

Figure 1: Ant-associating salticids (a) Cosmophasis bitaeniata, chemical, aggressive myrmecophile and (c) Myrmarachne smaragdina, myrmecomorphic Batesian mimic, and their common model ant species (b) Oecophylla smaragdina in an aggressive display posture.

For those spiders that do not live in, or enter into the ants' nests, there does not seem to be as much danger as of being killed by an ant. However, for the myrmecomorphs that are Batesian mimics, the premise is to be in the model ants' vicinity, which nevertheless poses a considerable danger [61]. As there is no known case of chemical mimicry in myrmecomorphic spiders [69], their defence strategies need to rely on different approaches, which are mainly behavioural. It has long been observed that ant-associating spiders such as Myrmarachne generally avoid contact with ants [45, 58, 61, 70], and this holds true not only for myrmecomorphs, but also for aggressive mimics such as C. bitaeniata, despite its chemical protection [35, 58]. Upon seeing an ant approach, myrmecomorphic spider species of the genus Myrmarachne actively move away from the ant, regardless of the ant species, and contact occurs in fewer than 3% of the cases when the spiders react to the presence of the ants [58]. These spiders are able to distinguish between ants and conspecifics, due to their remarkable visual acuity [71]. They also react differently to ants depending on whether the ants are facing them, side-on, moving, or stationary but generally do not let the ant get closer than approximately 2 cm [58]. At times, however, contact is unavoidable, and the spiders flee even upon contact, only very rarely reacting aggressively, perhaps as a last resort [72]. Active avoidance of ants is common in myrmecomorphic spiders, and the behavioural reactions of myrmecomorphs towards sympatric ant species are different depending on the species of spiders (as was shown with Myrmarachne). Innate behavioural traits are different between species due to selection (as is the case in morphological traits). Aversion to ants is innate in arthropods such as mantises [39], and avoidance of ants could therefore also be an innate trait in myrmecomorphic jumping spiders such as Myrmarachne. If that is the case, the fact that each species of Myrmarachne reacts differently to the presence of ants suggests that these behavioural traits are under selection pressure [58].

5. CONCLUSIONS

There are advantages and disadvantages for ant-associating spiders related to living near or inside ant colonies. When looking purely at ant mimicry, it is clear that there is an arms race between the parties involved in terms of evolutionary costs and benefits. Varied strategies have evolved in ant-mimicking spiders allowing them to reap the benefits of resembling ants. In addition, these spiders have innate and/or learned behaviours that reduce the costs of having models that are often aggressive and a real danger to the spiders themselves. Despite the considerable studies that have been carried out recently, ant mimicry in spiders is definitely a topic which deserves more attention and in-depth studies. In particular, with the increasing use of genomics, it is possible to carry out studies relating the underlying genetic mechanisms to phenotypic adaptations to ant mimicry, as have been carried out by D. Charlesworth and B. Charlesworth [72] which would give even more insight into the evolution of mimicry.

ACKNOWLEDGMENTS

The author thanks all the people who supported her during her time working on ant-mimicking spiders in North Queensland, in particular her supervisors Richard Rowe and Ross Crozier, and people from the Crozier Lab and the School of Marine and Tropical Biology, James Cook University, Townsville, for funding. The author also thanks two the anonymous referees for their comments on the paper.

REFERENCES

1. G. D. Ruxton, T. N. Sherratt, and M. Speed, Avoiding Attack: The Evolutionary Ecology of Crypsis, Warning Signals and Mimicry, Oxford University Press, Oxford, UK, 2004.
2. R. I. Vane-Wright, "A unified classification of mimetic resemblances," Biological Journal of the Linnean Society, vol. 8, no. 1, pp. 25–56, 1976.
3. H. W. Bates, "Contributions to an insect fauna of the Amazon valley. Lepidoptera: Heliconidae," Transactions of the Linnean Society London, vol. 23, no. 3, pp. 495–566, 1862.
4. J. A. Welbergen and N. B. Davies, "A parasite in wolf's clothing: hawk mimicry reduces mobbing of cuckoos by hosts," Behavioral Ecology, vol. 22, no. 3, pp. 574–579, 2011.
5. K. L. Cheney and N. J. Marshall, "Mimicry in coral reef fish: how accurate is this deception in terms of color and luminance?" Behavioral Ecology, vol. 20, part 3, pp. 459–468, 2009.
6. W. Wickler, Mimicry in Plants and Animals, McGraw-Hill, New York, NY, USA, 1968.
7. U. Nur, "Evolutionary rates of models and mimics in Batesian mimicry," The American Naturalist, vol. 104, no. 939, pp. 477–486, 1970.
8. M. Edmunds, Defence in Animals—A Survey of Anti-Predator Defences, Longman, Harlow, UK, 1974.
9. D. Charlesworth and B. Charlesworth, "Theoretical genetics of Batesian mimicry. I. Single locus models," Journal of Theoretical Biology, vol. 55, no. 2, pp. 283–303, 1975.
10. C. Matessi and R. Cori, "Models of population genetics of Batesian mimicry," Theoretical Population Biology, vol. 3, no. 1, pp. 41–68, 1972.
11. O. H. Holen and R. A. Johnstone, "The evolution of mimicry under constraints," The American Naturalist, vol. 164, no. 5, pp. 598–613, 2004.
12. O. H. Holen and R. A. Johnstone, "Context-dependent discrimination and the evolution of mimicry," The American Naturalist, vol. 167, no. 3, pp. 377–389, 2006.
13. W. Dittrich, F. Gilbert, P. Green, P. Mcgregor, and D. Grewcock, "Imperfect mimicry: a pigeon's perspective," Proceedings of the Royal Society B, vol. 251, no. 1332, pp. 195–200, 1993.
14. M. Edmunds, "Why are there good and poor mimics?" Biological Journal of the Linnean Society, vol. 70, no. 3, pp. 459–466, 2000.
15. T. N. Sherratt, "The evolution of imperfect mimicry," Behvioral Ecology, vol. 13, part 6, pp. 821–826, 2002.
16. S. Pekár and M. Jarab, "Assessment of color and behavioral resemblance to models by inaccurate myrmecomorphic spiders (Araneae)," Invertebrate Biology, vol. 130, no. 1, pp. 83–90, 2011.

17. R. A. Johnstone, "The evolution of inaccurate mimics," Nature, vol. 418, no. 6897, pp. 524–526, 2002.
18. M. P. Speed and G. D. Ruxton, "Imperfect Batesian mimicry and the conspicuousness costs of mimetic resemblance," The American Naturalist, vol. 176, no. 1, pp. E1–E14, 2010.
19. S. Pekár, M. Jarab, L. Fromhage, and M. E. Herberstein, "Is the evolution of inaccurate mimicry a result of selection by a suite of predators? A case study using myrmecomorphic spiders," The American Naturalist, vol. 178, no. 1, pp. 124–134, 2011.
20. J. E. Randall, "A review of mimicry in marine fishes," Zoological Studies, vol. 44, no. 3, pp. 299–328, 2005.
21. J. D. McIver and G. Stonedahl, "Myrmecomorphy: morphological and behavioral mimicry of ants," Annual Review of Entomology, vol. 38, pp. 351–379, 1993.
22. B. Hölldobler and E. O. Wilson, The Ants, The Belknap Press of Harvard University Press, Cambridge, Mass, USA, 1990.
23. R. F. Foelix, Biology of Spiders, Oxford University Press, Oxford, UK, 3rd edition, 2011.
24. P. E. Cushing, "Myrmecomorphy and myrmecophily in spiders: a review," Florida Entomologist, vol. 80, no. 2, pp. 165–193, 1997.
25. P.E. Cushing, "Spider-ant associations: an updated review of myrmecomorphy, myrmecophily, and myrmecophagy in spiders," Psyche, vol. 2012, Article ID 151989, 23 pages, 2012.
26. R. R. Jackson and X. J. Nelson, "Specialised exploitation of ants (Hymenoptera: Formicidae) by spiders (Araneae)," Myrmecological News, vol. 17, pp. 33–49, 2012.
27. X. J. Nelson and R. R. Jackson, "How spiders practice aggressive and Batesian mimicry," Current Zoology, vol. 58, part 4, pp. 620–629, 2012.
28. M. A. Fürst, M. Durey, and D. R. Nash, "Testing the adjustable threshold model for intruder recognition on Myrmica ants in the context of a social parasite," Proceedings of the Royal Society B, vol. 279, pp. 516–522, 2012.
29. J. Reiskind, "Ant-mimicry in Panamanian clubionid and salticid spiders (Araneae: Clubionidae, Salticidae)," Biotropica, vol. 9, no. 1, pp. 1–8, 1977.
30. F. S. Ceccarelli, "Behavioral mimicry in Myrmarachne species (Araneae, Salticidae) from North Queensland, Australia," Journal of Arachnology, vol. 36, no. 2, pp. 344–351, 2008.
31. R. R. Jackson, "The biology of ant-like jumping spiders: intraspecific interactions of Myrmarachne lupata(Araneae, Salticidae)," Zoological Journal of the Linnean Society, vol. 76, no. 4, pp. 293–319, 1982.
32. J. L. Mercier, A. Lenoir, and A. Dejean, "Ritualised versus aggressive behaviours displayed by Polyrhachis iaboriosa (F. Smith) during intraspecific competition," Behavioural Processes, vol. 41, no. 1, pp. 39–50, 1997.
33. M. C. Hedin and W. P. Maddison, "A combined molecular approach to phylogeny of the jumping spider subfamily Dendryphantinae (Araneae: Salticidae)," Molecular Phylogenetics and Evolution, vol. 18, no. 3, pp. 386–403, 2001.
34. J. Proszynski, "Monograph of Salticidae (Araneae) of the World," 2012, http://www.peckhamia.com/salticidae/.
35. B. Hölldobler, "Communication between ants and their guests," Scientific American, vol. 224, pp. 86–93, 1971.
36. P. E. Cushing, "Population structure of the ant nest symbiont Masoncus pogonophilus (Araneae: Linyphiidae)," Annals of the Entomological Society of America, vol. 91, part 5, pp. 626–631, 1998.
37. R. A. Allan and M. A. Elgar, "Exploitation of the green tree ant, Oecophylla smaragdina, by the salticid spider Cosmophasis bitaeniata," Australian Journal of Zoology, vol. 49, no. 2, pp. 129–137, 2001.
38. M. Edmunds, "Does mimicry of ants reduce predation by wasps on salticid spiders?" Memoirs—Queensland Museum, vol. 33, no. 2, pp. 507–512, 1993.

39. X. J. Nelson, R. R. Jackson, D. Li, A. T. Barrion, and G. B. Edwards, "Innate aversion to ants (Hymenoptera: Formicidae) and ant mimics: experimental findings from mantises (Mantodea)," Biological Journal of the Linnean Society, vol. 88, no. 1, pp. 23–32, 2006.
40. B. Cutler, "Reduced predation on the antlike jumping spider Synageles occidentalis (Araneae: Salticidae)," Journal of Insect Behavior, vol. 4, part 3, pp. 401–407, 1991.
41. C. A. Durkee, M. R. Weiss, and D. B. Uma, "Ant mimicry lessens predation on a North American jumping spider by larger salticid spiders," Environmental Entomology, vol. 40, part 5, pp. 1223–1231, 2011.
42. X. J. Nelson, "A predator's perspective of the accuracy of ant mimicry in spiders," Psyche, vol. 2012, Article ID 168549, 5 pages, 2012.
43. X. J. Nelson and R. R. Jackson, "Vision-based innate aversion to ants and ant mimics," Behavioral Ecology, vol. 17, no. 4, pp. 676–681, 2006.
44. J. N. Huang, R. C. Cheng, D. Li, and I. M. Tso, "Salticid predation as one potential driving force of ant mimicry in jumping spiders," Proceedings of the Royal Society B, vol. 278, no. 1710, pp. 1356–1364, 2011.
45. B. Hölldobler, "Multimodal signals in ant communication," Journal of Comparative Physiology A, vol. 184, no. 2, pp. 129–141, 1999.
46. K. Wing, "Tutelina similis (araneae: Salticidae): an ant mimic that feeds on ants," Journal of the Kansas Entomological Society, vol. 56, no. 1, pp. 55–58, 1983.
47. P. S. Oliveira and I. Sazima, "The adaptive bases of ant-mimicry in a neotropical aphantochilid spider (Araneae: Aphantochilidae)," Biological Journal of the Linnean Society, vol. 22, no. 2, pp. 145–155, 1984.
48. X. J. Nelson and R. R. Jackson, "Aggressive use of Batesian mimicry by an ant-like jumping spider," Biology Letters, vol. 5, no. 6, pp. 755–757, 2009.
49. A. P. Mathew, "Transformational deceptive resemblance as seen in the life history of a plant bug (Riptortus pedestris), and of a mantis (Evantissa pulchra)," Journal of the Bombay Natural History Society, vol. 37, pp. 803–813, 1935.
50. M. Edmunds, "On the association between Myrmarachne spp. (Salticidae) and ants," Bulletin of the British Arachnological Society, vol. 4, pp. 149–160, 1978.
51. S. D. Pollard, "Consequences of sexual selection on feeding in male jumping spiders (Araneae: Salticidae)," Journal of Zoology, vol. 234, no. 2, pp. 203–208, 1994.
52. X. J. Nelson and R. R. Jackson, "Compound mimicry and trading predators by the males of sexually dimorphic Batesian mimics," Proceedings of the Royal Society B, vol. 273, no. 1584, pp. 367–372, 2006.
53. X. J. Nelson and R. R. Jackson, "Collective Batesian mimicry of ant groups by aggregating spiders," Animal Behaviour, vol. 78, no. 1, pp. 123–129, 2009.
54. F. S. Ceccarelli and R. H. Crozier, "Dynamics of the evolution of Batesian mimicry: molecular phylogenetic analysis of ant-mimicking Myrmarachne (Araneae: Salticidae) species and their ant models," Journal of Evolutionary Biology, vol. 20, no. 1, pp. 286–295, 2007.
55. X. J. Nelson, "Polymorphism in an ant mimicking jumping spider," Journal of Arachnology, vol. 38, no. 1, pp. 139–141, 2010.
56. X. J. Nelson, D. Li, and R. R. Jackson, "Out of the frying pan and into the fire: a novel trade-off for batesian mimics," Ethology, vol. 112, no. 3, pp. 270–277, 2006.
57. X. J. Nelson, R. R. Jackson, and D. Li, "Conditional use of honest signaling by a Batesian mimic," Behavioral Ecology, vol. 17, no. 4, pp. 575–580, 2006.
58. F. S. Ceccarelli, Dynamics of Salticid-Ant Mimicry Systems—A Study of Behavioural Ecology and Evolution in Two Genera of Ant-Associating Jumping Spiders, Lambert Academic Publishing, Saarbrücken, Germany, 2010.
59. B. Hölldobler, "Territorial behavior in the green tree ant (Oecophylla smaragdina)," Biotropica, vol. 15, no. 4, pp. 241–250, 1983.
60. X. J. Nelson, R. R. Jackson, G. B. Edwards, and A. T. Barrion, "Living with the enemy: jumping spiders that mimic weaver ants," Journal of Arachnology, vol. 33, pp. 813–819, 2005.

61. X. J. Nelson and R. R. Jackson, "The influence of ants on the mating strategy of a myrmecophilic jumping spider (Araneae, Salticidae)," Journal of Natural History, vol. 43, no. 11-12, pp. 713–735, 2009.
62. R. A. Allan, R. J. Capon, W. V. Brown, and M. A. Elgar, "Mimicry of host cuticular hydrocarbons by salticid spider Cosmophasis bitaeniata that preys on larvae of tree ants Oecophylla smaragdina," Journal of Chemical Ecology, vol. 28, no. 4, pp. 835–848, 2002.
63. M. A. Elgar and R. A. Allan, "Chemical mimicry of the ant Oecophylla smaragdina by the myrmecophilous spider Cosmophasis bitaeniata: is it colony-specific?" Journal of Ethology, vol. 24, no. 3, pp. 239–246, 2006.
64. M. A. Elgar and R. A. Allan, "Predatory spider mimics acquire colony-specific cuticular hydrocarbons from their ant model prey," Naturwissenschaften, vol. 91, no. 3, pp. 143–147, 2004.
65. R. H. Crozier, P. S. Newey, E. A. Schlüns, and S. K. A. Robson, "A masterpiece of evolution Oecophyllaweaver ants (Hymenoptera: Formicidae)," Myrmecological News, vol. 13, pp. 57–71, 2009.
66. K. Dettner and C. Liepert, "Chemical mimicry and camouflage," Annual Review of Entomology, vol. 39, pp. 129–154, 1994.
67. C. von Beeren, R. Hashim, and V. Witte, "The social integration of a myrmecophilous spider does not depend exclusively on chemical mimicry," Journal of Chemical Ecology, vol. 38, no. 3, pp. 262–271, 2012.
68. S. Pekar and P. Jiros, "Do ant mimics imitate cuticular hydrocarbons of their models?" Animal Behaviour, vol. 82, no. 5, pp. 1193–1199, 2011.
69. A. P. Mathew, "Observations on the habits of two spider mimics of the red ant, Oecophylla smaragdina(Fabr.)," Journal of the Bombay Natural History Society, vol. 52, pp. 249–263, 1944.
70. X. J. Nelson, "Visual cues used by ant-like jumping spiders to distinguish conspecifics from their models," Journal of Arachnology, vol. 38, no. 1, pp. 27–34, 2010.
71. F. S. Ceccarelli, "Contact between Myrmarachne (Araneae: Salticidae) and ants," Bulletin of the British Arachnological Society, vol. 14, part 2, pp. 54–58, 2007.
72. D. Charlesworth and B. Charlesworth, "Mimicry: the hunting of the supergene," Current Biology, vol. 21, part 20, pp. R846–R848, 2011.

CHAPTER 6

Pimachrysa (Neuroptera: Chrysopidae: Nothochrysinae): Larval Description and Support for Generic Relationships

Catherine A. Tauber[1,2] and David K. Faulkner[3,4]

[1]Department of Entomology, Comstock Hall, Cornell University, Ithaca, NY 14853-2601, USA

[2]Department of Entomology & Nematology, University of California, Davis, CA 95616, USA

[3]Forensic Entomology Services, 2321 Gladwick Street, Rancho Dominguez, CA 90220, USA

[4]Entomology Department, San Diego Natural History Museum, 1788 El Prado, San Diego, CA 92101, USA

ABSTRACT

We describe the previously unknown larval characteristics of Pimachrysa (second and third instars), and we provide new comparative data on the Nothochrysa californica Banks larvae that were described earlier. The Pimachrysa larvae (identified as Pimachrysa fusca Adams) express the numerous features that characterize the chrysopid subfamily Nothochrysinae. They resemble Hypochrysa larvae in many respects, but several differences support retaining the two genera as separate. It now appears that the reputedly more archaic, small bodied, nothochrysine genera (e.g., Dictyochrysa, Hypochrysa, and Pimachrysa) have naked larvae that differ distinctly from the larvae of Nothochrysa species, all of which are debris-carriers. In addition, the pattern of variability in Nothochrysa larval features supports the proposal that this genus has a closer relationship with the other chrysopid subfamilies than do the small-bodied nothochrysine genera. The findings are consistent with the hypothesis that larval nakedness (as opposed to debris-carrying) was the ancestral life style for the extant chrysopid clade.

1. INTRODUCTION

Considerable effort has focused on elucidating the evolutionary history of the neuropteran family Chrysopidae, and yet significant challenges remain. The subfamily Nothochrysinae is of particular interest, largely because of its position at or very near the base of the chrysopid phylogenetic tree and also because of its substantial and ancient fossil record. Members of this subfamily have primitive patterns of wing venation and unique features in their larval morphology that have been used to explore evolutionary trends within the subfamily [1–6]. However, generic relationships among the extant taxa remain ambiguous. On-going molecular studies [4, 7, 8] thus far have yielded some interesting, but contradictory results concerning the origin of the subfamily, and they have not yet resolved the relationships among its constituent genera.

Ideally, the above issues would be met with systematic comparisons among all included taxa using a broad range of approaches (adult morphology, larval morphology, comparative biology, and molecular sequencing). In reality, a large barrier prevents such an analysis, that is, the rarity of specimens from relevant taxa. For example, except for a few species, specimens of Nothochrysinae and other basal chrysopid taxa are seldom encountered in the field; as a result the range of taxonomic coverage has been relatively narrow.

To continue efforts in overcoming this barrier, we describe the hitherto unknown larva (Semaphoront B) of the nothochrysine genus, Pimachrysa. Along with Nothochrysa, Dictyochrysa, and Hypochrysa, whose larvae were described earlier (Box 1), Pimachrysa is now the fourth genus of Nothochrysinae for which larvae are known. Also, we provide new data on the larva of Nothochrysa californica Banks that document a broader range of larval variation than previously reported for the genus. Finally, we discuss the pattern of variation among known nothochrysine larvae in relation to the current understanding of generic relationships and the phylogeny of the subfamily.

Nothochrysinae

- *Dictyochrysa* – **Naked**
  - *fulva* Esben-Petersen
    - New 1981 [14] [Desc-L1; Draw-L1; Note]
- *Hypochrysa* – **Naked**
  - *elegans* (Burmeister)
    - Brauer 1867 [15], as *Hypochrysa nobilis* Heyd. [Desc-L3; Draw-L3; Note]
    - Lacroix 1925, as *Hypochrysa nobilis* Schneider [Note]
    - Principi 1956b, as *H. nobilis* Schneider [Note: svelt, agile, green larva, no mention of debris]
    - Gepp 1983 [Photo-L3; Note]
    - Gepp 1988 [Photo-L3; Note]
    - Díaz-Aranda and Monserrat 1995 [18] [Draw-L1, L3; Note]
    - Gepp 1999 [Photo-L3]
    - Díaz-Aranda et al. 2001 [Draw-L1, L3; Note]
    - Monserrat and Díaz-Aranda 2012 [13] [Desc-L1, L3; Draw-L1, L3; Note]
  - *africana* (Kimmins)
    - Duelli et al. 2010 [4] [Photo-L3]
    - Tauber 2014 [6] [Desc-L2; Draw-L2; Photo-L2; Note: Synonymy of *Kimochrysa* with *Hypochrysa*, largely based on larval characters]
    - Tauber and Faulkner, this paper [Draw-L2; Note]
- *Nothochrysa* – **Varied, with both Debris-carrier & Occasional Debris-carrier**
  - *californica* Banks
    - Toschi 1965 [16] [Desc-L1–L3; Draw-L1–L3; Note; Mat: not specified; Beh]
    - Tauber and Tauber, unpublished data [Mat: bracts from *Quercus* catkins, frass, plant fragments]
    - Tauber and Faulkner, this paper [Redesc-L2; Draw-L2; Photo-L2]
  - *capitata* (Fabricius)'
    - Withycombe 1922 [Desc-L1–L3; Draw-L3; Note: a few large pieces; Mat: not specified]
    - Lacroix 1925 [Note: packet loosely attached, not elaborate; Mat: not specified]
    - Killington 1931 [Note: a few large pieces; Mat: not specified]
    - Killington 1937 [17], as *Nathanica* [Desc-L1–L3; Draw-L3; Note: information from Withycombe 1922]
    - Kimmins 1939, as *Nathanica* [Desc-L1; Draw-L1]
    - Fraser 1959, as *Nathanica* [Draw-L2 or L3; Note; Mat: not specified]
    - New 1969, as *Nathanica* [Note: L2, L3 – most carried some to much debris; Mat: mainly animal remains; Beh]
    - Gepp 1983 [Photo-L3; Note; Mat: frass, twisted plant material, lichens]
    - Barnard et al. 1986 [Desc: dorsal coloration]
    - Gepp 1988 [Note: packet loose, on central part of abdomen; Mat: bud scales, frass]
    - Monserrat and Díaz-Aranda 2012 [13] [Desc-L3; Draw-L3; Note; Mat: few large, diverse fragments]
  - *fulviceps* (Stephens)
    - Lacroix 1925 [Note: packet loosely attached, not elaborate; Mat: not specified]
    - Killington 1935, as *Nathanica* [Desc-L1; Draw-L1; Note; Mat: not specified]
    - Killington 1937 [17], as *Nathanica* [Desc-L1; Draw-L1; Note; Mat: not specified]
    - Gepp 1983 [Photos-L3, with and without debris; Note; Mat: large pieces of plant material, shed bark etc.]
    - Gepp 1984a [Photo-L3; Note; Mat: not specified]
    - Gepp 1988 [Draw-L3, Photo-L3; Note: packet on abdomen; Mat: bud scales, frass, coleophorid cases]
    - Díaz-Aranda & Monserrat 1995 [18] [Draw-L3; Note; Mat: not specified]
    - Díaz-Aranda et al. 2001 [Draw-L3; Note; Mat: not specified]
    - Monserrat & Díaz Aranda 2012 [13] [Desc-L3; Draw-L3; Note; Mat: few large pieces of different nature]
- *Pimachrysa* – **Naked**
  - *fusca* Adams
    - Tauber & Faulkner, this paper [Desc-L2, L3; Draw-L2; Photo-L2, L3; Note]
- *Asthenochrysa* – **Larvae Unknown**
- *Leptochrysa* – **Larvae Unknown**
- *Pamochrysa* – **Larvae Unknown**
- *Triplochrysa* – **Larvae Unknown**

Box 1: Catalog of larval descriptions for Nothochrysinae (updated from Tauber et al. [5], where the citations not listed in the references here can be found). See Monserrat and Díaz-Aranda 2012 [13] regarding problems in distinguishing N. capitata and N. fulviceps.

2. MATERIALS AND METHODS

2.1. Larval Material and Procedures

Our study is based on four specimens of Pimachrysa fusca Adams collected in late February and late April at Lower Covington Flat, Joshua Tree National Park (previously a national monument), Riverside Co., CA. Three of the specimens (two third instars and the single second instar) were used for morphological study, and one third instar was reared to confirm the specific identity of our larval specimens. The three specimens were treated in KAAD solution [9] for 24 hours and then transferred to 95% ethyl alcohol. In the laboratory the specimens were photographed, and the external gross features were described. Subsequently, one specimen from each instar was cleared in KOH and transferred to glycerin for examination of fine structures and setation. The second instar died soon after it was collected. We noted a globule of apparently sticky material on the labrum and base of the left mandible; presumably the material was exudate from aphids that we supplied the larva while we were in the field. We were able to remove the exudate from the preserved specimen without damage.

All voucher specimens are now in the Tauber Research Collection (Lot 2015: 01); a leg from one of the third instars was given to Shaun Winterton, California Department of Food and Agriculture, Sacramento, for DNA analysis.

To rear one of the third instars to the adult stage, we held it in a glass vial provided with frozen eggs of Ephestia kuehniella (Zeller), the Mediterranean flour moth. It was maintained at room temperature on a long-day light : dark cycle (~LD 16 : 8).

2.2. Terminology and Comparisons

The chrysopid larval stage typically includes three instars, the first of which (L1, Semaphoront A) differs markedly from the latter two (L2, L3), which, in turn, usually resemble each other (Semaphoront B). The main differences between second and third instars are in size, markings, and the number of thoracic and abdominal setae. In this paper, both instars of Semaphoront B are considered.

With minor exceptions, morphological terminology and chaetotaxy followed the usage of Rousset [10], Tsukaguchi [11], Tauber et al. [12], and Monserrat and Díaz-Aranda [13]. In one case there is a discrepancy. Some abdominal segments of Nothochrysa have a pair of laterodorsal tubercles (LDTs); in papers by Tauber and coauthors, the term tubercle refers to a protuberance bearing more than one seta. In Nothochrysa californica Banks, each LDT bears one long, filiform seta and one shorter seta. Thus, the tubercles of N. californica are consistent with those of other chrysopids. However, third instars of the two European species of Nothochrysa also were reported to have laterodorsal "tubercles" on abdominal segments A5 to A7; but in these species, the "tubercles" were reported to have only one, relatively long, seta [13]. If such a situation pertains (and no small seta is found on these "tubercles"), the homology of the structures should be examined and the usage of the term "tubercle" clarified. For convenience, in this paper, we refer to the structures reported for the European species as laterodorsal tubercles (LDTs).

For our comparisons, we used specimens of Nothochrysa californica Banks (all instars from Alameda and Marin Counties, California) and Hypochrysa africana Kimmins (second instar from South Africa, see [6]), as well as earlier descriptions in the literature, as follows: Dictyochrysa fulva Esben-Petersen (first instar) [14], Hypochrysa africana (Kimmins) (second instar) [6], Hypochrysa elegans (Burmeister) (first and third instars) [13, 15], N. californica (all instars) [16], Nothochrysa capitata (Fabricius) (third instar) [13, 17], and Nothochrysa fulviceps (Stephens) (first and third instars) [13, 17]. Our comparisons are summarized in Box 2.

3. BACKGROUND

3.1. Habitat

The collection site (northwestern corner of Joshua Tree National Park) is in a high elevation portion of the Mohave Desert (elevation: ~4670 ft). The area is open and arid, with scattered, low-growing trees and bushes amongst rocky outcroppings. Vegetation at the site includes Quercus cornelius-mulleri (Muller oak), Juniperus californica (California juniper), Pinus monophylla (single-leaf pinyon pine), Larrea tridentata (creosote bush), Bebbia juncea (Bebbia), Yucca brevifolia (Joshua tree), and Yucca schidigera (Mohave yucca). Our specimens were beaten from oak bushes, Q. cornelius-mulleri. The collections occurred on two days: the first (a single second instar, L2) on 26-2-2015 (cool and sunny), and the second (three third instars, L3) on 28-4-2015 (warm and sunny).

Pimachrysa (L2, L3)
Larvae naked, without a layer of camouflaging material.
Small to medium sized larvae (L2: ~5.5 mm; L3: 8.5–9.0 mm); shape fusiform, narrow, abdomen rounded, slender.
Cephalic markings obvious, lateral epicranial markings elongate; mesal epicranial and/or postfrontal markings reduced; with a conspicuous median stripe.
Cephalic setae: all primary setae except S10 and S12 present; Vx obscure, probably two setae and a pore.
Antennae with terminal segment (flagellum) stubby, short (0.07–0.08x length of pedicel).
Mouthparts elongate (mandible ~1.4–1.6x longer than head capsule).
Mandible with single seta on basolateral margin.
Thorax without lateral tubercles.
Paired prothoracic sclerites (Sc1) elongate.
Metanotum with three distinct folds, each bearing a transverse band of unhooked setae with acute tips.
Legs fairly long in relation to body; tarsi dark brown throughout.
Abdominal segments II–VII without lateral tubercles, with some lateral setae slightly elongated, but without distinct lateral setae (LS).
Abdominal segments I–VII dorsally with transverse bands of numerous short, acute setae (SMS).
Abdominal segments V–VII without laterodorsal tubercles, without LDS (but with some laterodorsal setae slightly elongated).

Nothochrysa (all instars, unless noted otherwise)
Larvae usually bearing a camouflaging layer of debris.
Large larvae (L3: 9–12 mm; L2: 6-7 mm; L1: 4-5 mm); shape very slightly fusiform, abdomen convex, almost flattened to globular.
Cephalic markings obvious, epicranial markings large, fused with postfrontal markings, occupying almost the entire cephalic dorsal surface.
Cephalic setae: all primary setae including S12 present; Vx = two short anterior setae, pore, single posterior seta, in longitudinally elongate row.
Antennae with terminal segment (flagellum) of variable length.
Pedicel length/flagellar length (= number of times flagellum shorter than pedicel):
N. capitata L3 = ~8.8 (from Figures 47-6 [13])
N. fulviceps L3 = ~13.0 (from Figures 47-12 [13])
N. californica L1 = ~3.6, L2 = 5.1–6.2, L3 = 9.4–9.5
Flagellar terminus and terminal bristle variable.
N. capitata, N. fulviceps L3 with rounded flagellar terminus, very short terminal bristle (from Figure 1b [18], Figures 47-6, 47-12 [13])
N. californica L3–L1 with flagellum tapered, slightly elongate seta (setal length/flagellar length: L3 = 0.18, L2 = 0.18, L1 = 0.25)
Mouthparts noticeably longer than the head capsule.
N. capitata, N. fulviceps L3 (from Figures 46-1, 46-5 [13])
N. californica L1–L3 (mandible length/head length: L1 = 1.36, L2 = ~1.60–1.65, L3 = 1.30–1.48)
Mandible with two setae on basolateral margin.
Thorax with lateral tubercles present, carrying 1-2 or few elongate lateral setae (LS).
Paired prothoracic sclerites (Sc1) triangular.
Metanotum with transverse row of filiform (hooked or straight) setae posteriorly.
Legs very long in relation to body size; tarsi pale to dark brown.
Abdominal segments II–VII with lateral tubercles well developed, hemispherical, bearing filiform or hooked ("uncinate") setae (LS).
Abdominal segments VI-VII (sometimes also V, VIII) with laterodorsal tubercles present, bearing single long seta (LDS?) (sometimes also a shorter seta).
Dorsum of abdominal segments I–VI with transverse rows or bands of hooked submedian setae (SMS).

Hypochrysa (L1, L3)
Larvae naked, without a layer of camouflaging material.
Small to medium sized larvae (L2: 6-7 mm); shape fusiform, narrow, abdomen rounded, slender.
Cephalic markings light or dark; lateral epicranial markings narrow, elongate; mesal epicranial and/or postfrontal markings faint, either with a conspicuous median stripe or with pair of small frontal markings.
Cephalic setae: all primary setae including S12 present; Vx obscure, probably two setae and a pore or three setae.
Antennae with terminal segment (flagellum) stubby, short (~0.05–0.06x length of pedicel).

Mouthparts elongate (mandible ~1.7–1.8x longer than head capsule).
Mandible with single seta on basolateral margin.
Labial palp with terminal segment bearing 4 lateral sensilla.
Thorax without lateral tubercles.
Paired prothoracic sclerites (Scl) elongate.
Metanotum with three distinct folds, each bearing a transverse band or row of unhooked, blunt to clavate (bacilliform) setae.
Legs of normal size in relation to body; tarsi dark brown thoughout.
Abdominal segments II–VII without lateral tubercles, without distinct lateral setae (LS).
Abdominal segments I–VII dorsally with transverse rows or bands of numerous short, blunt to clavate (bacilliform) setae (SMS).
Abdominal segments V–VII without laterodorsal tubercles, without LDS.

Dictyochrysa (L1)
Larvae naked, without a layer of camouflaging material.
Small to medium sized larvae (L1: ~2 mm); shape fusiform, narrow, abdomen rounded, slender.
Cephalic markings unknown.
Cephalic setae: S12 absent; S8–10 unknown; Vx short.
Antennae with terminal segment (flagellum) rounded apically, pedicel length/flagellar length = 9 (number of times flagellum shorter than pedicel) (from Figure 3 [14])
Mouthparts nearly twice as long as head capsule.
Number of setae on mandible unknown.
Thorax without lateral tubercles.
Paired prothoracic sclerites (Scl) triangular.
Metanotum with three distinct folds, each bearing a transverse band of unhooked setae with acute tips.
Legs fairly long in relation to body; tarsi dark brown throughout.
Abdominal segments II–VII without lateral tubercles.
Abdominal segments I–VII dorsally with transverse bands of blunt setae (SMS).
Abdominal segments V–VII without laterodorsal tubercles, without LDS.

Box 2: Summary of larval features of nothochrysine genera with described larvae. In addition to information in this paper; see [6, 13–18].

3.2. Larval Behavior

The specimens did not carry debris and they gave the overall impression of a Chrysoperla larva, but their bodies appeared thicker and more rounded (Figures 1(a)–1(c)). They remained motionless during much of the time we observed them; occasionally they moved quickly. The third instars, in the laboratory, fed readily on eggs of E. kuehniella; one specimen spun a delicate, oval shaped cocoon on the side of its vial 5-6 days after captivity; small pieces of debris were incorporated into the outer layers of the cocoon. The larva (within the cocoon) remained in diapause for 4 to 5 months, pupated ~155 days after spinning, and emerged as an adult 15 days later. The pharate adult exited the cocoon via a circular opening that it cut at the head end of the cocoon.

Figure 1: Pimachrysa fusca Adams, Joshua Tree National Park, Riverside Co., CA. (a, b, and c) Body, dorsal, illustrating the variation in the dorsal body markings among three third-instar specimens collected at the same locality on the same day. (d) Antenna (second instar), lateral, showing foreshortened flagellum typical of nothochrysine larvae. (e) Labial palps (second instar), lateral, showing row of five sensilla on terminal segment.

3.3. Generic Diagnosis

Our P. fusca larvae have the three diagnostic features that typify nothochrysine larvae. (i) The flagellum is very short relative to the pedicel (~1/14th the length of the pedicel) (Figure 1(d)). (ii) The terminus of the flagellum has a group of short apical setae. (iii) More than three lateral sensilla are present on the terminal segment of the labial palps (L2 = 5, L3 = 4-5) (Figure 1(e)).

Among the Nothochrysinae with known larvae, Pimachrysa (P. fusca) closely resembles Hypochrysa (two species) (Box 2). That is, the larvae (second and third instars) from both genera have elongate, naked bodies; the thorax and abdomen lack lateral tubercles; the abdomen has no laterodorsal tubercles. In addition, none of the setae are hooked, and the primary cephalic setae of the Pimachrysa larvae are slightly blunt and thus similar to those of Hypochrysa. Finally, the Pimachrysa head and body markings resemble those of H. elegans; that is, larvae of both species have dark brown, elongate, dorsal head markings, and dark brown, sublateral stripes on the thorax, and abdomen (most noticeable on the L3; see Figures 1(a), 1(c), 2(a), 2(b), and 6).

Figure 2: Pimachrysa fusca Adams, second instar (Joshua Tree National Park, Riverside Co., CA). (a) Body, dorsal. (b) Body, lateral. (c) Body, ventral. (d) Head, dorsal. (e) Head, ventral. (f) Head, lateral. (g) Thorax, dorsal. (h) Hind leg, anterior. (i) Tarsus, dorsal.

However, a number of distinguishing features differentiate the larvae of the two genera. First, other than the cephalic setae, all other setae on the Pimachrysa larvae, especially the abdominal submedian setae (SMS), are short and acutely tipped, whereas on Hypochrysa they are short and blunt or clavate. Second, the jaws of Hypochrysa are longer than those of the Pimachrysa (Hypochrysa: 1.7-1.8x head length; Pimachrysa 1.4–1.6x head length). Third, the white-to-tan body coloration of the Pimachrysa larvae differs markedly from the green-to-bright green coloration of both H. elegans and H. africana. In our opinion, the above differences offer significant supporting evidence for the separate identity of the two genera.

4. DESCRIPTION OF PIMACHRYSA FUSCA ADAMS

4.1. Second Instar

4.1.1. Body (Figures 2(a)–2(c))

Length ~5.5 mm (measured in lateral view through spiracles), height ~0.94 mm (thickest section of abdomen). Cream-colored-to-white with dark brown dorsolateral and lateral markings on each segment; lateral surface with brown dorsal band above lateral band of white and lower narrow band of dark brown; pronotal sclerites brown. All setae short, smooth, pale, of two types: "simple" = with acute tip (most setae) or "slightly blunt" = with slightly rounded tip (some primary cranial setae). Thoracic, abdominal setae often with swollen, lightly sclerotized setal bases (Figures 3(a) and 3(c)), these setal bases becoming increasingly larger and more sclerotized posteriorly.

4.1.2. Head (Figures 2(a)–2(c), 4(a), and 4(b))

Dorsum white, with dark brown markings as in figures; frontal marking elongate, without lateral arms; mesal epicranial and/or postfrontal markings reduced to pair of posterior spots below frontal marking; lateral epicranial marking elongate, extending to lateral side of antennal base; eyes with stemmata clear, integument between stemmata, surrounding all stemmata dark brown to black; cranium roughly triangular, with posterior margin (dorsal) rounded, ventral margin elongate; width (across eyes) ~0.79 mm, length (dorsum) ~0.64 mm, depth (midregion to top of eye) ~0.20 mm; base mostly exposed. Clypeolabral area elongated distally, extending over basal segment of labial palps; with large brown mark mesally. All dorsal primary cephalic setae present except S10 and S12; S11 long, S3 short and shifted mesally, others of medium length, most slightly blunt apically; Vx setae obscured by cervix, probably with three small setae (as drawn on Figure 4(a)) or two setae and pore; labrum with three pairs of setae (one mesally, two laterally); dorsum without secondary setae.

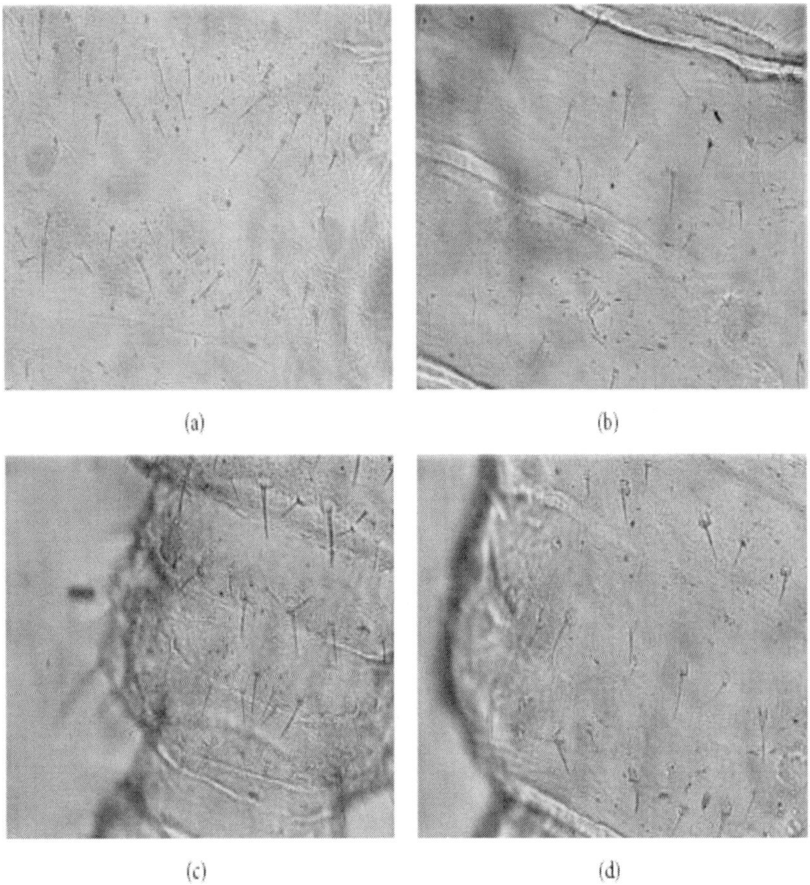

Figure 3: Pimachrysa fusca Adams (Joshua Tree National Park, Riverside Co., CA), thoracic (a, b) and abdominal (c, d) integument and setae, dorsal. Note slight differences in size of setae and chalazae between second (a, c) and third (b, d) instars. All images are the same magnification (×100).

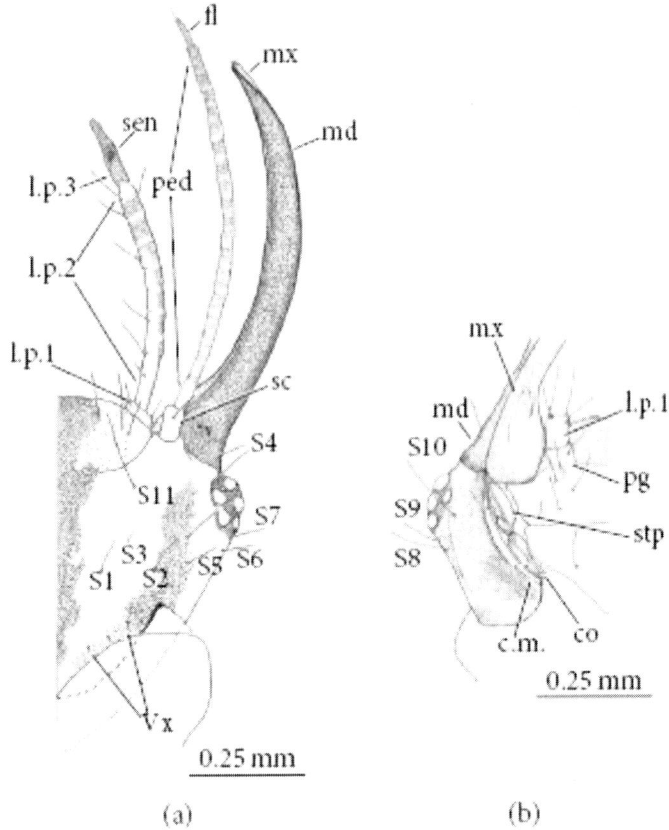

Figure 4: Head of Pimachrysa fusca Adams, second instar. (a) Dorsal. (b) Ventral. co: cardo; c.m.: cranial margin; fl: flagellum; l.p.x: labial palps, number of segment; md: mandible; mx: maxilla; ped: pedicel; pg: palpiger; sc: scape; stp: stipes; Sx: primary seta number; Vx: Vx setae.

Cervix full laterally, ventrally, slightly withdrawn from cranium dorsally; with row of 6–8 pairs of setae along anterior margin, dorsum to venter.

4.1.3. Cephalic Appendages (Figures 2(d)–2(f), 4(a), and 4(b))

Mandible long, thin, length ~0.88–0.90 mm (along curved exterior margin), length 0.74–0.78 mm (straight line), width (at base) 0.10–0.13 mm; ratio mandible length (curved margin) to head width = 1.11–1.13; ratio mandible length (curved margin) to head length (dorsal) = 1.37–1.40. Mandible very slightly upturned distally, with single acute basolateral seta; terminus sharp, with four very small teeth. Antennal length 0.94 mm, ~1.5x length of cranium; width ~0.04 mm (at widest part of pedicel); scape with relatively straight sides, two pairs of distal setae (one lateral, other dorsal); pedicel annulate, with ~18–22 medium-length annulations mesally, one elongate annulation basally, one distally, distal annulations with two small setae, one lateral, one mesal, distomesal margin concave, bearing small claw; flagellum short, stubby, ~0.08 times length of pedicel, terminus with one short apical seta, surrounded by three or four microsetae. Labial palp long, slender, ~0.85 times length of mandible (straight line); basal segment with four distal setae, two dorsally, two ventrally; middle segment long, with long, undivided basal subsegment bearing ~three setae mesally, with ~six mesal subsegments, one mesal seta; terminal subsegment elongate, with two long setae distally, one basally; terminal labial segment ~one-third length of middle segment, slightly tapered distally, heavily sclerotized, no setae observed, ventrolateral margin with five closely spaced, elongate sensilla. Palpiger erect, with relatively straight sides, one mesal seta, one lateral seta; prementum broad, with three pairs of setae; cardo, stipes elongate, narrow, longitudinally arranged, cardo behind stipes, stipes with long seta.

4.1.4. Thorax (Figures 2(g) and 5)

Lateral tubercles absent; dorsum with scattered, short, acute setae; pleurum with longitudinal, dorsal band of short bristles, spines; primary setae unidentified (except as noted below and in Figure 5). Legs (Figures 2(h) and 2(i)) white, with white-to-cream-colored setae; coxa dark brown; distal one-third of femur, basal one-fifth of tibia dark brown, with margins of all adjoining segments black distally; tarsus, claw, empodium dark brown.

Prothorax (T1) with two well-delineated subsegments separated by transverse depression. Paired Sc1 dark brown, elongate, extending full length of anterior subsegment, slightly narrow anteriorly, broader mesally, posteriorly, underlaid by dark brown, pigmented lateral stripe. Sc2 broadly ovate, extending from middle to posterior margin of anterior subsegment, with longitudinally elongate, brown marking posteromesally.

Mesothorax (T2) consisting of three well-delineated subsegments separated by distinct depressions. Anterior subsegment with pair of ovate sclerites (Sc1) on anterior margin, with single lateral seta (probably S1Sc1); spiracles simple, sessile, circular, brown, with cylindrical atrium. Second subsegment with pair of small sclerites (Sc2) on anterior margin, each with single seta (S1Sc2), juxtaposed much smaller seta (S2Sc2); subsegment separated mesally from posterior subsegment by transverse depression with pair of large, brown, circular sclerites (Sc3) laterally, each with small mesal seta (S1Sc3).

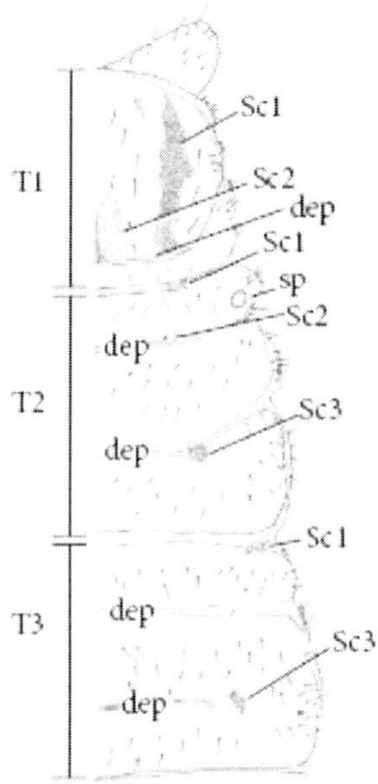

Figure 5: Thorax (dorsal) of Pimachrysa fusca Adams, second instar. dep: smooth-surfaced, intrasegmental depression between subsegments; Sc1, Sc2, and Sc3: first, second, and third primary sclerites of each segment; sp: spiracle; Tx: number of thoracic segment.

Metathorax (T3) consisting of three well-delineated subsegments separated by transverse depressions, each with pair of small setae laterally (probably S1Sc1, S1Sc2), but sclerites Sc1, Sc2 not identified. Sc3 medium-sized, brown, circular, with small mesal seta (S1Sc3).

4.1.5. Abdomen (Figures 2(a), 2(b), 2(c), and 6)

Lateral tubercles (LTs), laterodorsal tubercles (LDTs) absent; spiracles (A1–A8) circular, sessile, with simple atrium. Each segment (A1–A7) roughly quadrangular, divided into three subsegments (or raised folds) separated by transverse depressions, anterior one extending to margin of segment; spiracle located laterally on dorsum of second subsegment; dorsum of all subsegments with acute submedian setae (SMS) on chalazae of small to medium sizes, arranged in transverse rows on each subsegment; pleura with patches of small, mostly acute, sometimes blunt setae, spines. A1: anterior subsegment short, dorsum with four pairs of SMS; second,

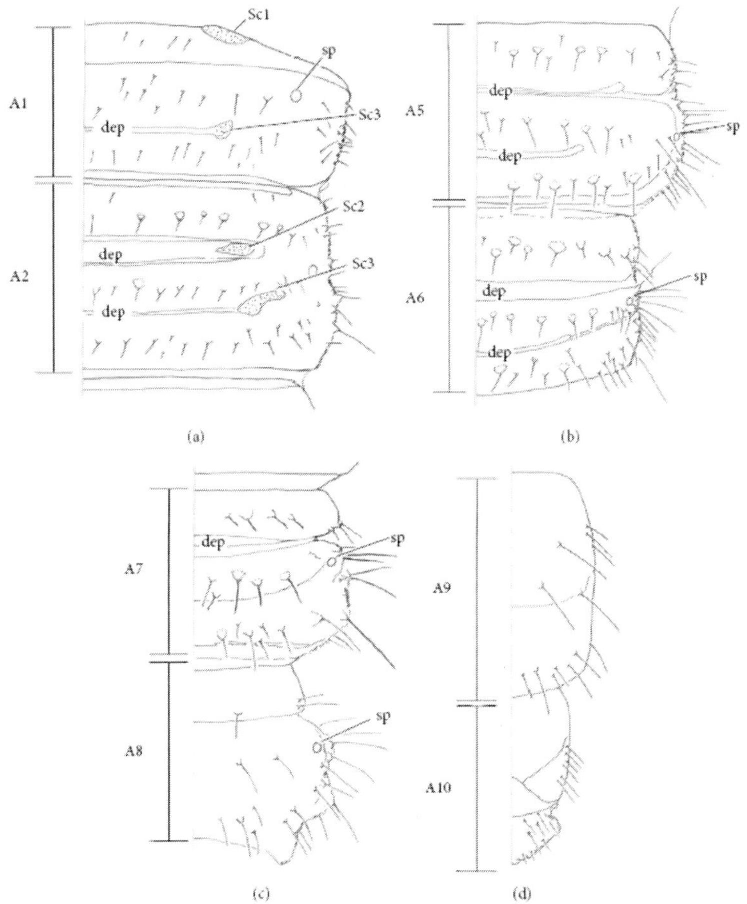

Figure 6: Abdomen (dorsal) of Pimachrysa fusca Adams, second instar. (a) First and second segments. (b) Fifth and sixth segments. (c) Seventh and eighth segments. (d) Ninth and tenth segments. Ax: number of abdominal segment; dep: smooth-surfaced, intrasegmental depression between subsegments; Sc1, Sc2, and Sc3: first, second, and third primary sclerites of each segment; sp: spiracle.

third subsegments longer, similar in size to each other, each with row of ~ten to twelve pairs of SMS, most stemming from relatively small chalazae. A2–A7: dorsum of each subsegment with ~nine to twelve pairs of SMS; number and length of SMS, size (plumpness) of chalazae becoming progressively larger posteriorly. A7: first subsegment with four pairs of short SMS stemming from relatively large, plump chalazae; second and third subsegments each with five to seven pairs of SMS, most with plump chalazae. A8: segment divided into three subsegments; anterior subsegment narrow, well-delineated from subsequent subsegment; second, third subsegments not well-delineated from each other; dorsum of anterior subsegment with one pair of SMS, middle subsegment with two pairs of SMS; posterior subsegment with ~seven pairs of SMS; pleural region of segment with two, unsclerotized, lateral swellings, each bearing ~five medium length setae. A9: cylindrical, with two visible subsegments; anterior subsegment without setae anteriorly, with ~three pairs of robust dorsal setae posteriorly, lateral-most pair on swelling; posterior subsegment with row of ~seven pairs of medium length setae on robust chalazae posteriorly. A10: dorsum with ~six pairs of setae laterally, ~eight pairs of setae extending from posterior membrane on terminus. Venter with subsegmentation poorly demarcated; each segment (A1–A8) with two transverse bands of acute setae and spines (microsetae), pair of large, sublateral patches of relatively robust setae and spines (microsetae); setae often longer than those on dorsum, but without enlarged chalazae.

4.2. Third Instar

Similar to second instar except for measurements and noteworthy features, as follows.

4.2.1. Body (Figures 1(a)–1(c), 7, 8(a), and 8(b))

Body length (measured in lateral view through spiracles) 8.5–9.0 mm (L2: 5.5 mm). Body depth (thickest) 1.5-1.6 mm (L2: ~0.94 mm). Subsegmentation, setation, coloration as in L2. Setal size approximately same as L2. Thoracic, abdominal setal bases sometimes slightly swollen, sclerotized (Figures 3(b) and 3(d)).

4.2.2. Head (Figures 7(a) and 8(c))

Head width (across eyes) 0.77-0.78 mm (L2: 0.79 mm). Head length (dorsum) 0.63-0.64 mm (L2: ~0.64 mm). Head depth (midregion to top of eye) 0.18-0.19 mm (L2: ~0.20 mm).

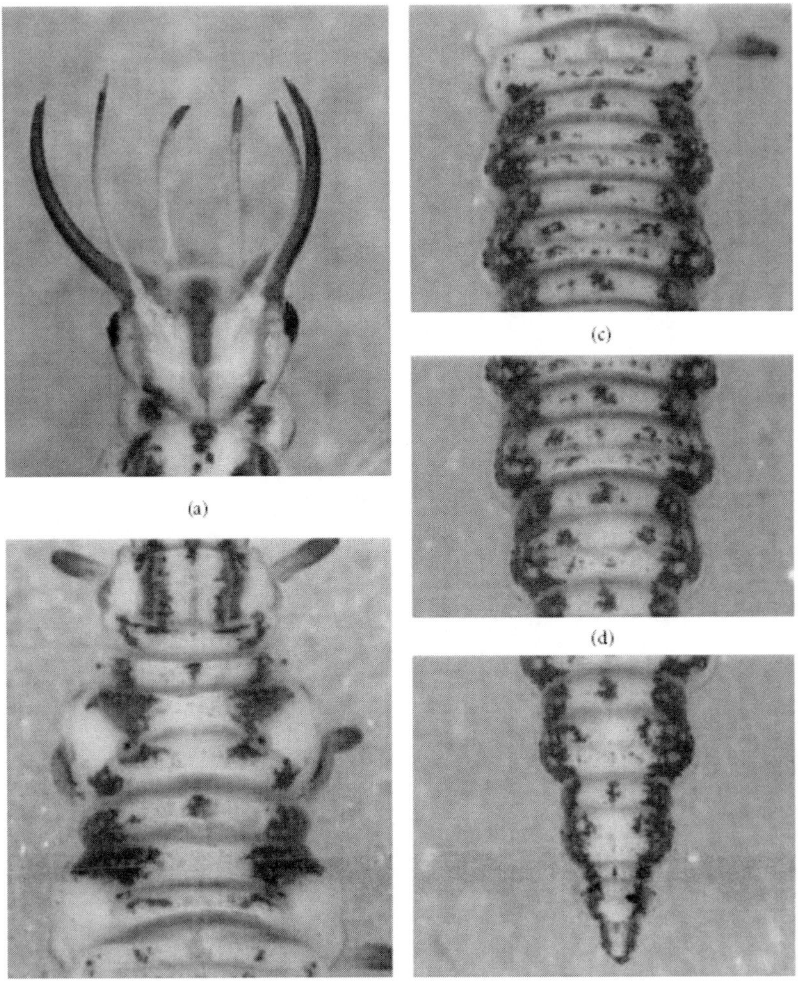

Figure 7: Pimachrysa fusca Adams, third instar (Joshua Tree National Park, Riverside Co., CA). (a) Head, dorsal. (b) Thorax, dorsal. (c) Abdominal segments 1 through 3, dorsal. (d) Abdominal segments 4 and 5, dorsal. (e) Abdominal segments 6 through 10, dorsal.

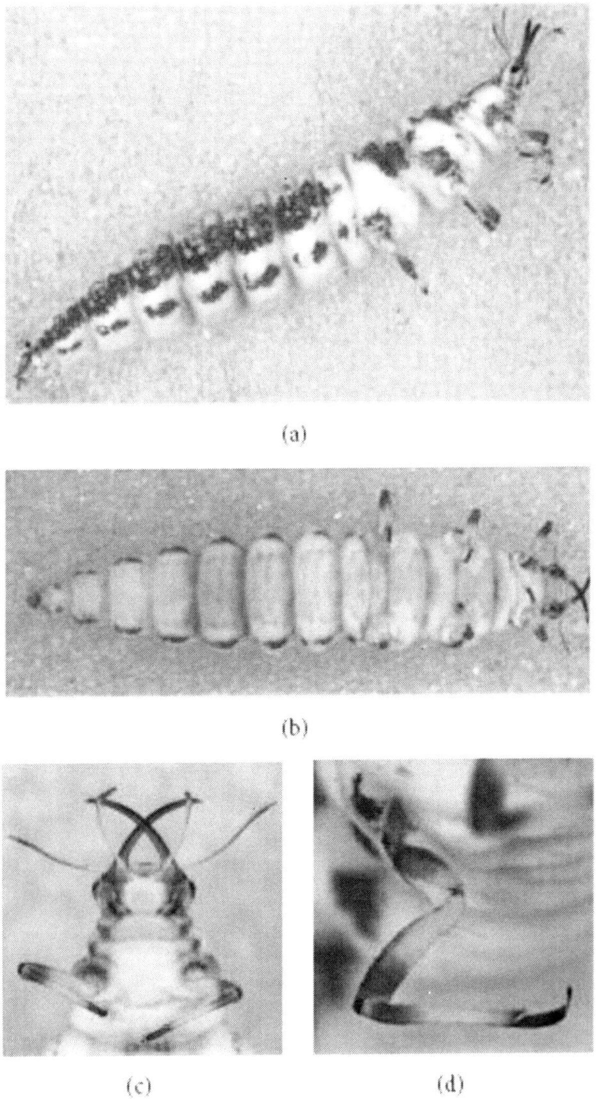

Figure 8: Pimachrysa fusca Adams, third instar (Joshua Tree National Park, Riverside Co., CA). (a) Body, lateral. (b) Body, ventral. (c) Head, prothorax, ventral. (d) Hind leg, anterior.

4.2.3. Cephalic Appendages (Figures 7(a) and 8(c))

Mandible length (along curved exterior margin) 0.93–0.97 mm (L2: ~0.88–0.90 mm). Ratio, mandible length to head width = 1.19–1.29 : 1 (L2: 1.11–1.13 : 1). Ratio mandible length to head length (dorsal) = 1.46–1.57 : 1 (L2: 1.37–1.40). Antennal length ~0.96 mm, ~1.5x length of cranium (L2: ~0.94 mm, ~1.5x length of cranium). Flagellar length 0.06-0.07 mm (L2: 0.06 mm); terminus with single very short apical seta, few, if any, surrounding microsetae. Ratio, length of flagellum : length of pedicel 0.08 : 1 (L2: 0.07 : 1). Ratio, length of labial palp : length of mandible (straight line) ~0.78 : 1 (L2: ~0.85 : 1). Labial palp: middle segment with ~seven subsegments; ventrolateral margin of terminal segment with four to five closely spaced, elongate sensilla.

4.2.4. Thorax (Figures 7(b) and 8)

Legs as in L2, except coxa white distally; basal one-third of tibia dark brown.

4.2.5. Abdomen (Figures 7(c)–7(e), 8(a), and 8(b))

As in L2, with following exceptions: A1: anterior subsegment dorsally with eight pairs of SMS; second, third subsegments each with row of ~twelve to fifteen pairs of SMS, most stemming from relatively small chalazae. A2–A7: dorsum of each subsegment with ~nine to fifteen pairs of SMS—one to three more than L2. A8: dorsum of anterior subsegment with two pairs of SMS; pleural region with single pair of broad, unsclerotized, lateral swellings, each bearing ~ten setae. A10: basal portion of segment withdrawn and shorter than on L2. Venter similar to L2.

4.3. Size Differences between Second and Third Instars

Among our specimens, the L2 () and L3 () express some unusual variation in the relative sizes of body parts. The L3s, as expected, have much larger bodies than the L2, and yet the head size is almost identical among our specimens of the two instars. Also, while the L3 mandibles appear somewhat larger than those of the L2, the antennae and labial palps do not. In addition, the L3 setal size is approximately the same as in the L2, but the degree of enlargement and sclerotization of the abdominal setal bases is less than on L2 (Figures 3(a) and 3(c) versus 3(b) and 3(d)). Thus, the L3 setae seem less prominent than those of the L2 (Figure 3).

Given the small number of specimens, we are puzzled by the above variation in the relative sizes of larval body parts. First, we do not know whether our observations represent the normal situation for this species. Second, if it does, we suggest that the variation could be related to some unusual pattern of annual or seasonal variation in the availability of food or perhaps seasonal or generational differences in physical factors.

4.4. Cocoon (Figures 16(a) and 16(b))

White, oblong longitudinally, round transversely, with basal side spun directly against substrate; attached base with single layer of thin, transparent silk; upper surface with two layers of silk: inner layer: dense, relatively smooth and outer layer: sparse, mixed with debris.

5. COMPARATIVE INFORMATION ON NOTHOCHRYSA CALIFORNICA BANKS

Previously, the three instars and cocoon of N. californica were described in detail [16]. However, full comparisons with other described nothochrysine immatures are hampered because the descriptions lack adequate illustrations and several larval characters were not included. Here, we provide images and drawings of the N. californica second instar (Figures 9–13) and the following notes on all instars for use in comparisons. All observations and measurements below are based on one to two specimens per instar: Mandible length (along curved exterior margin) = L3: 1.63–1.66 mm; L2: 1.15 mm; L1: 0.80 mm. Antennal length = L3: ~1.65 mm, 1.5x length of cranium; L2: 1.24 mm, 1.8x length of cranium; L1: 0.84 mm, 1.6x length of cranium. Pedicel with claw at tip = L3: yes; L2: yes; L1: yes. Pedicel length = L3: 2.16 mm; L2: 0.97 mm; L1: 0.58 mm. Flagellar length = L3: 0.18 mm; L2: 0.17 mm; L1: 0.17 mm. Length of flagellar seta = L3: 0.03 mm; L2: 0.03 mm; L1: 0.04 mm. Labial palp, number of lateral sensilla on terminal segment = L3: 4; L2: 4-5; L1: 0.

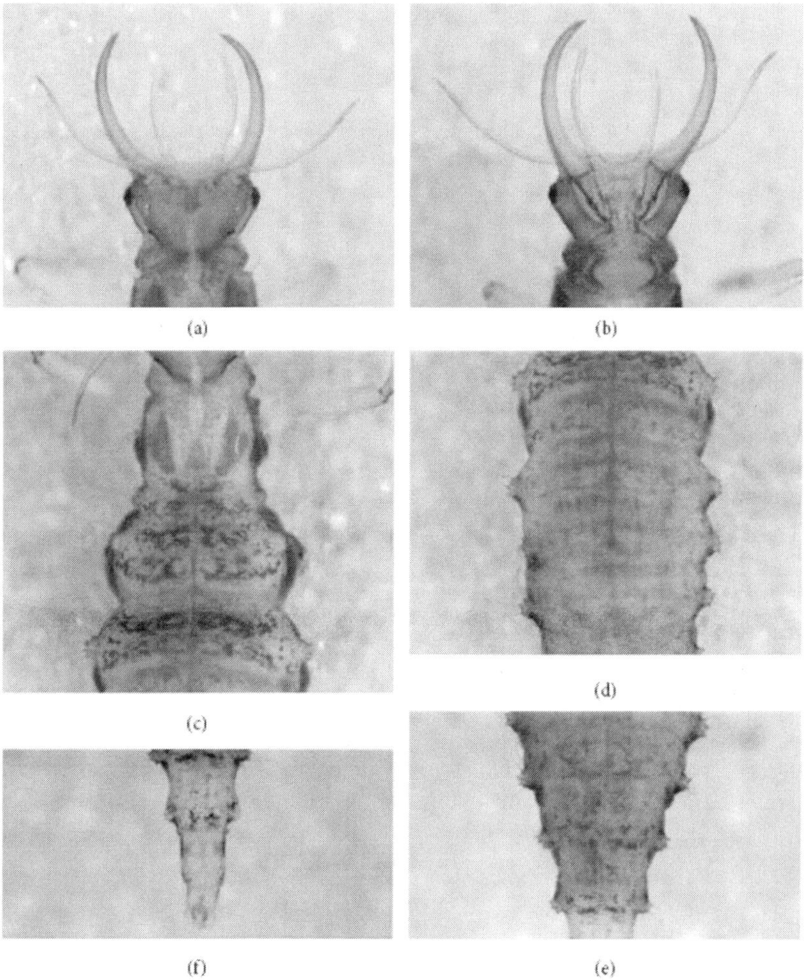

Figure 9: Nothochrysa californica Banks, second instar (Strawberry Canyon, University of California Campus, Alameda Co., CA). (a) Head, dorsal. (b) Head, ventral. (c) Thorax, dorsal. (d) Abdominal segments 1 through 4, dorsal. (e) Abdominal segments 5 through 7, dorsal. (f) Abdominal segments 8 through 10, dorsal.

Figure 10: Nothochrysa californica Banks, second instar (Strawberry Canyon, University of California Campus, Alameda Co., CA). (a) Body, dorsal. (b) Body, lateral. (c) Body, ventral.

6. UNIQUE LARVAL CHARACTERS FOR NOTHOCHRYSINAE

Three larval features have been proposed as unique, and perhaps apomorphic, for the subfamily Nothochrysinae [5, 13, 18]. Given the recent data from Hypochrysa and now Pimachrysa, as well as our reexamination of the N. californica larvae, we were led to reevaluate the diagnostic and phylogenetic value of these characters. In addition, we propose an addition larval character that may be unique to the Nothochrysinae. For our larval comparisons, we used our own measured values and estimated others from published drawings (Dictyochrysa: Figure by New [14]; N. capitata, N. fulviceps: Figures 46 and 47 by Monserrat and Díaz-Aranda [13]).

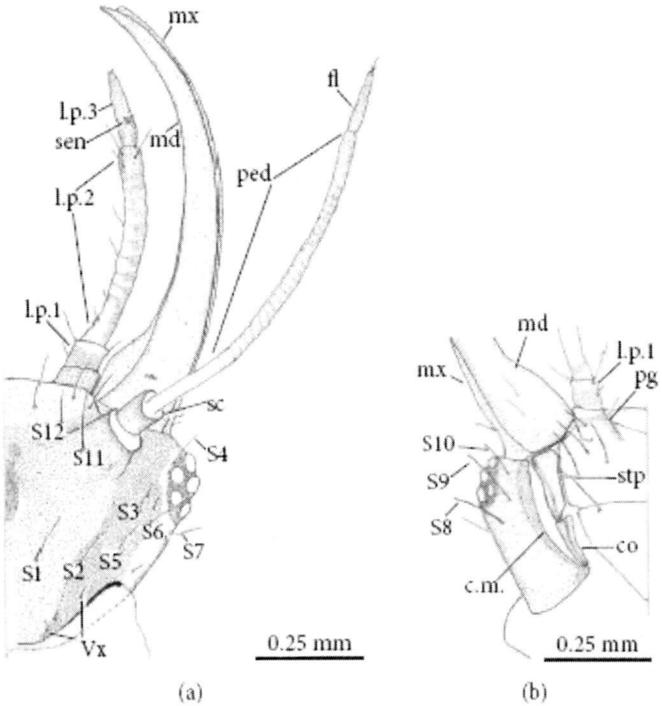

Figure 11: Head of Nothochrysa californica Banks, second instar. (a) Dorsal. (b) Ventral. co: cardo; c.m.: cranial margin; fl: flagellum; l.p.1–3: labial palps, number of segment; md: mandible; mx: maxilla; ped: pedicel; pg: palpiger; sc: scape; sen: sensilla; stp: stipes; Sx: primary seta number; Vx: Vx setae.

6.1. Length of Flagellum

The nothochrysine antenna is characterized as having a terminal segment (flagellum) that typically is short and stubby, that is, approximately 10 or more times shorter than the pedicel; in larvae of Apochrysinae and Chrysopinae, the flagellum is considerably longer—only about 3–6 times shorter than the pedicel. For Hypochrysa and Pimachrysa second and third instars, as well as N. fulviceps and N. californica third instars, this character is consistent (their flagella are ~12–20 times shorter than the pedicel). Also, for the Dictyochrysa first instar, it is close at ~9 times shorter than the pedicel. However, there are interesting examples of interspecific and developmental variation in Nothochrysa. For example, the third instar of N.

fulviceps, with a flagellum that is 12.5 times shorter than the pedicel, is consistent with the character, and the third instars of both N. capitata and N. californica (with flagella 8.3–9.1 times shorter than the pedicel) are near the 10x designation (Box 2). However, the flagella of the first and second instars of N. californica are noticeably longer, with the first instar being the most deviant (L3: flagellum 9.4-9.5 times shorter, L2: 5–7 times shorter, L1: only 4 times shorter than the pedicel; see Figure 14). Whether this pattern of developmental variation applies to the first instars of the European Nothochrysa species is unknown. However, if the old maxim "ontogeny recapitulates phylogeny" holds for this genus, the pattern could have phylogenetic importance.

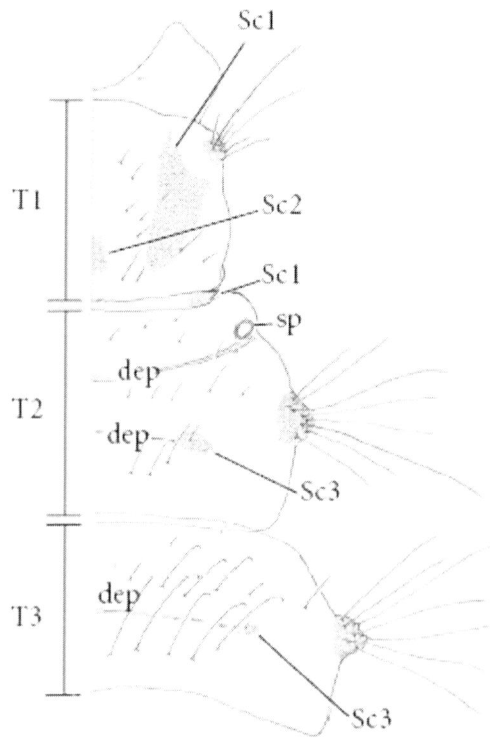

Figure 12: Thorax (dorsal) of Nothochrysa californica Banks, second instar. dep: smooth-surfaced, intrasegmental depression between subsegments; Sc1, Sc2, and Sc3: first, second, and third primary sclerites of each segment; sp: spiracle; Tx: number of thoracic segment.

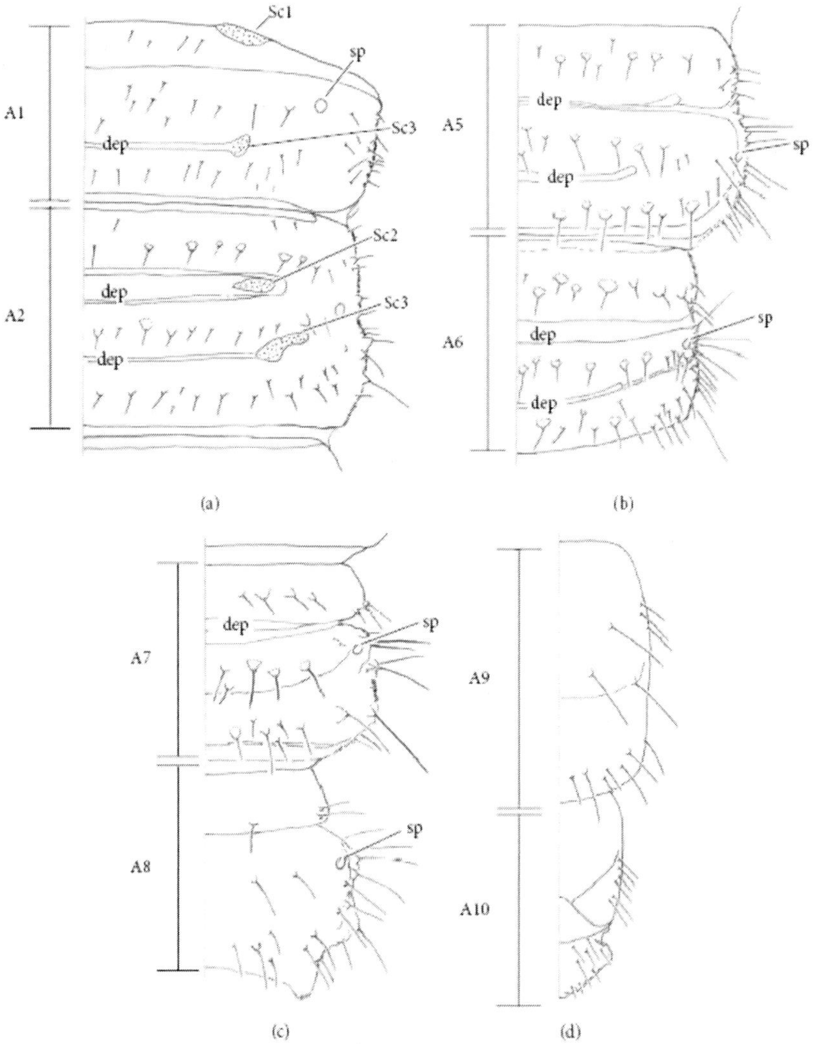

Figure 13: Abdomen (dorsal) of Nothochrysa californica Banks, second instar. (a) First and second segments. (b) Fifth and sixth segments. (c) Seventh and eighth segments. (d) Ninth and tenth segments. Ax: number of abdominal segment; dep: smooth-surfaced, intrasegmental depression between subsegments; Sc1, Sc2, and Sc3: first, second, and third primary sclerites of each segment; sp: spiracle.

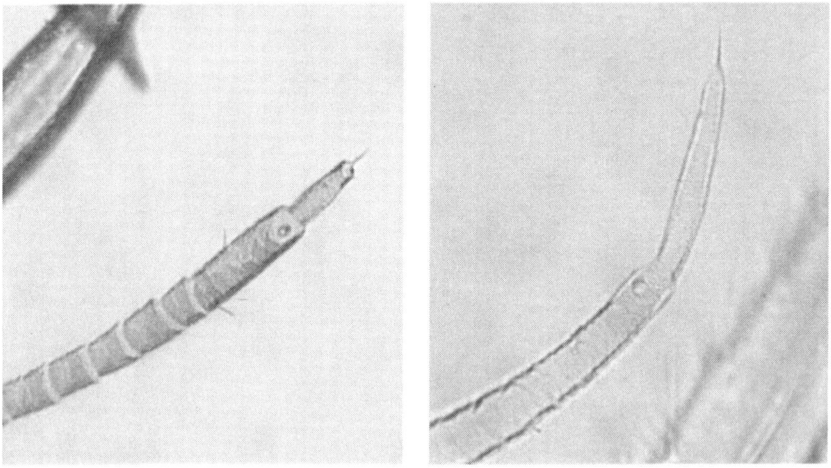

(a) (b)

Figure 14: Flagellum and distal annulations of pedicel on (a) Pimachrysa fusca Adams and (b) Nothochrysa californica Banks, second instar antenna. Note: stubby flagellum and short terminal setae on P. fusca and elongate flagellum and single, more robust terminal seta on N. californica.

6.2. Terminal Flagellar Setae

The nothochrysine flagellum is typified as bearing several, very small terminal setae; in contrast, the tip of the flagellum in Chrysopinae and Apochrysinae bears a single long seta with a robust base. Upon reevaluation, this trait holds for most of the known nothochrysine larvae—Dictyochrysa (L1, one species), Hypochrysa (L1, L2, and L3, one species each), Pimachrysa (L2, L3, one species, Figure 14(a)), and Nothochrysa (L3, two European species). However, the structure differs in N. californica; in this species the flagellum terminates with a single elongate seta having a broad base (Figure 14(b)). The length of the terminal seta varies from 0.18 to 0.25 times the length of the flagellum. Again, the deviation from the typical nothochrysine pattern is most pronounced in the N. californica first instar. Unfortunately, the situation with the first and second instars of the European species is unknown.

6.3. Number of Sensilla on Terminal Segment of Labial Palp

Thus far, this character appears to be a consistent means of distinguishing nothochrysine larvae (second and third instars, Semaphoront B; there are no sensilla on the palps of the first instar). A number of four or more sensilla is typical of nothochrysine larvae. Three or fewer sensilla are present on larvae of Apochrysinae and Chrysopinae (note: this character is difficult to examine; it requires careful placement and often manipulation of a cleared specimen with labial palps in ventrolateral position and under sufficient lighting.)

6.4. Cocoon Structure

As described here, the Pimachrysa cocoon is sessile and strongly oblong; the basal membrane of the cocoon is single, thin, and highly transparent, whereas the upper layer is double, thick, and opaque. Among the Nothochrysinae, this cocoon structure is shared with Nothochrysa (Toschi 1965: N. californica); however, the situation is not clear for Hypochrysa. Hypochrysa africana shares the oblong ("barrel") shape of Pimachrysa and Nothochrysa (Duelli et al. 2010, as Kimochrysa), but it is unknown if its cocoon is sessile or entire. In any case, the Pimachrysa and Nothochrysa cocoons appear unlike the entire, more spherical cocoons spun by larvae of many Chrysopinae (Chrysopini, Leucochrysini) that we have observed.

6.5. Synapomorphies?

Whether the above characters constitute apomorphies or symplesiomorphies is unconfirmed, and available evidence from the hemerobiids (the presumed sister group of Chrysopidae) is inconclusive. For example, among the hemerobiids, (1) the length of the flagellum exhibits intergeneric variability: in Boriomyia fidelis (Banks), the flagellum is short relative to the pedicel [19]; in Hemerobius species, it is elongate [20]; in Micromus species, it is indistinguishable from the pedicel [20]. (2) The length of the terminal flagellar seta also shows variability among hemerobiid genera: in B. fidelis and Drepanacra binocula (Newman) the flagellum is elongate [19, 21]; in Micromus it is short and accompanied by a very small seta [20]; in Hemerobius it is also short, but without a small seta [20]. To our knowledge, no hemerobiids have been shown to have a cluster of very short setae on the terminal antennal segment as found in Hypochrysa, Pimachrysa, and Dictyochrysa. (3) To our knowledge, lateral sensilla have not been reported from the terminal segment of hemerobiid labial palpi. They appear to be unique to Chrysopidae; however, the plesiomorphic number is unknown. (4) At this point, reports on the structural features of Neuropteran cocoons are not sufficient for meaningful phylogenetic comparisons.

From the above, it is obvious that the choice of an out-group is of crucial importance for the interpretation of an apomorphic versus plesiomorphic state for the various larval characters. Unfortunately, there has been relatively little comparative systematic work that compares hemerobiid and chrysopid larvae (see [22]).

7. NOTHOCHRYSINE PHYLOGENY

When Adams [1] revised the subfamily Nothochrysinae, it comprised seven extant genera and four fossil genera attributed to the Miocene. Later, two extant and eight fossil genera were added, and one extant genus was synonymized [3, 6, 23]. So, now the subfamily includes eight extant genera and twelve extinct fossil genera from the early Eocene (Ypresian) to the Miocene (Aquitanian, Burdigalian?).

Although Adams [1] included Nothochrysa within the subfamily because it retained many archaic and/or unusual features that are not expressed in the Chrysopinae, he pointed out that it has other advanced features that are more typical of Chrysopinae and Apochrysinae than Nothochrysinae (e.g., in its development of the pseudomedia and pseudocubitus, its wing folds). And recently, an extensive study of all known fossil chrysopids identified patterns of evolutionary progression in certain venational features among the nothochrysine fossils that support a distinct position for Nothochrysa [3]. But, the relationships among the extant genera remain unresolved.

Based on his morphological studies, Brooks [2] proposed that the extant genera of the subfamily fall into two groups. In his scheme Nothochrysa was included with Dictochrysa and Hypochrysa, separate from Pimachrysa and Kimochrysa. These groupings subsequently were found to be weak; indeed Kimochrysa was synonymized with Hypochrysa [6]. Finally, in another study that used molecular and other data, Duelli et al. [4] proposed that the small bodied, pollen-feeding nothochrysine genera (e.g., Hypochrysa sen. lat. (including Kimochrysa), Pamochrysa, Pimachrysa, and Asthenochrysa) are very similar to each other. These authors concurred with the opinion that although Nothochrysa shows some affinities with these nothochrysine genera, they also resemble Apochrysinae and Chrysopinae in other respects. Given their findings, they conjectured that the small bodied, pollen-feeding nothochrysines represent the most ancient group of extant chrysopids. Moreover, they concluded that pollen-feeding is the likely ancestral condition for chrysopid adults.

It appears from the descriptions of now four of the eight nothochrysine genera (Box 1) that larval characters support the conclusions of Adams, Archibald and Makarkin, and Duelli et al. [1, 3, 4]. The pattern of larval differences between Nothochrysa and the other known nothochrysine genera is consistent with the pattern of variation expressed in adult morphology. Most notable among the larval traits is the difference in larval life style between the small-bodied genera and Nothochrysa. Three of the small-bodied genera—Hypochrysa, Pimachrysa, and Dictyochyrsa—are now known to have naked larvae; they have narrow, fusiform bodies that lack lateral tubercles on the thorax and abdomen and are without laterodorsal tubercles on all abdominal segments. In marked contrast, the larvae of the European Nothochrysa species and the North American N. californica are debris-carriers; they have broad, somewhat flattened bodies, small but distinct, sclerotized, lateral tubercles on the thorax and abdominal segments A2 through A7, as well as other morphological adaptations for holding debris (Box 1).

It is noteworthy that within the family Chrysopidae, life style (larval nakedness versus debris-carrying) is associated with numerous morphological and behavioral adaptations, and yet it is an evolutionarily very labile trait. Debris carrying occurred very early in chrysopid evolutionary history [24], and both naked and debris-carrying forms have evolved numerous times within the family [5, 24]. Thus, it could be argued that the characters above are redundant—all an expression of the naked versus debris-carrying life styles. However, it is important that, in addition to features associated with life style, the larvae of Nothochrysa species also differ from other nothochrysines in features that appear unrelated to their debris-carrying habit (Box 2). For example, as discussed above, larvae of Nothochrysa species show a broad range of interspecific and developmental variation in flagellar length and in the length and structure of the terminal seta(e) on the flagellum. They also differ in their head markings and in the shape and size of the pronotal sclerites. Each of these characters provides independent support that Nothochrysa is distinct from the smaller-bodied nothochrysine genera and that species within Nothochrysa present a series of intermediate conditions between the more archaic extant nothochrysines (e.g., Dictyochrysa, Hypochrysa, and Pimachrysa) and the Apochrysinae-Chrysopinae lineage(s). If this hypothesis proves true, not only would pollen-feeding likely be an ancestral adult feature of the extant Chrysopidae (as proposed by Duelli et al. [4]), but also nakedness would likely be their ancestral larval life style.

(Note: a previous description [6] of the H. africana second instar had some errors in the designation of setal numbers on the dorsum of the head. For accuracy in comparisons with Pimachrysa and other nothochrysine larvae, Figure 15 here corrects those errors.)

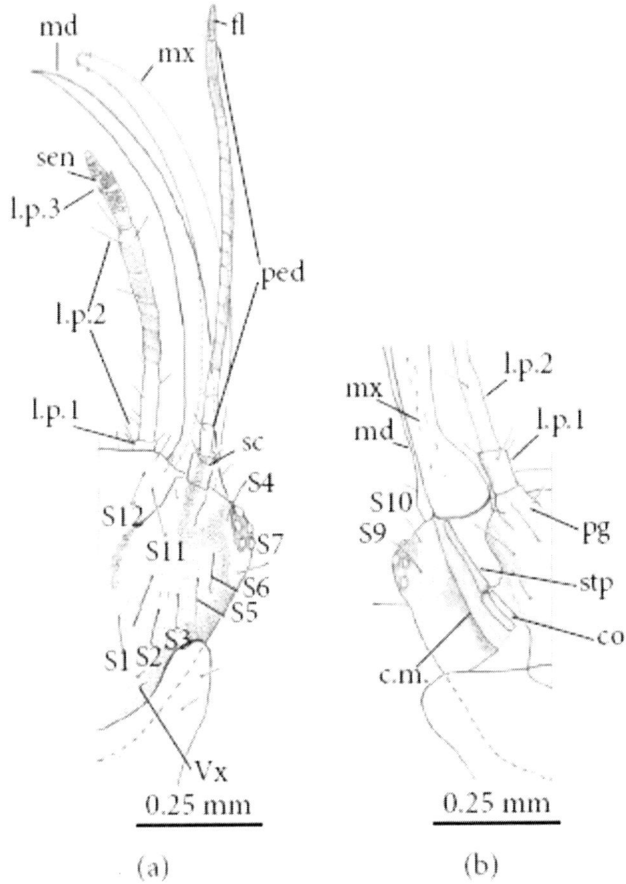

Figure 15: Head of Hypochrysa africana Kimmins, second instar. (a) Dorsal. (b) co: cardo; c.m.: cranial margin; fl: flagellum; l.p.x: labial palps, number of segment; md: mandible; mx: maxilla; ped: pedicel; pg: palpiger; sc: scape; sen: sensilla; stp: stipes; Sx: primary seta number; Vx: Vx setae.

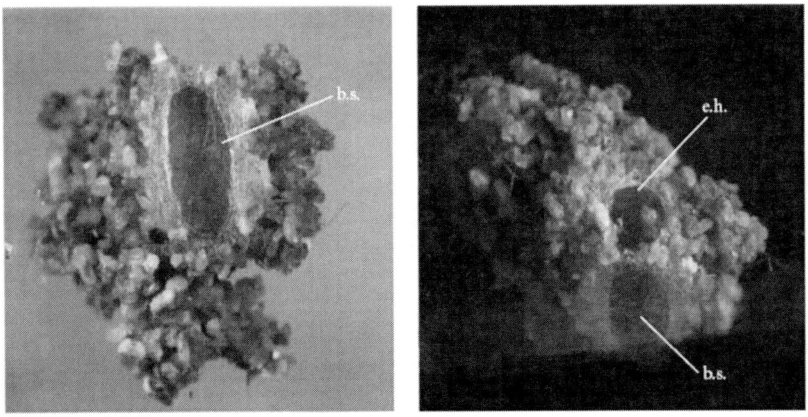

(a) (b)

Figure 16: Pimachrysa fusca Adams, cocoon (Joshua Tree National Park, Riverside Co., CA). (a) Basal surface attached to glass vial. (b) Round exit hole on upper surface (head end) of cocoon. b.s.: basal surface; e.h.: exit hole.

DISCLOSURE

This study is part of Regional Project W-3185.

CONFLICT OF INTERESTS

The authors declare that there is no conflict of interests regarding the publication of this paper.

ACKNOWLEDGMENTS

The authors thank Michael J. Tauber for his enthusiastic help collecting the second batch of larvae, Gilberto S. Albuquerque and Agatha J. Tauber for their comments on the paper, and Sheila and David Brodovsky as well as Jeanne Bellemin, Bob Allen, Michelle Rivers, and Marcus Hughes for logistical help and encouragement during the first field trip. The project benefitted from earlier funding by the National Science Foundation, the

USDA Competitive Grants Program, the National Geographic Society, and Cornell University (to Maurice J. Tauber and CAT).

REFERENCES

1. P. A. Adams, "A review of the mesochrysinae and nothochrysinae (Neuroptera: Chrysopidae)," Bulletin of the Museum of Comparative Zoology, vol. 135, no. 4, pp. 215–238, 1967.
2. S. J. Brooks, "An overview of the current status of Chrysopidae (Neuroptera) systematics," Deutsche Entomologische Zeitschrift, vol. 44, no. 2, pp. 267–275, 1997.
3. S. B. Archibald and V. N. Makarkin, "A new species of Archaeochrysa Adams (Neuroptera: Chrysopidae) from the early Eocene of Driftwood Canyon, British Columbia, Canada," Canadian Entomologist, vol. 147, no. 4, pp. 359–369, 2015.
4. P. Duelli, H. Hölzel, and M. W. Mansell, "Habitat and larvae of the enigmatic genus Kimochrysa Tjeder (Neuroptera: Chrysopidae) in South Africa," in Proceedings of the Tenth International Symposium on Neuropterology (22–25 June 2008, Piran, Slovenia), D. Devetak, S. Lipovsek, and A. E. Arnett, Eds., pp. 153–158, University of Maribor, Maribor, Slovenia, 2010.
5. C. A. Tauber, M. J. Tauber, and G. S. Albuquerque, "Debris-carrying in larval Chrysopidae: unraveling its evolutionary history," Annals of the Entomological Society of America, vol. 107, no. 2, pp. 295–314, 2014.
6. C. A. Tauber, "Nothochrysinae (Neuroptera: Chrysopidae): new larval description and generic synonymy, with a consideration of generic relationships," Psyche, vol. 2014, Article ID 839261, 10 pages, 2014.
7. S. Winterton and S. de Freitas, "Molecular phylogeny of the green lacewings (Neuroptera: Chrysopidae)," Australian Journal of Entomology, vol. 45, no. 3, pp. 235–243, 2006.
8. N. Haruyama, A. Mochizuki, P. Duelli, H. Naka, and M. Nomura, "Green lacewing phylogeny, based on three nuclear genes (Chrysopidae, Neuroptera)," Systematic Entomology, vol. 33, no. 2, pp. 275–288, 2008.
9. F. W. Stehr, "Techniques for collecting, rearing, preserving, and studying immature insects," in Immature Insects, F. W. Stehr, Ed., pp. 7–18, Kendall/Hunt Publishing Company, Dubuque, Iowa, USA, 1987.
10. A. Rousset, Morphologie Céphalique des Larves de Planipennes (Insectes Névroptéroïdes), vol. 42 of Memoires du Museum National d'Histoire Naturelle Series A: Zoology, Éditions du Muséum, Paris, France, 1966.
11. S. Tsukaguchi, Chrysopidae of Japan (Insecta, Neuroptera), S. Tsukaguchi, Nishinomiya-shi, Japan, 1995.
12. C. A. Tauber, M. J. Tauber, and G. S. Albuquerque, "Berchmansus elegans (Neuroptera: Chrysopidae): larval and adult characteristics and new tribal affiliation," European Journal of Entomology, vol. 103, no. 1, pp. 221–231, 2006.
13. V. J. Monserrat and L. M. Díaz-Aranda, "Los estadios larvarios de los Crisópidos ibéricos (Insecta, Neuroptera, Chrysopidae), nuevos elementos sobre la morfología larvaria aplicables a la sistemática de la familia," Graellsia, vol. 68, no. 1, pp. 31–158, 2012.
14. T. R. New, "Some early stages of Dictyochrysa Esben-Petersen (Neuroptera, Chrysopidae)," Neuroptera International, vol. 1, no. 3, pp. 136–140, 1981.
15. F. Brauer, "Larve von Hypochrysa nobilis Heyd," in Verhandlungen der Kaiserlich-Königlichen Zoologisch-Botanischen Gesellschaft in Wien, vol. 17, pp. 27–30, Zoologisch-Botanische Gesellschaft in Wien, 1867.

16. C. A. Toschi, "The taxonomy, life histories, and mating behavior of the green lacewings of Strawberry Canyon (Neuroptera: Chrysopidae)," Hilgardia, vol. 36, no. 11, pp. 391–433, 1965.
17. F. J. Killington, A Monograph of the British Neuroptera, vol. 2, Ray Society, London, UK, 1937.
18. L. M. Díaz-Aranda and V. J. Monserrat, "Aphidophagous predator diagnosis: key to genera of European chrysopid larvae (Neur.: Chrysopidae)," Entomophaga, vol. 40, no. 2, pp. 169–181, 1995.
19. E. G. MacLeod, "The immature stages of Boriomyia fidelis (Banks) with taxonomic notes on the affinities of the genus Boriomyia (Neuroptera: Hemerobiidae)," Psyche, vol. 67, no. 1-2, pp. 26–40, 1960.
20. A. H. Krakauer and C. A. Tauber, "Larvae of Micromus: generic characteristics and a description of Micromus subanticus (Neuroptera: Hemerobiidae)," Annals of the Entomological Society of America, vol. 89, no. 2, pp. 203–211, 1996.
21. T. R. New, "The immature stages of Drepanacra binocula (Neuroptera: Hemerobiidae), with notes on the relationships of the genus," Australian Journal of Entomology, vol. 14, no. 3, pp. 247–250, 1975.
22. J. D. Oswald and C. A. Tauber, "Recognition of larval Neuroptera. Preimaginal stages of the family Hemerobiidae," in Lacewings in the Crop Environment, P. K. McEwen, T. R. New, and A. E. Whittington, Eds., pp. 50–60, Cambridge University Press, Cambridge, UK, 2001.
23. P. A. Adams and N. D. Penny, "Review of the South American genera of Nothochrysinae (Insecta: Neuroptera: Chrysopidae)," in Current Research in Neuropterology, Proceedings of the Fourth International Symposium on Neuropterology (24–27 June 1991, Bagnères-de-Luchon, Haute-Garonne, France), M. Canard, H. Aspöck, and M. W. Mansell, Eds., pp. 35–41, M. Canard, Toulouse, France, 1992.
24. R. Pérez-de la Fuente, X. Delclòs, E. Peñalver et al., "Early evolution and ecology of camouflage in insects," Proceedings of the National Academy of Sciences of the United States of America, vol. 109, no. 52, pp. 21414–21419, 2012.

CHAPTER 7

Queen Quality and the Impact of Honey Bee Diseases on Queen Health: Potential for Interactions between Two Major Threats to Colony Health

Esmaeil Amiri [1,2,*], Micheline K. Strand [3], Olav Rueppell [1] and David R. Tarpy [2]

[1]*Department of Biology, University of North Carolina at Greensboro, Greensboro, NC 27402, USA*

[2]*Department of Entomology & Plant Pathology, North Carolina State University, Raleigh, NC 27695, USA*

[3]*Life Science Division, U.S. Army Research Office, Research Triangle Park, Durham, NC 27709, USA*

ABSTRACT

Western honey bees, *Apis mellifera*, live in highly eusocial colonies that are each typically headed by a single queen. The queen is the sole reproductive female in a healthy colony, and because long-term colony survival depends on her ability to produce a large number of offspring, queen health is essential for colony success. Honey bees have recently been experiencing considerable declines in colony health. Among a number of biotic and abiotic factors known to impact colony health, disease and queen failure are repeatedly reported as important factors underlying colony losses. Surprisingly, there are relatively few studies on the relationship and interaction between honey bee diseases and queen quality. It is critical to understand the negative impacts of pests and pathogens on queen health, how queen problems might enable disease, and how both factors influence colony health. Here, we review the current literature on queen reproductive potential and the impacts of honey bee parasites and pathogens on queens. We conclude by highlighting gaps in our knowledge on the combination of disease and queen failure to provide a perspective and prioritize further research to mitigate disease, improve queen quality, and ensure colony health.

Keywords: honey bee queen; queen quality; honey bee disease; colony health; colony fitness; queen fitness; pollinators

1. INTRODUCTION

Pollination is required for approximately 35% of global food production, and honey bees (*Apis mellifera*) are the only pollinators successfully managed on a large scale [1,2]. Forty percent of invertebrate pollinators are at risk of extinction, and even managed honey bee colonies have been dying at high rates for several decades [3,4,5,6,7]. Continued declines in pollinator health imperil global food security and economic stability.

The death of honey bee colonies is a complex phenomenon that can be attributed to multiple biotic and abiotic factors, as well as interactions among them [6,8,9,10,11]. Pathogens and parasites have been linked to poor colony health and can increase colony losses [12,13,14,15,16]. The ectoparasitic mite, *Varroa destructor*, is considered the main candidate involved in winter colony losses in most managed honey bee populations [15,17,18,19,20,21]. Furthermore, pathogenic viruses are a major threat to honey bee colonies [12], especially those associated with *V. destructor*, including Deformed wing virus (DWV), Israeli acute bee paralysis virus (IAPV), and Acute bee paralysis virus (ABPV) [12,15,19,22,23,24,25,26]. Honey bee mortality may increase when virulent forms of associated viruses are vectored by the varroa mite [27]. The microsporidian *Nosema ceranae* has also been associated with collapse of honey bee colonies in some countries, although its role remains controversial [28,29,30,31]. *Melissococcus plutonius*, the etiological agent of European Foulbrood, is a serious disease of honey bee brood and can lead to the loss of the host colony [32], which has been increasingly observed in

Switzerland and the UK [33,34]. *Paenibacillus larvae* causes American Foulbrood, which has historically been the most noxious honey bee larval disease and is also causing economic losses to beekeepers worldwide [35].

In addition to parasites and pathogens, failure or loss of queens has been considered one of the most important factors leading to colony losses [5,15,36,37,38], especially outside of the typical queen-rearing season. "Poor queens" are reported as the primary problem in a number of beekeeping operations, and it consistently ranks among the top reasons for colony failure [38,39]. The queen is the sole reproductive female in a colony, and the presence of a healthy, high-quality queen is essential for colony survival not only because of her ability to lay large numbers of female and male eggs but also because the social coherence of a colony depends on her pheromones [40]. Although queens have a 3–4 year adult lifespan [40], diminished longevity of queens (<1 year) is commonly and increasingly observed [7,39,41,42]. As such, commercial beekeepers typically replace their queens every 1–2 years [43,44,45,46] because of the critical importance of a vigorous queen to colony survival and productivity.

Here, we review the literature on queen quality and the impacts that honey bee parasites and pathogens have on queen health. We aim to highlight the gaps in this area of research to give perspective and priority for further investigations in order to improve queen quality and colony health. Ultimately, we seek to help beekeepers produce and maintain healthy and high-quality queens, in order to promote colony survival and to maintain the economic viability of the apiculture industry.

2. HONEY BEE QUEEN REPRODUCTIVE POTENTIAL

The growth, productivity, and survival of a colony, with tens of thousands worker bees, depends in large part on the health and reproductive capacity of its queen and the number of drones with which she has mated [41,46,47,48]. The quantitative and qualitative reproductive potential of a queen represents her "quality", which results from her genome, her developmental conditions, mating success, and adult environment (including beekeeper management) [49,50,51]. Here, we chose to focus on several fitness-related criteria of queen quality, including physical characteristics and mating success that define reproductive quality in queens and have been empirically linked to colony health and productivity.

2.1. Physical Quality

2.1.1. Body Size

Queen weight is a physical characteristic that is critical for evaluating the quality of honey bee queens [52,53,54,55]. Specifically, common measures have been weight at emergence (which is presumably associated with queen's ovaries; but see [56,57]), weight after mating, and the onset of oviposition [52,58]. Most often, queen weight at emergence is used as an indicator of queen quality, a measure that varies drastically [52,58,59,60]. This variation is influenced by a variety of factors, including the genetic background, the age at which the larva is initially reared as a queen, time of the year, and rearing conditions of the colony [49,52,58,61,62,63,64,65].

The weight of a queen varies at different periods of her adult life [52,54,58]. In virgin queens, it decreases gradually from emergence until mating, with the most rapid loss during the first 36 h. Heavy queens decrease their weight more than moderate and lighter queens [52,58]. After mating, queens start to recover their weights to post-emergence levels [52,58,66]. This seems reasonable because the mating flight(s) requires a lighter body for adequate lift and flight duration, otherwise it can decrease her mating success [67]. After queen mating and the onset of oviposition, a queen's ovaries start to develop, significantly contributing to an increase in body weight [68]. Queen weight, either at emergence or after becoming reproductive, is significantly correlated with queen attractiveness to worker bees, mating, and reproductive success, although different empirical studies are equivocal on the magnitude of this effect [53,55,60,61,69,70]. Weight is also positively associated with greater acceptance when a new queen is introduced into a foreign colony [70,71,72,73]. Mated and older virgin queens are more attractive to workers and may be more readily accepted than younger virgin queens, because they produce a full queen mandibular pheromone (QMP) profile [74,75], but see [76].

Most reports demonstrate a positive correlation between adult queen weight, the number of mating flights, and overall mating success [52,53,55,77]. While weight at emergence does not seem to be associated with mating [52,69], heavier queens at emergence tend to initiate oviposition later than queens with lower weight after being artificially inseminated [78]. Queen weight also has been shown to correlate with ovary weight, the size and number of ovarioles, the diameter of the spermatheca, and the number of stored spermatozoa [52,53,60,61,77].

Weight is an integrative measure of size and physiological condition. Thus, it could be argued that body weight is the most informative indicator of queen quality. However, external measurements of queen size, such as thorax width, head width, and wing lengths, have also been investigated in correlation with queen reproductive organs in different studies. Results of these studies are equivocal. For example, thorax width was positively correlated with both stored sperm number and mating frequency [53], but no correlation was found between thorax width and ovariole number, ovary weight, or mating number [55,57,77].

2.1.2. Internal Reproductive Organs

In an egg-laying queen, the ovaries occupy the vast majority of the abdominal cavity [40]. The weight of ovaries has been investigated as one of the physical criteria to assess the reproductive potential of honey bee queens [52,79,80]. Ovaries of virgin queens are morphologically different and smaller compared to egg-laying queens [68,81]; because well-developed ovaries are required for egg production, the ovaries of egg-laying queens are about eight times larger than those in virgins [68]. Ovary development occurs soon after mating and is associated with distinct gene-expression patterns in the brain and ovaries, physiological, and behavioral changes in the queen [75,82,83]. The weight of ovaries in a mature egg-laying queen not only depends on the number of ovarioles but also on the number and developmental stage of eggs in them. During winter, egg-laying activity by queens decreases or stops, which results in the queen having smaller and less-developed ovaries [68]. Ovary size and fertility are usually positively correlated [62]. However, under certain circumstances, perhaps due to stress or disease, this relation may not hold [52]. The number of ovarioles can be evaluated at any time during the life of a queen, but it is most reliably scored a few months after mating [84]. In addition to ovariole number, the size of a honey bee queen ovary is determined by the length of the ovarioles, which is more flexible during a queen's lifetime and reflects her physiological status. Symmetry between the left and right ovaries may [84] or may not [57] be consistent within and among queens or honey bee races [54,57]. Queen ovary size and symmetry are affected by larval nutrition, and during artificial queen production the age of larvae that are transferred into queen cells is critical [60,85], but see [56,62].

After mating, spermatozoa are stored in the spermatheca for the duration of the queen's remaining life [40]. The number of stored sperm and their viability (i.e., the percentage of live stored sperm) are two critical measures of the reproductive capacity of the queen and her lifespan [59,86]. Therefore, the spermatheca size is another valid measure of physical queen quality, because larger spermatheca can hold more sperm. The spermatheca wall is composed of a single layer of columnar epithelium, lined internally with a thick mucinous cuticle and covered with an extensive tracheal network [84]. Spermatheca size can be measured with or without the tracheal net, and its diameter should be larger than 1.2 mm for high-quality queens [54,84]. The size of the spermatheca in a queen is also influenced by rearing conditions and the genetic source of the queen and is inversely proportional to the age at which the queen is initially reared as larvae [55,62]. Queens raised from newly hatched larvae have larger spermathecae [56,62,80]. This measurement has been used as a direct estimation of the volume and as an indirect estimation of the theoretical maximum number of spermatozoa stored in spermatheca [55,61,77,84]. However, spermathecae are rarely filled completely, as the occupied volume in experimental queens is only an average of 47% [55,77].

2.2. Mating Quality

2.2.1. Mating Number

Very early in their lives (1–2 weeks of age), honey bee queens take one or several mating flights [87] to mate with drones. Mating usually occurs at a considerable distance from their natal colonies at so-called drone congregation areas (DCAs) [88,89,90,91], presumably to avoid inbreeding. Honey bee queens are highly polyandrous, with the typical mating number of queens varying between 6 and 26 with an average of 12–14 [53,87,89,92]. Such estimates have typically been obtained using genetic marker analysis (normally microsatellites) applied to the worker offspring of a queen to determine the number of subfamilies (patrilines) and therefore the number of drones with which she mated [53,89,92]. There have been dozens of theories to explain polyandry among insects [93,94]. While there are certainly fitness costs of multiple mating at the individual level (e.g., longer exposure to predation, increased exposure to diseases [95]), several counterbalancing benefits to polyandry exist. Perhaps most importantly, multiple mating increases the genetic diversity within the colony, leading to an increased likelihood that the genetic resources exist within the colony to withstand new biotic or abiotic threats [96]. More specifically, multiple mating has been shown to offer benefits including an increase in the number of stored sperm [53,91], enhanced queen attractiveness [75], enhanced division of labor within the colony [97,98,99], stabilization of brood nest temperature [100,101], improved communication among the workers [102,103,104], reduced incidence of disease, and improved colony fitness [48,105,106,107,108], all of which positively affect colony growth and survival [41,48]. It should be noted that mating number is generally used as a proxy for genetic diversity, if the drones are too inbred or exhibit low genetic diversity, the advantage of multiple mating for the queen and the colony will be muted. Overall, however, increased mating number and the resultant intra-colony genetic diversity of nest-mates is another important criterion for determining the reproductive quality of queens [48,106].

2.2.2. Insemination Success

Queen longevity is inextricably linked to an adequate number of viable stored sperm in the spermatheca, because queens are often replaced once they begin to lay unfertilized drone eggs inside the worker brood nest as a result of sperm depletion [40]. The critical stage of sperm storage occurs immediately after mating [109] when a queen returns to her hive with an average of 10–20 μL of semen (containing 100 million spermatozoa) in the median and lateral oviducts within her genital tract [110]. The vast majority of this semen is discharged, with only 3%–5% of each drone's sperm actively migrating into the spermatheca where it is stored [110]. The estimated number of stored sperm in a queen's spermatheca has been used to assess her reproductive quality [53,55]. Woyke [110] considered queens that carried fewer than 3 million sperm to be "inadequately mated". Different studies have shown that 13.6%–19.0% of commercially produced queens are below this threshold which indicates a serious problem, although significant variation among commercial sources exist [53,55,79,111]. Insufficient sperm supply clearly constitutes a fitness disadvantage for the queen herself as well as her colony, as the total number of sperm decreases with the age of the queen through use and abiotic stressors such as pesticides [112,113].

The viability of spermatozoa is also a crucial parameter of mating and reproductive success, which should be maintained at a high level throughout mating and storage [114]. After sperm storage, sperm cells remain quiescent and the queen provides them with secretions from the spermathecal gland to keep the sperm viable over several years [115,116]. Sperm viability in commercially produced queens showed significant differences across queen producers with an average of 83.7% [55]. Generally, higher sperm mortality is observed in the spermatheca of older queens [112]. Even the movement patterns of the spermatozoa stored in the spermathecae of older queens are slower than those from younger queens [86]. Low sperm viability due to temperature spikes has also been directly linked to colony failure [117].

3. PARASITES AND QUEEN QUALITY

Queens are considered to be less susceptible to infections than workers [118]. Queens are continuously attended to and fed by young worker bees [40], which may provide physical and social barriers to protect them from infection [119,120]. In addition to such social immune mechanisms, queens may also be inherently better equipped with individual immune defenses [121]. Nonetheless, queens are susceptible to most of the diseases that infect worker bees, including *Nosema* spp. [28,122,123,124,125], the tracheal mite *Acarapis woodi* [111,126,127,128], and numerous viruses [53,129,130,131,132,133]. Although studies have measured parasite and pathogen levels in queens, only a few studies have investigated the impact of these threats on queen health, performance, and lifespan [119,130,134,135]. Even fewer studies have investigated transmission routes of these pathogens and parasites to the queen [95,136,137,138] and the vertical transmission to offspring throughout the colony [138,139,140]. Due to modern apicultural practices, queens represent an important mode of inter-colonial disease spread even over long geographic distances [141]. Thus, the physiological quality of queens may be affected by diseases but the vector capacity of the queen for diseases is another measure of quality [54,142].

3.1. Varroa Mites

The parasitic varroa mite, *V. destructor*, is an obligate ectoparasite of honey bees and generally considered to be the most serious threat for managed honey bee populations [11,16,17,18,143,144]. The impact of varroa mites on honey bee health is particularly severe because varroa can also act as a vector for a number of bee viruses [145,146,147,148]. The act of feeding on bees by varroa directly injects a large number of viral particles into the host [145,147,149], selecting for certain (more virulent) virus strains and causing viral pathologies, such as immunosuppression, weight loss, decreased flight ability, and reduced lifespan [19,150,151,152]. Population increases of varroa mites lead to varroosis in the colony [20], which ultimately leads to colony loss. The reproductive success of a mite is positively correlated with its host's post-capping developmental duration. Consequently, mites prefer to parasitize drone brood over the workers and queens [153,154] presumably because the drones' longer development time increases mite fitness. Drone brood suffers from higher varroa parasitism than worker brood because of active choices by the mites based on brood or food odors, or because a slower drone development and more nurse bee visits translate into more opportunities to infest drone cells [155,156].

Contrary to their preference for drones, mites are rarely observed in queen cells. This aversion is likely an adaptive response to the shorter post-capping stage of queen brood (8–8.5 days) and may be mediated by intrinsic differences of larval scents and presence of octanoic acid in royal jelly [154,157]. Because of the short post-capping period of queens, varroa mites cannot successfully complete their reproductive cycle in a queen cell. Therefore, the damage of varroa to queen brood has not been considered important. Similarly, phoretic varroa mites are mainly seen on adult drones and workers but not on the queen, probably due to constant attendance of the queen by workers. Queen infestation by varroa occurs only at extremely high mite prevalence in the colony and the near absence of drone or worker brood [158]. Thus, varroa is generally not a direct concern for queen health or quality but their impact on viruses (see Section 3.4) within a colony poses an indirect threat to queens.

3.2. Tracheal Mites

The tracheal mite, *Acarapis woodi*, is an obligate endoparasite of honey bees and was discovered following extensive colony mortality in the UK [159]. This microscopic parasite mostly dwells in the respiratory airways and causes acarapisosis or acariosis disease in adult worker bees [160]. The pathogenic effects of *A. woodi* on individual bees depend on the number of parasites within the tracheae, which can be attributed to mechanical injury, physiological disorders stemming from nutritional loss, damage to tracheal systems, reduction of tracheal airflow and paralysis of flight muscles [161,162]. Colonies with high mite infestation tend to have poor winter survival [163].

Although the concerns about *A. woodi* have decreased recently, studies have demonstrated a wide range of mite prevalence [111,126,127]. Similar to workers, queen infestation is age dependent, with 1-day-old queens carrying on average 6.5 mites and 10-day-old queens only 1.0 mite [128]. The infestation of queens tends to increase with the level of worker infestation around the queen [127,128]. In commercial queen breeding operations, young queens may become infested in mating nuclei if worker bees are already infested, although the results from different surveys vary [53,111]. It is not yet known whether tracheal mite infestation impairs the performance of queens, both in terms of mating or long term productivity and survival. Breeding selection programs have tried to develop resistance to *A. woodi* by improving worker autogrooming, but experimental studies comparing queens from resistant and susceptible lines indicate that the queens from resistant colonies do not have reduced tracheal mite infestation [127].

3.3. Nosema Species

Nosema apis and *N. ceranae* are two common intestinal parasites, causing Nosemosis in European honey bees [164] by attacking the epithelial cells in the midgut [30,165]. While *N. apis* shares a co-evolutionary history with the European honey bee, *N. ceranae* has more recently been transmitted from its original host, *Apis cerana* to Western honey bees [166]. *Nosema* spp. are obligate parasites and their spores are transmitted horizontally through oral and fecal pathways [167,168]. Recent evidence also suggests that *Nosema* may be transmitted sexually but not vertically [122,168,169,170]. Both *Nosema* species are limited to reproducing in the midgut [165,171], although *N. ceranae* seems to be less tissue-specific and has been found in queen ovaries [122,170]. Symptoms caused by *N. apis* are a large number of dead adult bees in the colony and diarrhea spotting at the hive entrance early in the spring. For *N. ceranae* infections, however, no conclusive evidence of strong seasonal patterns exists [28,30]. Its symptoms are increased foraging duration, decreased flight frequencies, decreased immune functions, and general stress in worker bees that reduces longevity, leading to colony depopulation and collapse [30,172,173,174].

Queens, like other members of a colony, can be infected by both *N. apis* and *N. ceranae* [122,124], and *N. ceranae* has even been detected in larval queens [122]. Most transmission of *Nosema* spp., presumably occurs during the adult stage, including mating, although antimicrobial molecules in drones' semen are able to kill *N. apis* spores and reduce the risk of disease transmission during mating [169]. Nosemosis in queens causes aberrant physiology, as well as similar gut lesions and metabolic costs as in workers [135,167]. In addition, queens infected by *N. apis* start oviposition later than healthy ones [175,176], display changed pheromone production [135] and in extreme cases their oocytes degenerate leading to infertility [134]. *N. apis* infection may severely reduce queen lifespan to an average of nearly 50 days, resulting in queen supersedure [176], but not in all cases [127]. Compensatory increases in the level of vitellogenin and other antioxidant enzymes occur in infected queens [135]. These counterintuitive changes may be protective mechanisms that are too costly in the long-term for the infected queen to survive. Recent surveys indicate that infection levels in commercialized produced queens are far less than during past decades [53,55,111].

3.4. Viruses

Viruses in honey bees, like in all organisms, are non-living, opportunistic, and obligate intracellular pathogens that require the host cellular machinery for transcription, translation, and replication [177]. Viral infection may cause direct effects on the morphology, physiology, and behavior of honey bees [24,25,178]. Thus far, 23 positive-strand RNA viruses have been reported to infect honey bees [179,180,181]. The most commonly studied honey bee viruses can be grouped into the families *Dicistroviridae* (Black queen cell virus, BQCV; ABPV; IAPV; and Kashmir bee virus, KBV), *Iflaviridae* (DWV; Slow bee paralysis virus, SBPV; and Sacbrood virus, SBV), and an uncategorized family (Chronic bee paralysis virus, CBPV) [179].

Deformed wing virus (DWV) is a nearly omnipresent and persistent virus, associated with varroa mite infestation [24]. However, prior to or in the absence of varroa, DWV can also be transmitted through other pathways between and within castes [137,140,146]. In the absence of varroa, sexual and vertical transmission of DWV may have been more important. In the post-varroa era of apiculture, drones from colonies with high mite infestations show high levels of DWV infection [182]. Infected drones are able to fly and frequently reach drone congregation areas to mate with young queens [95,136]. DWV has been detected in drone samples from DCAs, the queen "mating sign" (the remaining endophallus of her last mating partner), and semen collected from queen spermathecae, all of which indicate venereal transmission of DWV [95,136,137]. However, it is yet unclear if this is the main mechanism by which uninfected queens become newly infected. It has been also found that the virus is able to transmit vertically from an infected queen to her offspring [138,139,140,183]. Heavily infected workers usually emerge with the symptomatic crippled-wing syndrome and shortened abdomens, resulting in premature mortality [24,184]. The virus has been detected in older queens as well as young queens. The latter have usually a lower virus prevalence [53,129,133], suggesting that the virus replicates in adult queens. DWV has been found to infect the head, fat body, gut, and ovaries of queens [95,130,182,183], although crippled wings have seldom been reported in queens as a consequence of DWV infections [185]. A high virus titer in reproductive tissues can lead to ovarian degeneration or possibly affect stored sperm viability [130]. Therefore, a DWV-induced decline of reproduction quality may seriously affect colony performance, productivity, and queen supersedure.

Chronic bee paralysis virus (CBPV) was the first isolated and described virus causing chronic bee paralysis in adult worker bees [186]. CBPV can be transmitted through the fecal–oral pathway or via contact between adult bees when healthy bees are crowded together with infected individuals [187]. An outbreak of CBPV can lead to severe mortality of the workers and eventually to a collapse of the colony even within a single active season [188]. Often, the queen and a few workers are the only remaining individuals in collapsed colonies [189]. Several studies have detected a low prevalence of CBPV in surveyed queens [131,133,183]. Experimentally infected queens show the same symptoms as worker bees with trembling of legs, spread and disjunct wings, and a bloated abdomen full of hemolymph with a dilated honey sac. Queens are as susceptible to CBPV as workers [119].

Acute bee paralysis virus (ABPV) and Israeli Acute Paralysis Virus (IAPV) are closely related viruses, commonly existing as covert low-titer infections [25]. In association with varroa mites, they are additional factors that promote colony collapse [23,132,147]. ABPV and IAPV are virulent pathogens that cause paralysis, trembling, and rapid death of workers in 1–2 days after infection [25]. These viruses have been found in different developmental stages and castes [132], although high titers have been rarely detected in queen surveys [129,131,133,183]. As such, the impact of ABPV and IAPV on queen health has not yet been well studied. The only investigation by in situ hybridization of IAPV showed specific detection of the IAPV in the egg, gut, ovaries, and spermatheca of infected queens [132].

Sacbrood virus (SBV) causes brood disease of the honey bee, but it has also been reported from adult honey bees without any obvious sign of disease [190]. The virus mostly transmits through the oral-oral pathway from adult worker bees to the larvae, and clear symptoms appear a few days after capping [178]. Infected larvae fail to pupate, and ecdysial fluid accumulates beneath their unshed cuticle, the larvae change color from pearly white to pale yellow, and shortly after the larvae die [190]. The virus also has been detected in queens, mostly in the ovary and eviscerated body (i.e., non-specific tissues) but how it affects queen health or whether it can be vertically transmitted is not well studied [131,139,183].

Black queen cell virus (BQCV) was first isolated from dead queen pre-pupae and pupae sealed in their cells that had turned dark brown to black along the walls of the cell [191]. It is consistently one of the most common viruses detected using laboratory techniques in adult honey bees worldwide [177]. BQCV infection is often observed in queen-rearing colonies, attacking queen larvae and pupae. Coinfection of workers with BQCV and *N. apis* within a colony results in increased worker bee mortality [192], although data to support synergistic interaction between BQCV and *N. ceranae* are controversial [193,194]. The virus has been detected with high titers in collapsed colonies [13], and it has been detected in honey bee queens mostly in gut, feces, and ovaries [131,183].

In sum, it is clear that several viruses cause overt problems for queen health but more commonly viruses remain asymptomatic and can be efficiently transmitted by queens into the next generation [133,139]. Thus, viruses should be under selection for decreased virulence towards queens, but little data are available to test this prediction. Practically, it is unknown how the presence of covert viruses in queens affects them because non-invasive methods for virus detection in honey bees remain to be developed. Many other viruses are also known to exist in honey bee colonies, however they remain relatively poorly characterized and little or nothing is known about their effects on queens. This lack of even simple descriptive information makes them important targets for further empirical investigations, especially with regards to their connection to queen and colony health.

4. CONCLUSIONS

We have explained above how to quantify queen quality and how different diseases can affect queen reproductive potential. In turn, high quality queens with high reproductive potential produce colonies that exhibit high growth and survival [46]. Young and healthy queens produce a solid brood pattern, facilitating brood care that can prevent disease, and a greater proportion of this brood will develop into healthy workers to replenish dying workers [44,195]. These colonies are able to store more honey and pollen throughout the year in comparison to the colonies headed by low quality queens [46], which translates into better winter survival. Moreover, reproductive quality influences queen mandibular gland pheromone profile and influences colony cohesion [196]. Replacing the queen in *Nosema*- and CBPV-infected colonies with a young, healthy, and productive queen has been recommended to maintain the colony homeostasis [188,197] and presumably is also advisable when other diseases are detected. Through direct and indirect effects a healthy, high-quality queen can mitigate disease effects on colony fitness or even make colonies more resistant to disease [105]. However, more research to demonstrate these interaction effects is clearly needed.

Our review focuses on the intersection between queen reproductive potential, and honey bee diseases. How these interactions manifest themselves will be significant for understanding colony health problems and the means to mitigate them. Based on our review, we believe the following questions should be prioritized for future investigation to help elucidate these interactions and improve colony health. (1) How do queens initially become infected? There are multiple routes that have been identified (orally during juvenile and adult life stages, as well as sexually), but we lack in many cases a comprehensive understanding of exposure, within-queen pathways, and key stages of infection in queens. If we can identify the means by which queens become infected, then we may be able to identify the means to avoid infection of queens and their colonies. (2) What triggers viral replication so that a previously chronic low level (asymptomatic) infection becomes harmful? Several studies have shown that viruses can be detected as a low, covert level in honey bee queens [129,133], but the effect of different stressors on the queen to trigger the viruses to become overt and harm the queen is not clear yet. (3) What is the collective decision-making process of workers to replace sick queens through supersedure? Researchers have developed a comprehensive model for the internal and external conditions that trigger swarming behavior in colonies [198], but we lack a similar theoretical context for non-swarming queen replacement. Developing such a model is a top priority for research, because early or failed supersedure attempts are an increasingly common problem experienced by beekeepers, and such events lead to loss of colony productivity and increased mortality. (4) What are the specific, various phenotypes of "poor queens"? There are many "queen problems" experienced by beekeepers, but the mechanistic basis is unclear. It will be critical to tease apart the various phenotypes of "poor queens" and explain their underlying causes, as it is possible if not likely that different symptoms of queen problems can be attributed to different factors. (5) What is the effect of a queen's microbiome on her health and disease resistance? The typical microbiomes of workers and hive substrates have been described (e.g., [199]; reviewed in [200]), but our understanding of queen microbiota is relatively lacking [201,202]. One particularly interesting question is the effect of the queen on the colony microbiota and vice versa (the effect of diverse colony parameters on the queen's microbiota). In the light of the increased interest in probiotics, this area will also have direct implications for hive management. (6) How does disease mechanistically affect the queen's pheromones, chemical signaling, and reproductive physiology? Investigating the interactions between queen physiology and disease is critical for understanding which resistance factors are more important than others. Functionally, these interactions are likely going to be expressed through the pheromonal signaling of the queen and their perception by the workers. Determining how infection may influence a queen's signals, the workers' perception of them, or both will demonstrate how colony cohesion is ultimately affected by disease. Finally, (7) How does a potential shift to microbreeding affect queen quality and spread of disease? Currently, the majority of queens in the US are produced by relatively few commercial queen producers [203], which raises concerns about a lack of genetic diversity and the spread of certain diseases [141,203], particularly those that are transmitted vertically. There are currently many efforts to promote smaller-scale, localized production of queens ("microbreeding") to address supply and favor locally adapted genotypes (see [204]). However, there is little if any information about whether such efforts may help bolster genetic diversity, reduce disease prevalence or intensity, or otherwise have a positive effect on overall queen quality. Potentially, new metrics of queen quality that include disease resistance or immune measures are needed to guide breeding efforts.

In sum, the interactions between queen health and honey bee diseases have not been sufficiently studied because their importance was underestimated. However, many unanswered questions in this area provide fertile grounds for fundamental biological research and the possibility to apply insights to improve honey bee health.

ACKNOWLEDGMENTS

The authors would like to thank the guest editors for the invitation to write this review, as well as for anonymous referees for their helpful and constructive comments. Funding was provided by the U.S. Army Research Laboratory (Grant: W911NF1520045) in support of Esmaeil Amiri, as well as a USDA-AFRI grant 2016-07962 awarded to David R. Tarpy and Project ApisM and USDA-APHIS support (agreement #16-8130-0636-CA) to Olav Rueppell.

AUTHOR CONTRIBUTIONS

Esmaeil Amiri, Micheline K. Strand, Olav Rueppell and David R. Tarpy wrote the manuscript and all authors reviewed the manuscript.

CONFLICTS OF INTEREST

The authors declare no conflict of interest. The founding sponsors had no role in the writing of the manuscript, and in the decision to publish this manuscript.

REFERENCES

1. Breeze, T.D.; Bailey, A.P.; Balcombe, K.G.; Potts, S.G. Pollination services in the UK: How important are honeybees? Agric. Ecosyst. Environ. 2011, 142, 137–143.
2. Morse, R.A.; Calderone, N.W. The value of honey bees as pollinators of US crops in 2000. Bee Cult. 2000, 128, 1–15.
3. Potts, S.; Roberts, S.; Dean, R.; Marris, G.; Brown, M.; Jones, R.; Neumann, P.; Settele, J. Declines of managed honey bees and beekeepers in Europe. J. Apic. Res. 2010, 49, 15.
4. Pirk, C.W.W.; Human, H.; Crewe, R.M.; VanEngelsdorp, D. A survey of managed honey bee colony losses in the Republic of South Africa—2009 to 2011. J. Apic. Res. 2014, 53, 35–42.
5. Liu, Z.; Chen, C.; Niu, Q.; Qi, W.; Yuan, C.; Su, S.; Liu, S.; Zhang, Y.; Zhang, X.; Ji, T.; et al. Survey results of honey bee (Apis mellifera) colony losses in China (2010–2013). J. Apic. Res. 2016, 55, 29–37.
6. Vanengelsdorp, D.; Meixner, M.D. A historical review of managed honey bee populations in Europe and the United States and the factors that may affect them. J. Invertebr. Pathol. 2010, 103, 80–95.
7. Seitz, N.; Traynor, K.S.; Steinhauer, N.; Rennich, K.; Wilson, M.E.; Ellis, J.D.; Rose, R.; Tarpy, D.R.; Sagili, R.R.; Caron, D.M.; et al. A national survey of managed honey bee 2014–2015 annual colony losses in the USA. J. Apic. Res. 2015, 54, 292–304.
8. Meixner, M.D.; Francis, R.M.; Gajda, A.; Kryger, P.; Andonov, S.; Uzunov, A.; Topolska, G.; Costa, C.; Amiri, E.; Berg, S. Occurrence of parasites and pathogens in honey bee colonies used in a European genotype-environment interactions experiment. J. Apic. Res. 2014, 53, 215–219.
9. Goulson, D.; Nicholls, E.; Botías, C.; Rotheray, E.L. Bee declines driven by combined stress from parasites, pesticides, and lack of flowers. Science 2015, 347, 1255957.
10. Potts, S.G.; Biesmeijer, J.C.; Kremen, C.; Neumann, P.; Schweiger, O.; Kunin, W.E. Global pollinator declines: Trends, impacts and drivers. Trends Ecol. Evol. 2010, 25, 345–353.
11. Smith, K.M.; Loh, E.H.; Rostal, M.K.; Zambrana-Torrelio, C.M.; Mendiola, L.; Daszak, P. Pathogens, pests, and economics: Drivers of honey bee colony declines and losses. EcoHealth 2013, 10, 434–445.
12. McMenamin, A.J.; Genersch, E. Honey bee colony losses and associated viruses. Curr. Opin. Insect Sci. 2015, 8, 121–129.
13. Cornman, R.S.; Tarpy, D.R.; Chen, Y.; Jeffreys, L.; Lopez, D.; Pettis, J.S.; van Engelsdorp, D.; Evans, J.D. Pathogen webs in collapsing honey bee colonies. PLoS ONE 2012, 7, e43562.
14. Dainat, B.; Evans, J.D.; Chen, Y.; Gauthier, L.; Neumann, P. Predictive markers of honey bee colony collapse. PLoS ONE 2012, 7, e32151.
15. Genersch, E.; von der Ohe, W.; Kaatz, H.; Schroeder, A.; Otten, C.; Büchler, R.; Berg, S.; Ritter, W.; Mühlen, W.; Gisder, S.; et al. The German bee monitoring project: A long term study to understand periodically high winter losses of honey bee colonies. Apidologie 2010, 41, 332–352.
16. Genersch, E. Honey bee pathology: Current threats to honey bees and beekeeping. Appl. Microbiol. Biotechnol. 2010, 87, 87–97.
17. Le Conte, Y.; Ellis, M.; Ritter, W. Varroa mites and honey bee health: Can Varroa explain part of the colony losses? Apidologie 2010, 41, 353–363.
18. Guzmán-Novoa, E.; Eccles, L.; Calvete, Y.; Mcgowan, J.; Kelly, P.G.; Correa-Benítez, A. Varroa destructor is the main culprit for the death and reduced populations of overwintered honey bee (Apis mellifera) colonies in Ontario, Canada. Apidologie 2010, 41, 443–450.
19. Nazzi, F.; Brown, S.P.; Annoscia, D.; Del Piccolo, F.; Di Prisco, G.; Varricchio, P.; Della Vedova, G.; Cattonaro, F.; Caprio, E.; Pennacchio, F. Synergistic parasite-pathogen

interactions mediated by host immunity can drive the collapse of honeybee colonies. PLoS Pathog. 2012, 8, e1002735.
20. Boecking, O.; Genersch, E. Varroosis—The ongoing crisis in bee keeping. Journal für Verbraucherschutz und Lebensmittelsicherheit 2008, 3, 221–228.
21. Di Prisco, G.; Annoscia, D.; Margiotta, M.; Ferrara, R.; Varricchio, P.; Zanni, V.; Caprio, E.; Nazzi, F.; Pennacchio, F. A mutualistic symbiosis between a parasitic mite and a pathogenic virus undermines honey bee immunity and health. Proc. Natl. Acad. Sci. USA 2016, 113, 3203–3208.
22. Francis, R.M.; Nielsen, S.L.; Kryger, P. Varroa-Virus interaction in collapsing honey bee colonies. PLoS ONE 2013, 8, e57540.
23. Cox-Foster, D.L.; Conlan, S.; Holmes, E.C.; Palacios, G.; Evans, J.D.; Moran, N.A.; Quan, P.-L.; Briese, T.; Hornig, M.; Geiser, D.M.; et al. A metagenomic survey of microbes in honey bee colony collapse disorder. Science 2007, 318, 283–287.
24. De Miranda, J.R.; Genersch, E. Deformed wing virus. J. Invertebr. Pathol. 2010, 103, S48–S61.
25. De Miranda, J.R.; Cordoni, G.; Budge, G. The Acute bee paralysis virus—Kashmir bee virus—Israeli acute paralysis virus complex. J. Invertebr. Pathol. 2010, 103, S30–S47.
26. Highfield, A.C.; El Nagar, A.; Mackinder, L.C.M.; Noël, L.M.-L.J.; Hall, M.J.; Martin, S.J.; Schroeder, D.C. Deformed wing virus implicated in overwintering honeybee colony losses. Appl. Environ. Microbiol. 2009, 75, 7212–7220.
27. McMahon, D.P.; Natsopoulou, M.E.; Doublet, V.; Fürst, M.; Weging, S.; Brown, M.J.F.; Gogol-Döring, A.; Paxton, R.J. Elevated virulence of an emerging viral genotype as a driver of honeybee loss. Proc. R. Soc. B Sci. 2016, 283.
28. Higes, M.; Martin-Hernandez, R.; Botias, C.; Bailon, E.G.; Gonzalez-Porto, A.V.; Barrios, L.; del Nozal, M.J.; Bernal, J.L.; Jimenez, J.J.; Palencia, P.G.; et al. How natural infection by Nosema ceranae causes honeybee colony collapse. Environ. Microbiol. 2008, 10, 2659–2669.
29. Higes, M.; Martin-Hernandez, R.; Garrido-Bailon, E.; Gonzalez-Porto, A.V.; Garcia-Palencia, P.; Meana, A.; del Nozal, M.J.; Mayo, R.; Bernal, J.L. Honeybee colony collapse due to Nosema ceranae in professional apiaries. Environ. Microbiol. Rep. 2009, 1, 110–113.
30. Higes, M.; Meana, A.; Bartolomé, C.; Botías, C.; Martín-Hernández, R. Nosema ceranae (Microsporidia), a controversial 21st century honey bee pathogen. Environ. Microbiol. Rep. 2013, 5, 17–29.
31. Soroker, V.; Hetzroni, A.; Yakobson, B.; David, D.; David, A.; Voet, H.; Slabezki, Y.; Efrat, H.; Levski, S.; Kamer, Y.; et al. Evaluation of colony losses in Israel in relation to the incidence of pathogens and pests. Apidologie 2011, 42, 192–199.
32. Forsgren, E.; Budge, G.E.; Charrière, J.-D.; Hornitzky, M.A.Z. Standard methods for European foulbrood research. J. Apic. Res. 2013.
33. Roetschi, A.; Berthoud, H.; Kuhn, R.; Imdorf, A. Infection rate based on quantitative real-time PCR of Melissococcus plutonius, the causal agent of European foulbrood, in honeybee colonies before and after apiary sanitation. Apidologie2008, 39, 362–371.
34. Wilkins, S.; Brown, M.A.; Cuthbertson, A.G. The incidence of honey bee pests and diseases in England and Wales. Pest Manag. Sci. 2007, 63, 1062–1068.
35. Genersch, E. American Foulbrood in honeybees and its causative agent, Paenibacillus larvae. J. Invertebr. Pathol. 2010, 103, 10–19.
36. Brodschneider, R.; Moosbeckhofer, R.; Crailsheim, K. Surveys as a tool to record winter losses of honey bee colonies: A two year case study in Austria and South Tyrol. J. Apic. Res. 2010, 49, 23–30.
37. Spleen, A.M.; Lengerich, E.J.; Rennich, K.; Caron, D.; Rose, R.; Pettis, J.S.; Henson, M.; Wilkes, J.T.; Wilson, M.; Stitzinger, J.; et al. A national survey of managed honey bee 2011–2012 winter colony losses in the United States: Results from the Bee Informed Partnership. J. Apic. Res. 2013, 52, 44–53.
38. Vanengelsdorp, D.; Tarpy, D.R.; Lengerich, E.J.; Pettis, J.S. Idiopathic brood disease syndrome and queen events as precursors of colony mortality in migratory

beekeeping operations in the eastern United States. Prev. Vet. Med. 2013, 108, 225–233.
39. Vanengelsdorp, D.; Hayes, J., Jr.; Underwood, R.M.; Pettis, J. A Survey of honey bee colony losses in the U.S., Fall 2007 to Spring 2008. PLoS ONE 2009, 3, e4071.
40. Winston, M.L. The Biology of the Honey Bee; Harvard University Press: London, UK, 1987.
41. Tarpy, D.R.; vanEngelsdorp, D.; Pettis, J.S. Genetic diversity affects colony survivorship in commercial honey bee colonies. Naturwissenschaften 2013, 100, 723–728.
42. Tarpy, D.R.; Summers, J.; Keller, J.J. Comparison of parasitic mites in Russian-hybrid and Italian honey bee (Hymenoptera: Apidae) colonies across three different locations in North Carolina. J. Econ. Entomol. 2007, 100, 258–266.
43. Cobey, S.W. Comparison studies of instrumentally inseminated and naturally mated honey bee queens and factors affecting their performance. Apidologie 2007, 38, 390–410.
44. Akyol, E.; Yeninar, H.; Korkmaz, A.; Çakmak, İ. An observation study on the effects of queen age on some characteristics of honey bee colonies. Ital. J. Anim. Sci. 2008, 7, 19–25.
45. Kostarelou-Damianidou, M.; Thrasyvoulou, A.; Tselios, D.; Bladenopoulos, K. Brood and honey production of honey bee colonies requeened at various frequencies. J. Apic. Res. 1995, 34, 9–14.
46. Rangel, J.; Keller, J.J.; Tarpy, D.R. The effects of honey bee (Apis mellifera L.) queen reproductive potential on colony growth. Insect Soc. 2013, 60, 65–73.
47. Nelson, D.L.; Gary, N.E. Honey productivity of honeybee colonies in relation to body weight, attractiveness and fecundity of the queen. J. Apic. Res. 1983, 22, 209–213.
48. Tarpy, D.R. Genetic diversity within honeybee colonies prevents severe infections and promotes colony growth. Proc. R. Soc. B Sci. 2003, 270, 99–103.
49. Hoopingarner, R.; Farrar, C. Genetic control of size in queen honey bees. J. Econ. Entomol. 1959, 52, 547–548.
50. Dodologlu, A.; Gene, F. Comparison of some features of queens reared from different honeybee (Apis mellifera L.) genotypes. J. Appl. Anim. Res. 2003, 24, 105–109.
51. Oldroyd, B.P.; Goodman, R.D.; Allaway, M.A. On the relative importance of queens and workers to honey production. Apidologie 1990, 21, 153–159.
52. Kahya, Y.; Gençer, H.V.; Woyke, J. Weight at emergence of honey bee (Apis mellifera caucasica) queens and its effect on live weights at the pre and post mating periods. J. Apic. Res. 2008, 47, 118–125.
53. Delaney, D.A.; Keller, J.J.; Caren, J.R.; Tarpy, D.R. The physical, insemination, and reproductive quality of honey bee queens (Apis mellifera L.). Apidologie 2011, 42, 1–13.
54. Hatjina, F.; Bieńkowska, M.; Charistos, L.; Chlebo, R.; Costa, C.; Dražić, M.M.; Filipi, J.; Gregorc, A.; Ivanova, E.N.; Kezić, N.; et al. A review of methods used in some European countries for assessing the quality of honey bee queens through their physical characters and the performance of their colonies. J. Apic. Res. 2014, 53, 337–363.
55. Tarpy, D.R.; Keller, J.J.; Caren, J.R.; Delaney, D.A. Assessing the mating 'health'of commercial honey bee queens. J. Econ. Entomol. 2012, 105, 20–25.
56. Hatch, S.; Tarpy, R.D.; Fletcher, C.D.J. Worker regulation of emergency queen rearing in honey bee colonies and the resultant variation in queen quality. Insect Soc. 1999, 46, 372–377.
57. Jackson, T.J.; Tarpy, R.D.; Fahrbach, E.S. Histological estimates of ovariole number in honey bee queens, Apis mellifera, reveal lack of correlation with other queen quality measures. J. Insect Sci. 2011, 11, 1–11.
58. Skowronek, W.; Bieñkowska, M.G.; Kruk, C. Changes in body weight of honey bee queens during their maturation. J. Apic. Res. 2004, 48, 61–68.
59. Akyol, E.; Yeninar, H.; Kaftanoglu, O. Live weight of queen honey bees (Apis Mellifera L.) predicts reproductive characteristics. J. Kans. Entomol. Soc. 2008, 81, 92–100.

60. Woyke, J. Correlations between the age at which honeybee brood was grafted, characteristics of the resultant queens, and results of insemination. J. Apic. Res. 1971, 10, 45–55.
61. Collins, A.M.; Pettis, J.S. Correlation of queen size and spermathecal contents and effects of miticide exposure during development. Apidologie 2013, 44, 351–356.
62. Tarpy, D.R.; Hatch, S.; Fletcher, D.J.C. The influence of queen age and quality during queen replacement in honeybee colonies. Anim. Behav. 2000, 59, 97–101.
63. Gençer, H.V.; Shah, S.Q.; Firatli, Ç. Effects of supplemental feeding of queen rearing colonies and larval age on the acceptance of grafted larvae and queen traits. Pak. J. Biol. Sci. 2000, 3, 1319–1322.
64. Koç, A.U.; Karacaoglu, M. Effects of rearing season on the quality of queen honeybees (Apis mellifera L.) raised under the conditions of aegean region. Mellifera 2004, 4, 34–37.
65. Masry, S.; El-Wahab, T.A.; Hassona, N.M. Origin, weight at emergence of virgin honey bee queens and its effect on acceptance during introduction. Acad. J. Entomol. 2015, 8, 174–182.
66. Harano, K.-I.; Sasaki, M.; Sasaki, K. Effects of reproductive state on rhythmicity, locomotor activity and body weight in European honeybee, Apis mellifera (Hymenoptera, Apini) queens. Sociobiology 2007, 50, 189–200.
67. Hayworth, M.K.; Johnson, N.G.; Wilhelm, M.E.; Gove, R.P.; Metheny, J.D.; Rueppell, O. Added weights lead to reduced flight behavior and mating success in polyandrous honey bee queens (Apis mellifera). Ethology 2009, 115, 698–706.
68. Shehata, S.M.; Townsend, G.F.; Shuel, R.W. Seasonal physiological changes in queen and worker honeybees. J. Apic. Res. 1981, 20, 69–78.
69. Szabo, T.I.; Mills, P.F.; Heikel, D.T. Effects of honeybee queen weight and air temperature on the initiation of oviposition. J. Apic. Res. 1987, 26, 73–78.
70. Szabo, T.I. Relationship between weight of honey-bee queens (Apis mellifera L.) at emergence and at the cessation of egg laying. Am. Bee J. 1973, 13, 127–135.
71. Yadava, R.P.S.; Smith, M.V. Aggressive behavior of Apis mellifera L. workers towards introduced queens. III. Relationship between the attractiveness of the queen and worker aggression. Can. J. Zool. 1971, 49, 1359–1362.
72. Moretto, G.; Guerra, J.; Kalvelage, H.; Espindola, E. Maternal influence on the acceptance of virgin queens introduced into Africanized honey bee (Apis mellifera) colonies. Genet. Mol. Res. 2004, 3, 441–445. [PubMed]
73. Szabo, T.I.; Townsend, G.F. Behavioural studies on queen introduction in the honeybee 1. effect of the age of workers (from a colony with a laying queen) on their behaviour towards an introduced virgin queen. J. Apic. Res. 1974, 13, 19–25.
74. Szabo, T.I. Behavioural studies of queen introduction in the honeybee 6. multiple queen introduction. J. Apic. Res.1977, 16, 65–83.
75. Richard, F.J.; Tarpy, D.R.; Grozinger, C.M. Effects of insemination quantity on honey bee queen physiology. PLoS ONE 2007, 2, e980.
76. Pettis, J.S.; Westcott, L.C.; Winston, M.L. Balling behaviour in the honey bee in response to exogenous queen mandibular gland pheromone. J. Apic. Res. 1998, 37, 125–131.
77. Tarpy, D.R.; Keller, J.J.; Caren, J.R.; Delaney, D.A. Experimentally induced variation in the physical reproductive potential and mating success in honey bee queens. Insect Soc. 2011, 58, 569–574.
78. Skowronek, W.; Kruk, C.; Klopot, J. Factors affecting oviposition of artificially inseminated honeybee queens. J. Apic. Res. 2002, 2, 85–95.
79. Gregorc, A.; Smodiš Škerl, M.I. Characteristics of honey bee (Apis Mellifera Carnica Pollman, 1879) queens reared in Slovenian commercial breeding stations. J. Apic. Sci. 2015, 59, 5–12.
80. Gilley, D.C.; Tarpy, D.R.; Land, B.B. Effect of queen quality on interactions between workers and dueling queens in honeybee (Apis mellifera L.) colonies. Behav. Ecol. Sociobiol. 2003, 55, 190–196.

81. Patricio, K.; Cruz-Landim, C. Mating influence in the ovary differentiation in adult queens of Apis mellifera L. (Hymenoptera, Apidae). Braz. J. Biol. 2002, 62, 641–649.
82. Kocher, S.D.; Richard, F.-J.; Tarpy, D.R.; Grozinger, C.M. Genomic analysis of post-mating changes in the honey bee queen (Apis mellifera). BMC Genom. 2008, 9, 232.
83. Niño, E.L.; Tarpy, D.R.; Grozinger, C.M. Differential effects of insemination volume and substance on reproductive changes in honey bee queens (Apis mellifera L.). Insect Mol. Biol. 2013, 22, 233–244.
84. Carreck, N.L.; Andree, M.; Brent, C.S.; Cox-Foster, D.; Dade, H.A.; Ellis, J.D.; Hatjina, F.; vanEnglesdorp, D. Standard methods for Apis mellifera anatomy and dissection. J. Apic. Res. 2013, 52, 1–40.
85. Dedej, S.; Hartfelder, K.; Aumeier, P.; Rosenkranz, P.; Engels, W. Caste determination is a sequential process: Effect of larval age at grafting on ovariole number, hind leg size and cephalic volatiles in the honey bee (Apis mellifera carnica). J. Apic. Res. 1998, 37, 183–190.
86. Al-Lawati, H.; Kamp, G.; Bienefeld, K. Characteristics of the spermathecal contents of old and young honeybee queens. J. Insect Physiol. 2009, 55, 117–122.
87. Kraus, F.B.; Neumann, P.; Moritz, R.F.A. Genetic variance of mating frequency in the honeybee (Apis mellifera L.). Insect Soc. 2005, 52, 1–5.
88. Koeniger, G.; Koeniger, N.; Ellis, J.; Connor, L.J. Mating Biology of Honey Bees (Apis mellifera); Wicwas Press: Kalamazoo, MI, USA, 2014.
89. Tarpy, D.R.; Nielsen, D.I. Sampling error, effective paternity, and estimating the genetic structure of honey bee colonies (Hymenoptera: Apidae). Ann. Entomol. Soc. Am. 2002, 95, 513–528.
90. Tarpy, D.R.; Nielsen, R.; Nielsen, D.I. A scientific note on the revised estimates of effective paternity frequency in Apis. Insect Soc. 2004, 51, 203–204.
91. Schlüns, H.; Moritz, R.F.A.; Neumann, P.; Kryger, P.; Koeniger, G. Multiple nuptial flights, sperm transfer and the evolution of extreme polyandry in honeybee queens. Anim. Behav. 2005, 70, 125–131.
92. Neumann, P.; Moritz, R.F.A.; van Praagh, J. Queen mating frequency in different types of honey bee mating apiaries. J. Apic. Res. 1999, 38, 11–18.
93. Boomsma, J.J.; Ratnieks, F.L.W. Paternity in eusocial Hymenoptera. Philos. Trans. R. Soc. B 1996, 351, 947–975.
94. Palmer, K.A.; Oldroyd, B.P. Evolution of multiple mating in the genus Apis. Apidologie 2000, 31, 235–248.
95. Amiri, E.; Meixner, M.D.; Kryger, P. Deformed wing virus can be transmitted during natural mating in honey bees and infect the queens. Sci. Rep. 2016, 6, 33065.
96. Rueppell, O.; Johnson, N.; Rychtář, J. Variance-based selection may explain general mating patterns in social insects. Biol. Lett. 2008, 4, 270–273.
97. Robinson, G.E. Regulation of division of labor in insect societies. Annu. Rev. Entomol. 1992, 37, 637–665.
98. Mattila, H.R.; Seeley, T.D. Extreme polyandry improves a honey bee colony's ability to track dynamic foraging opportunities via greater activity of inspecting bees. Apidologie 2014, 45, 347–363.
99. Kolmes, S.A.; Winston, M.L.; Fergusson, L.A. The division of labor among worker honey bees (Hymenoptera: Apidae): The effects of multiple patrilines. J. Kans. Entomol. Soc. 1989, 62, 80–95.
100. Jones, J.C.; Myerscough, M.R.; Graham, S.; Oldroyd, B.P. Honey bee nest thermoregulation: Diversity promotes stability. Science 2004, 305, 402–404.
101. Oldroyd, B.P.; Fewell, J.H. Genetic diversity promotes homeostasis in insect colonies. Trend Ecol. Evol. 2007, 22, 408–413.
102. Mattila, H.R.; Burke, K.M.; Seeley, T.D. Genetic diversity within honeybee colonies increases signal production by waggle-dancing foragers. Proc. R. Soc. B Sci. 2008, 275, 809–816.
103. Mattila, H.R.; Seeley, T.D. Does a polyandrous honeybee queen improve through patriline diversity the activity of her colony's scouting foragers? Behav. Ecol. Sociobiol. 2011, 65, 799–811.

104. Carr-Markell, M.K.; McDonald, K.M.; Mattila, H.R. Intracolonial genetic diversity increases chemical signaling by waggle-dancing honey bees, Apis mellifera. Insect Soc. 2013, 60, 485–496.
105. Tarpy, D.R.; Seeley, T.D. Lower disease infections in honeybee (Apis mellifera) colonies headed by polyandrous vs monandrous queens. Naturwissenschaften 2006, 93, 195–199.
106. Seeley, T.D.; Tarpy, D.R. Queen promiscuity lowers disease within honeybee colonies. Proc. R. Soc. B Sci. 2007, 274, 67–72.
107. Palmer, K.A.; Oldroyd, B.P. Evidence for intra-colonial genetic variance in resistance to American foulbrood of honey bees (Apis mellifera): Further support for the parasite/pathogen hypothesis for the evolution of polyandry. Naturwissenschaften 2003, 90, 265–268. [PubMed]
108. Mattila, H.R.; Seeley, T.D. Genetic diversity in honey bee colonies enhances productivity and fitness. Science 2007, 317, 362–364.
109. Woyke, J. Dynamics of entry of spermatozoa into the spermatheca of instrumentally inseminated queen honeybees. J. Apic. Res. 1983, 22, 150–154.
110. Woyke, J. Natural and artificial insemination of queen honeybees. Bee World 1962, 43, 21–25.
111. Camazine, S.; Cakmak, I.; Cramp, K.; Finley, J.; Fisher, J.; Frazier, M.; Rozo, A. How healthy are commercially-produced US honey bee queens? Am. Bee J. 1998, 138, 677–680.
112. Lodesani, M.; Balduzzi, D.; Galli, A. A study on spermatozoa viability over time in honey bee (Apis mellifera ligustica) queen spermathecae. J. Apic. Res. 2004, 43, 27–28.
113. Williams, G.R.; Troxler, A.; Retschnig, G.; Roth, K.; Yañez, O.; Shutler, D.; Neumann, P.; Gauthier, L. Neonicotinoid pesticides severely affect honey bee queens. Sci. Rep. 2015, 5, 14621.
114. Den Boer, S.P.A.; Boomsma, J.J.; Baer, B. Honey bee males and queens use glandular secretions to enhance sperm viability before and after storage. J. Insect Physiol. 2009, 55, 538–543.
115. Zareie, R.; Eubel, H.; Millar, A.H.; Baer, B. Long-term survival of high quality sperm: Insights into the sperm proteome of the honeybee Apis mellifera. J. Proteome Res. 2013, 12, 5180–5188.
116. King, M.; Eubel, H.; Millar, A.H.; Baer, B. Proteins within the seminal fluid are crucial to keep sperm viable in the honeybee Apis mellifera. J. Insect Physiol. 2011, 57, 409–414.
117. Pettis, J.S.; Rice, N.; Joselow, K.; vanEngelsdorp, D.; Chaimanee, V. Colony failure linked to low sperm viability in honey bee (Apis mellifera) queens and an exploration of potential causative factors. PLoS ONE 2016, 11, e0147220.
118. Rueppell, O.; Aumer, D.; Moritz, R.F.A. Ties between ageing plasticity and reproductive physiology in honey bees (Apis mellifera) reveal a positive relation between fecundity and longevity as consequence of advanced social evolution. Curr. Opin. Insect Sci. 2016, 16, 64–68.
119. Amiri, E.; Meixner, M.; Büchler, R.; Kryger, P. Chronic bee paralysis virus in honeybee queens: Evaluating susceptibility and infection routes. Viruses 2014, 6, 1188–1201.
120. Kulincevic, J.M.; Rothenbuhler, W.C. The effects of artificial infection with Chronic bee paralysis virus on queens from strains of honeybee resistant or susceptible to Hairless-black syndrome. J. Apic. Res. 1989, 28, 79–80.
121. Chen, X.; Hu, Y.; Zheng, H.; Cao, L.; Niu, D.; Yu, D.; Sun, Y.; Hu, S.; Hu, F. Transcriptome comparison between honey bee queen- and worker-destined larvae. Insect Biochem. Mol. 2012, 42, 665–673.
122. Traver, B.E.; Fell, R.D. Low natural levels of Nosema ceranae in Apis mellifera queens. J. Invertebr. Pathol. 2012, 110, 408–410.
123. Czekonska, K. The influence of Nosema apis on young honeybee queens and transmission of the disease from queens to workers. Apidologie 2000, 31, 701–706.

124. Webster, T.C.; Pomper, K.W.; Hunt, G.; Thacker, E.M.; Jones, S.C. Nosema apis infection in worker and queen Apis mellifera. Apidologie 2004, 35, 49–54.
125. Porporato, M.; Grillone, G.; Patetta, A.; Manino, A.; Laurino, D. Survey of the health status of some honey bee queens in Italy. J. Apic. Res. 2015, 59, 27–36.
126. Burgett, M.; Kitprasert, C. Tracheal mite infestation of queen honey bees. J. Apic. Res. 1992, 31, 110–111.
127. Villa, J.D.; Danka, R.G. Caste, sex and strain of honey bees (Apis mellifera) affect infestation with tracheal mites (Acarapis woodi). Exp. Appl. Acarol. 2005, 37, 157–164.
128. Pettis, J.S.; Dietz, A.; Eischen, F.A. Incidence rates of Acarapis woodi (Rennie) in queen honey bees of various ages. Apidologie 1989, 20, 69–75.
129. Francis, R.M.; Nielsen, S.L.; Kryger, P. Patterns of viral infection in honey bee queens. J. Gen. Virol. 2013, 94, 668–676.
130. Gauthier, L.; Ravallec, M.; Tournaire, M.; Cousserans, F.; Bergoin, M.; Dainat, B.; de Miranda, J.R. Viruses associated with ovarian degeneration in Apis mellifera L. queens. PLoS ONE 2011, 6, e16217.
131. Chen, Y.; Pettis, J.S.; Feldlaufer, M.F. Detection of multiple viruses in queens of the honey bee Apis mellifera L. J. Invertebr. Pathol. 2005, 90, 118–121.
132. Chen, Y.; Pettis, J.S.; Corona, M.; Chen, W.P.; Li, C.J.; Spivak, M.; Visscher, P.K.; DeGrandi-Hoffman, G.; Boncristiani, H.; Zhao, Y.; et al. Israeli acute paralysis virus: Epidemiology, pathogenesis and implications for honey bee health. PLoS Pathog. 2014, 10, e1004261.
133. Gregorc, A.; Bakonyi, T. Viral infections in queen bees (Apis mellifera carnica) from rearing apiaries. Acta Vet. Brno2012, 81, 15–19.
134. Liu, T.P. Oocytes degeneration in the queen honey bee after infection by Nosema apis. Tissue Cell 1992, 24, 131–138.
135. Alaux, C.; Folschweiller, M.; McDonnell, C.; Beslay, D.; Cousin, M.; Dussaubat, C.; Brunet, J.-L.; Le Conte, Y. Pathological effects of the microsporidium Nosema ceranae on honey bee queen physiology (Apis mellifera). J. Invertebr. Pathol. 2011, 106, 380–385.
136. Yañez, O.; Jaffé, R.; Jarosch, A.; Fries, I.; Moritz, R.F.A.; Paxton, R.J.; Miranda, J.R. Deformed wing virus and drone mating flights in the honey bee (Apis mellifera): Implications for sexual transmission of a major honey bee virus. Apidologie 2012, 43, 17–30.
137. Yue, C.; Schroder, M.; Bienefeld, K.; Genersch, E. Detection of viral sequences in semen of honeybees (Apis mellifera): Evidence for vertical transmission of viruses through drones. J. Invertebr. Pathol. 2006, 92, 105–108.
138. De Miranda, J.R.; Fries, I. Venereal and vertical transmission of Deformed wing virus in honeybees (Apis mellifera L.). J. Invertebr. Pathol. 2008, 98, 184–189.
139. Ravoet, J.; De Smet, L.; Wenseleers, T.; de Graaf, D.C. Vertical transmission of honey bee viruses in a Belgian queen breeding program. BMC Vet. Res. 2015, 11, 1–6.
140. Yue, C.; Schroder, M.; Gisder, S.; Genersch, E. Vertical-transmission routes for deformed wing virus of honeybees (Apis mellifera). J. Gen. Virol. 2007, 88, 2329–2336.
141. Mutinelli, F. The spread of pathogens through trade in honey bees and their products (including queen bees and semen): Overview and recent developments. Rev. Sci. Tech. Int. Epiz. 2011, 30, 257–271.
142. Büchler, R.; Andonov, S.; Bienefeld, K.; Costa, C.; Hatjina, F.; Kezic, N.; Kryger, P.; Spivak, M.; Uzunov, A.; Wilde, J. Standard methods for rearing and selection of Apis mellifera queens. J. Apic. Res. 2013, 52, 1–30.
143. Anderson, D.L.; Trueman, J.W. Varroa jacobsoni (Acari: Varroidae) is more than one species. Exp. Appl. Acarol. 2000, 24, 165–189.
144. Evans, J.D.; Schwarz, R.S. Bees brought to their knees: Microbes affecting honey bee health. Trends Microbiol. 2011, 19, 614–620.

145. Bowen-Walker, P.L.; Martin, S.J.; Gunn, A. The transmission of Deformed wing virus between honeybees (Apis mellifera L.) by the ectoparasitic mite Varroa jacobsoni Oud. J. Invertebr. Pathol. 1999, 73, 101–106.
146. Mockel, N.; Gisder, S.; Genersch, E. Horizontal transmission of Deformed wing virus: Pathological consequences in adult bees (Apis mellifera) depend on the transmission route. J. Gen. Virol. 2011, 92, 370–377.
147. Di Prisco, G.; Pennacchio, F.; Caprio, E.; Boncristiani, H.F.; Evans, J.D.; Chen, Y. Varroa destructor is an effective vector of Israeli acute paralysis virus in the honeybee, Apis mellifera. J. Gen. Virol. 2011, 92, 151–155.
148. Erban, T.; Harant, K.; Hubalek, M.; Vitamvas, P.; Kamler, M.; Poltronieri, P.; Tyl, J.; Markovic, M.; Titera, D. In-depth proteomic analysis of Varroa destructor: Detection of DWV-complex, ABPV, VdMLV and honeybee proteins in the mite. Sci. Rep. 2015, 5, 13907.
149. Shen, M.; Yang, X.; Cox-Foster, D.; Cui, L. The role of varroa mites in infections of Kashmir bee virus (KBV) and Deformed wing virus (DWV) in honey bees. Virology 2005, 342, 141–149.
150. Duay, P.; De Jong, D.; Engels, W. Weight loss in drone pupae (Apis mellifera) multiply infested by Varroa destructor mites. Apidologie 2003, 34, 61–66.
151. Duay, P.; De Jong, D.; Engels, W. Decreased flight performance and sperm production in drones of the honey bee (Apis mellifera) slightly infested by Varroa destructor mites during pupal development. Genet. Mol. Res. 2002, 1, 227–232. [PubMed]
152. Martin, S.J.; Highfield, A.C.; Brettell, L.; Villalobos, E.M.; Budge, G.E.; Powell, M.; Nikaido, S.; Schroeder, D.C. Global honey bee viral landscape altered by a parasitic mite. Science 2012, 336, 1304–1306.
153. Fuchs, S. Preference for drone brood cells by Varroa jacobsoni Oud in colonies of Apis mellifera carnica. Apidologie 1990, 21, 193–199.
154. Calderone, N.W.; Lin, S.; Kuenen, L.P.S. Differential infestation of honey bee, Apis mellifera, worker and queen brood by the parasitic mite Varroa destructor. Apidologie 2002, 33, 389–398.
155. Le Conte, Y.; Arnold, G.; Trouiller, J.; Masson, C.; Chappe, B.; Ourisson, G. Attraction of the parasitic mite Varroa to the drone larvae of honey bees by simple aliphatic esters. Science 1989, 245, 638–639.
156. Calderone, N.W.; Kuenen, L.P.S. Differential tending of worker and drone larvae of the honey bee, Apis mellifera, during the 60 h prior to cell capping. Apidologie 2003, 34, 543–552.
157. Nazzi, F.; Bortolomeazzi, R.; Della Vedova, G.; Del Piccolo, F.; Annoscia, D.; Milani, N. Octanoic acid confers to royal jelly varroa-repellent properties. Naturwissenschaften 2009, 96, 309–314.
158. Harizanis, P.C. Infestation of queen cells by the mite Varroa jacobsoni. Apidologie 1991, 22, 533–538.
159. Rennie, J. Isle of wight disease in hive bees—Acarine disease: The organism associated with the disease—Tarsonemus woodi, n. sp. Trans. R. Soc. Edinb. 1921, 52, 768–779.
160. Sammataro, D.; de Guzman, L.D.; George, S.; Ochoa, R.; Otis, G. Standard methods for tracheal mite research. J. Apic. Res. 2013, 52, 1–20.
161. Harrison, J.F.; Camazine, S.; Marden, J.H.; Kirkton, S.D.; Rozo, A.; Yang, X. Mite not make it home: Tracheal mites reduce the safety margin for oxygen delivery of flying honeybees. J. Exp. Biol. 2001, 204, 805–814.
162. Eischen, F.A.; Cardoso-Tamez, D.; Wilson, W.T.; Dietz, A. Honey production of honey bee colonies infested with Acarapis woodi (Rennie). Apidologie 1989, 20, 1–8.
163. Eischen, F.A. Overwintering performance of honey bee colonies heavily infested with Acarapis woodi (Rennie). Apidologie 1987, 18, 293–304.
164. Fries, I.; Chauzat, M.P.; Chen, Y.P.; Doublet, V.; Genersch, E.; Gisder, S.; Higes, M.; McMahon, D.P.; Martín-Hernández, R.; Natsopoulou, M.; et al. Standard methods for Nosema research. J. Apic. Res. 2013, 52, 1–28.

165. De Graaf, D.C.; Raes, H.; Sabbe, G.; Derycke, P.H.; Jacobs, F.J. Early development of Nosema apis (Microspora, Nosematidae) in the midgut epithelium of the honeybee (Apis mellifera). J. Invertebr. Pathol. 1994, 63, 74–81.
166. Natsopoulou, M.E.; McMahon, D.P.; Doublet, V.; Bryden, J.; Paxton, R.J. Interspecific competition in honeybee intracellular gut parasites is asymmetric and favours the spread of an emerging infectious disease. Proc. R. Soc. B Sci.2015.
167. Higes, M.; Martin-Hernandez, R.; Garcia-Palencia, P.; Marin, P.; Meana, A. Horizontal transmission of Nosema ceranae (Microsporidia) from worker honeybees to queens (Apis mellifera). Environ. Microbiol. Rep. 2009, 1, 495–498.
168. Webster, T.C.; Thacker, E.M.; Pomper, K.; Lowe, J.; Hunt, G. Nosema apis infection in honey bee (Apis mellifera) queens. J. Apic. Res. 2008, 47, 53–57.
169. Peng, Y.; Grassl, J.; Millar, A.H.; Baer, B. Seminal fluid of honeybees contains multiple mechanisms to combat infections of the sexually transmitted pathogen Nosema apis. Proc. R. Soc. B Sci. 2016, 283, 20151785.
170. Roberts, K.E.; Evison, S.E.F.; Baer, B.; Hughes, W.O.H. The cost of promiscuity: Sexual transmission of Nosema microsporidian parasites in polyandrous honey bees. Sci. Rep. 2015, 5, 10982.
171. Huang, W.-F.; Solter, L.F. Comparative development and tissue tropism of Nosema apis and Nosema ceranae. J. Invertebr. Pathol. 2013, 113, 35–41.
172. Higes, M.; Martin, R.; Meana, A. Nosema ceranae, a new microsporidian parasite in honeybees in Europe. J. Invertebr. Pathol. 2006, 92, 93–95.
173. Antunez, K.; Martin-Hernandez, R.; Prieto, L.; Meana, A.; Zunino, P.; Higes, M. Immune suppression in the honey bee (Apis mellifera) following infection by Nosema ceranae (Microsporidia). Environ. Microbiol. 2009, 11, 2284–2290.
174. Mayack, C.; Naug, D. Energetic stress in the honeybee Apis mellifera from Nosema ceranae infection. J. Invertebr. Pathol.2009, 100, 185–188.
175. Hassanein, M.H. Studies on the effect of infection with Nosema apis on the physiology of the queen honey bee. Q. J. Microsc. Sci. 1951, 92, 225–231.
176. Loskotova, J.; Peroutka, M.; Vesely, V. Nosema disease of honeybee queens (Apis mellifica L.). Apidologie 1980, 11, 153–161.
177. De Miranda, J.R.; Bailey, L.; Ball, B.V.; Blanchard, P.; Budge, G.E.; Chejanovsky, N.; Chen, Y.; Gauthier, L.; Genersch, E.; de Graaf, D.C.; et al. Standard methods for virus research in Apis mellifera. J. Apic. Res. 2013, 52, 1–56.
178. Ball, B.V.; Bailey, L. Viruses. In Honey Bee Pests, Predators, and Diseases; Morse, R.F.K., Ed.; AI Root Company: Medina, OH, USA, 1997; pp. 11–32.
179. Chen, Y.; Siede, R. Honey bee viruses. In Advances in Virus Research; Maramorosch, K., Shabalina, S.A., Murphy, F.A., Eds.; Elsevier Academic Press Inc.: San Diego, CA, USA, 2007; Volume 70, pp. 33–80.
180. Runckel, C.; Flenniken, M.L.; Engel, J.C.; Ruby, J.G.; Ganem, D.; Andino, R.; DeRisi, J.L. Temporal analysis of the honey bee microbiome reveals four novel viruses and seasonal prevalence of known viruses, Nosema, and Crithidia. PLoS ONE 2011, 6, e20656.
181. Maori, E.; Lavi, S.; Mozes-Koch, R.; Gantman, Y.; Peretz, Y.; Edelbaum, O.; Tanne, E.; Sela, I. Isolation and characterization of Israeli acute paralysis virus, a dicistrovirus affecting honeybees in Israel: Evidence for diversity due to intra- and inter-species recombination. J. Gen. Virol. 2007, 88, 3428–3438.
182. Fievet, J.; Tentcheva, D.; Gauthier, L.; de Miranda, J.; Cousserans, F.; Colin, M.E.; Bergoin, M. Localization of Deformed wing virus infection in queen and drone Apis mellifera L. Virol. J. 2006, 3, 16.
183. Chen, Y.; Pettis, J.S.; Collins, A.; Feldlaufer, M.F. Prevalence and transmission of honeybee viruses. Appl. Environ. Microbiol. 2006, 72, 606–611.
184. Yang, X.; Cox-Foster, D. Effects of parasitization by Varroa destructor on survivorship and physiological traits of Apis mellifera in correlation with viral incidence and microbial challenge. Parasitology 2007, 134, 405–412.
185. Williams, G.R.; Rogers, R.E.L.; Kalkstein, A.L.; Taylor, B.A.; Shutler, D.; Ostiguy, N. Deformed wing virus in western honey bees (Apis mellifera) from Atlantic Canada

and the first description of an overtly-infected emerging queen. J. Invertebr. Pathol. 2009, 101, 77–79.

186. Bailey, L.; Gibbs, A.J.; Woods, R.D. Two viruses from adult honey bees (Apis mellifera Linnaeus). Virology 1963, 21, 390–395.

187. Ribiere, M.; Lallemand, P.; Iscache, A.L.; Schurr, F.; Celle, O.; Blanchard, P.; Olivier, V.; Faucon, J.P. Spread of infectious Chronic bee paralysis virus by honeybee (Apis mellifera L.) feces. Appl. Environ. Microbiol. 2007, 73, 7711–7716.

188. Ribiere, M.; Olivier, V.; Blanchard, P. Chronic bee paralysis: A disease and a virus like no other? J. Invertebr. Pathol.2010, 103, 120–131.

189. Ribière, M.; Faucon, J.P.; Pépin, M. Detection of Chronic honey bee (Apis mellifera L.) paralysis virus infection: Application to a field survey. Apidologie 2000, 31, 567–577.

190. Bailey, L. The multiplication and spread of sacbrood virus of bees. Ann. Appl. Biol. 1969, 63, 483–491.

191. Bailey, L.; Woods, R.D. Two more small RNA viruses from honey bees and further observations on Sacbrood and Acute bee-paralysis viruses. J. Gen. Virol. 1977, 37, 175–182.

192. Bailey, L.; Ball, B.V.; Perry, J.N. Association of viruses with two protozoal pathogens of the honey bee. Ann. Appl. Biol.1983, 103, 13–20.

193. Retschnig, G.; Williams, G.R.; Mehmann, M.M.; Yañez, O.; de Miranda, J.R.; Neumann, P. Sex-specific differences in pathogen susceptibility in honey bees (Apis mellifera). PLoS ONE 2014, 9, e85261.

194. Doublet, V.; Labarussias, M.; de Miranda, J.R.; Moritz, R.F.A.; Paxton, R.J. Bees under stress: Sublethal doses of a neonicotinoid pesticide and pathogens interact to elevate honey bee mortality across the life cycle. Environ. Microbiol.2015, 17, 969–983.

195. Simeunovic, P.; Stevanovic, J.; Cirkovic, D.; Radojicic, S.; Lakic, N.; Stanisic, L.; Stanimirovic, Z. Nosema ceranae and queen age influence the reproduction and productivity of the honey bee colony. J. Apic. Res. 2014, 53, 545–554.

196. Rangel, J.; Böröczky, K.; Schal, C.; Tarpy, D.R. Honey bee (Apis mellifera) queen reproductive potential affects queen mandibular gland pheromone composition and worker retinue response. PLoS ONE 2016, 11, e0156027.

197. Botías, C.; Martín-Hernández, R.; Días, J.; García-Palencia, P.; Matabuena, M.; Juarranz, Á.; Barrios, L.; Meana, A.; Nanetti, A.; Higes, M. The effect of induced queen replacement on Nosema spp. infection in honey bee (Apis mellifera Iberiensis) colonies. Environ. Microbiol. 2012, 14, 845–859.

198. Grozinger, C.M.; Richards, J.; Mattila, H.R. From molecules to societies: Mechanisms regulating swarming behavior in honey bees (Apis spp.). Apidologie 2014, 45, 327–346.

199. Engel, P.; Martinson, V.G.; Moran, N.A. Functional diversity within the simple gut microbiota of the honey bee. Proc. Natl. Acad. Sci. USA 2012, 109, 11002–11007.

200. Kwong, W.K.; Moran, N.A. Gut microbial communities of social bees. Nat. Rev. Microbiol. 2016, 14, 374–384.

201. Tarpy, D.R.; Mattila, H.R.; Newton, I.L.G. Development of the honey bee gut microbiome throughout the queen-rearing process. Appl. Environ. Microbiol. 2015, 81, 3182–3191.

202. Kapheim, K.M.; Rao, V.D.; Yeoman, C.J.; Wilson, B.A.; White, B.A.; Goldenfeld, N.; Robinson, G.E. Caste-specific differences in hindgut microbial communities of honey bees (Apis mellifera). PLoS ONE 2015, 10, e0123911.

203. Cobey, S.W.; Sheppard, W.S.; Tarpy, D.R. Status of breeding practices and genetic diversity in domestic U.S. honey bee. In Honey Bee Colony Health; CRC Press: Boca Raton, FL, USA, 2011; pp. 25–36.

204. Büchler, R.; Costa, C.; Hatjina, F.; Andonov, S.; Meixner, M.D.; Conte, Y.L.; Uzunov, A.; Berg, S.; Bienkowska, M.; Bouga, M.; et al. The influence of genetic origin and its interaction with environmental effects on the survival of Apis mellifera L. colonies in Europe. J. Apic. Res. 2014, 53, 205–214.

CHAPTER 8

Understanding the Mechanisms Involved in the Regulation of Cytochrome P450 Gene Expression in Drosophila melanogaster (Diptera: Drosophilidae)

Mohammed BR[1,2*], Kawe MS[1], Nwachukwu OM[1], Jegede OC[1], Agbede RIS[1] and Finn RD[3]

[1]*School of Science Engineering and Technology, Abertay University, Dundee, DD1 1HG, UK*

[2]*Department of Parasitology and Entomology, Faculty of Veterinary Medicine, University of Abuja, Nigeria*

[3]*Department of Applied Sciences, Northumbria University, Newcastle Upon Tyne, NE1 8ST, UK*

ABSTRACT

Cytochrome P450 monooxygenases (P450s) are known to play a central role in the adaptive response of insects and other animals to chemicals in the environment. Drosophila spp. P450s are known to be regulated by Cap 'n' collar isoform C (CnCC) and/ Spineless (ss) which are orthologs to Nuclear factor erythroid-2 factor 2 (Nrf2)/Aryl hydrocarbon receptor (AhR) in higher mammals. However, the mechanism underpinning this regulation in insects including fruit fly, Drosophila melanogaster is poorly understood. Understanding the constitutive and inducible patterns of expression requires knowledge about the signalling pathways that control insect P450 expression, which is still lacking for most identified insect P450s. D. melanogaster, because of its longstanding use as a genetic model insect, is a powerful tool for identifying possible regulatory mechanisms and for following expression through to function. Here, we describe the roles played by the cis-acting elements and the Transcription Factors (TFIF) mechanisms involved in the regulation of cytochrome P450 genes in D. melanogaster in response to xenobiotic compounds. These cis-acting elements include; promoters, enhancers, repressors, silencers and insulators. The regulatory mechanisms involved in the regulation of the P450s by the spineless (ss)/tango (Tgo) and CnCC/dKeap1 (Drosophila Kelch-like ECH-associated protein 1) signalling pathways in insecticide resistance were also extensively discussed. This review increases our understanding of the regulatory mechanisms involved in the insecticide metabolism in Drosophila melanogaster.

Keywords: CnCC; dKeap 1; *Drosophila melanogaster*; p450s; Spineless; Tango

INTRODUCTION

The cytochrome P450 superfamily consists of a considerable amount of heme-containing monooxygenases and is established in all living organisms including insects [1,2]. In the fruit fly, *Drosophila melanogaster* alone there are 83 P450 genes with a few of these enzymes playing key roles in the metabolism or activation of xenobiotics [3-6]. These P450s include; *Cyp6a2* (substrate-DDT, aldrin, dieldrin and diazinon) [7-9] *CYP6G1* (substrate-DDT, imidacloprid) [10,11]. *Cyp12a4* (substrate-Lufeneron) and Cyp12d1 (substrate-DDT and carbaryl) [12-14]. In addition, P450s also partake in the biosynthesis of cuticular hydrocarbons, ecdysteroids, juvenile hormone, and pheromones [15,16]. In higher mammals, some of these xenobiotic metabolizing cytochrome P450 genes are established to be upregulated by the transcription factors such as the Aryl hydrocarbon receptor (AhR)/Aryl hydrocarbon receptor nuclear translocator (ARNT) and/or Nuclear factor erythroid-2 related factor-2 (Nrf2)/Kelchlike ECH-associated protein 1 (Keap 1) [17-21]. Interestingly, AhR/ ARNT and Nrf2/Keap 1 display a multilevel crosstalk where the latter is also a target of the former [22-24]. Previous studies also revealed the orthologs to Aryl hydrocarbon receptor/(AhR) Aryl hydrocarbon nuclear receptor (ARNT) and Nuclear factor erythroid factor 2 (Nrf2)/ Kelch-like associated protein 1(Keap 1) in *Drosophila melanogaster* to be Spineless

(Ss)/Tango (Tgo) and Cap 'n' collar isoform C (CnCC)/ *Drosophila* Kelch-like ECH-associated protein 1 (dKeap 1) respectively [25-28]. Therefore, to understand the mechanism by which these P450s are regulated in *Drosophila melanogaster* (an insect model) the basics underpinning this regulation are described below:

Regulation of Gene Expression

The fundamental component of the control of gene expression is attained at the transcriptional level [29]. This level of regulation consolidates the contribution of various types of cis-acting genomic elements, which are vital molecular switches involved in the transcriptional regulation of a productive chain of gene activities regulating numerous biological processes, including abiotic stress responses, developmental processes and hormone responses [30,31]. This transcriptional regulation transpires within a complex genomic milieu in which promoters, enhancers, and insulators are intimately connected both along the one-dimensional linear chromosome and within the three-dimensional nuclear chromatin environment [32-34].

PROMOTER

Regulation of gene expression at the promoter level is chiefly regulated by the cis-acting elements restricted upstream of the transcriptional start site [35]. The physical interplay between regulatory proteins and the basic transcriptional machinery is straight forward during initiation of transcription owing to the location of proximal elements to the core promoter [34].

ENHANCERS

In variance with promoters, enhancers are typically located far away from the genes they regulate [31,32]. Albeit a promoter is utterly vital for gene transcription, a significant part of metazoan transcriptional regulation emanates via the action of distal cisregulatory modules [33,36].

REPRESSORS

Repressors appear to function by blocking the binding of a nearby activator, or by directly competing for the same site [36,37]. It has been proposed that the difference between the two may be associated to the recruitment of distinctive cofactors [36].

SILENCERS

Silencers are binding sites for negative transcription factors termed repressors [36]. Silencers are sequence-specific elements that confer a negative (i.e., silencing or repressing) effect on the transcription of a target gene [38]. Repressor function can require the recruitment of negative cofactors, also termed co-repressors, and in some cases, an activator can switch to a repressor by differential cofactor recruitment [36]. In *Drosophila*, two classes of silencers have been observed: shortrange silencers, which generally must reside within ~100 bp of their target gene to have a repressive effect and long-range silencers, which can repress multiple enhancers or promoters over a span of a few kilo base pairs [36].

INSULATORS

A third critical component contributing to gene expression is the insulator. Originally defined in *Drosophila*, and still best understood in that organism, insulators were so named due to their ability to "insulate" genes from position effects in transgenic assays. Historically, two major roles have been ascribed to insulator elements: the ability to serve as boundary elements preventing the spread of heterochromatin, and the ability to prevent enhancer activity when interposed between an enhancer and promoter [33].

REGULATORY MECHANISM IN RESISTANCE

Drosophila melanogaster has been used broadly as a model system to understand the molecular mechanisms underlying insecticide resistance [25]. Further studies also addressed the mechanisms that underlie this regulation, mapping critical promoter elements that are required for P450 gene induction in response to pesticides or the wellstudied xenobiotic phenobarbital (PB) [39-45]. The Spineless/Tango and CnCC/dKeap1 signaling pathways are hereby described below.

THE SPINELESS (SS) GENE

The *Drosophila* gene spineless (Ss) is the ortholog of vertebrate AhR [26,46]. These proteins share extensive sequence identity, especially in their Basic Helix-Loop-Helix (bHLH) regions, and must share common ancestry, as several of the splice sites in the Ss and AhR genes are precisely conserved [47].

Function and structure of spineless: Like other invertebrate homologs of AhR, spineless does not bind prototypical xenobiotic ligands of the vertebrate receptor such as 2,3,7,8-Tetrachlorodibenzo-p- Dioxin (TCDD) [48]. Studies revealed that spineless, functions as a heterodimer with Tango, the *D. melanogaster* ortholog of ARNT and both appear to recognise the same DNA sequence, the xenobiotic response element (XRE), a core nucleotide sequence at the upstream of inducible target genes for the transcription factor Aryl Hydrocarbon receptor (AhR) that is responsible for recognition of exogenous environmental pollutants in eukaryotic cells [49,50]. Furthermore, Tango heterodimerize with Trachealess and Single minded (both bHLH-PAS family members) and regulates transcription in the trachea and central midline, respectively [51].

Ligands of spineless (AhR): Spineless (Ss) in *D. melanogaster* does not show the ability to bind to toxic agonists such as dioxin congener (TCDD) and Polycyclic Aromatic Hydrocarbons (PAHs) [52]. It is therefore plausible that, as a consequence of their toxic activity, an endogenous ligand competent of triggering the Ss protein is generated. Such a role is often played by one of the endogenous AhR ligands–a toxic tryptophan derivative Formyl-Indolo-Carbazole (FICZ). Consequently, cellular concentrations of FICZ levels are elevated considerably in response to ionizing radiation and this inevitably triggers elevated expression of the Ahr gene thereby leading to cellular reaction to toxin exposure, and, in particular, to ionizing radiation. This induces expression of detoxification-related genes *Cyp6g1* and Cg1681 hence ss gene is necessary for this induction [47]. This protein can however bind to the XRE and stimulate transcription from genes containing this cis-acting element [49,53]. Moreover, it regulates normal morphogenesis of the leg or antenna and bristles, all of which are dominant *Drosophila* sensor organs or tissues that operate in response to environmental chemicals [54].

Molecular Mechanisms of AhR Functions in the Regulation of Cytochrome P450S in *Drosophila*

SPINELESS (SS) TANGO (TGO) SIGNALLING PATHWAY

The spineless (Ss) protein remains predominantly cytoplasmic as part of a protein complex with the molecular chaperone heat shock protein 90 (HSP90), p23, and XAP2. It is known to interact with Tango (Tgo), the fly homolog of mammalian ARNT and through this interaction the protein is transported to the nucleus from the cytoplasm where it binds another bHLH-PAS-protein, the Aryl hydrocarbon receptor nuclear translocator (ARNT). The Ss: Tgo heterodimer binds to a specific motif, XRE in the promoters of its target genes and regulates their transcription [47,55]. The Ss: Tgo heterodimer can both repress and activate explicit genes, demonstrating the heterodimer's interplay with other transcription or nucleosome assembly factors [47]. AhR gene is exceedingly conserved between vertebrates and invertebrates (**FIGURE 1**) [47].

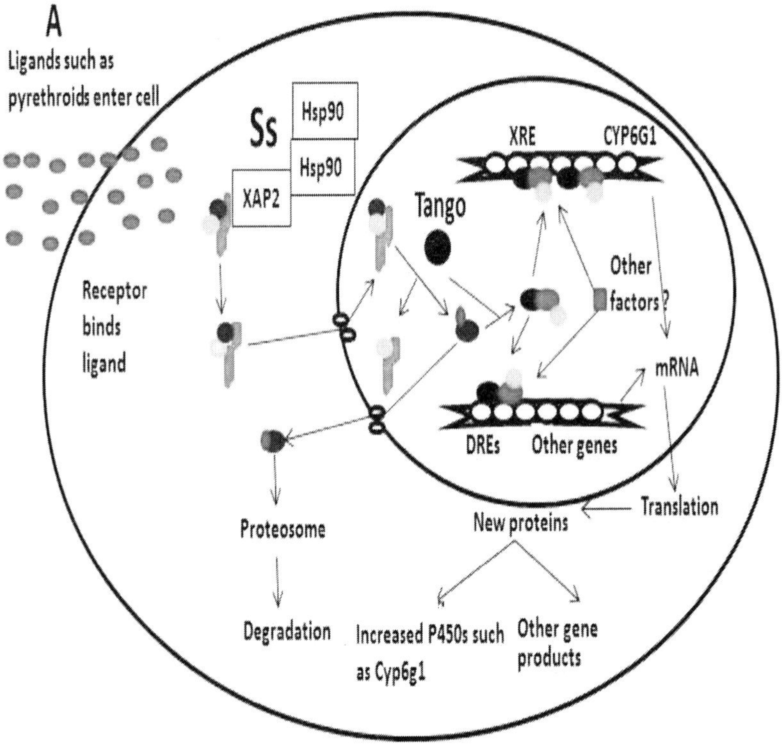

FIGURE 1: Schematic representation of Ss/tgo signalling pathway. The cytosolic ss are complexed of two molecules of heat shock protein (Hsp90), X associated protein (XAP2) and the co-chaperone p23 for the HSP90. Binding of a ligand, e.g. TCDD, leads to a conformational change, thereby allowing nuclear translocation of the spineless complex. In the nucleus, Spineless dissociates from the complex and dimerizes with Tango. The Ss–Tgo heterodimer then binds to xenobiotic response elements (XREs) in the promoters of genes encoding for several phase I and phase II metabolizing enzymes but also several other genes, e.g. *Cyp6g1* (Adopted and modified from Denison and Nagy) [64]. Schematic presentation of the sequence conservation between AhR and Ss, LBD, dioxin binding domain. (Adapted and modified from Céspedes et al. [52].

THE CAP 'N' COLLAR ISOFORM C (CNCC) (NAMED AFTER THE CNC GENE OF *Drosophila*)

Studies of the *Drosophila* orthologs to Nrf2 and dKeap1 have provided insights into the functions of this protein. The *Drosophila*Cap 'n' collar locus encodes CnCC, which contains a bZIP domain homologous to that of Nrf2, N-terminal DTG (Asp-Thr-Gly), a low-affinity motif and a high-affinity ETGE (Glu-Thr-Gly-Glu) motif, separated by a central lysine-rich α-helix and are homologous to those that mediate Nrf2 interaction with Keap1 (a member of the Kelch family of actin binding proteins, named after the fruit fly's Kelch protein (a component of the egg chambers) [56].

Function of the CnCC: CnCC regulates the transcriptional responses to xenobiotic compounds whilst the CnCC-Keap1 (dKeap 1) protein complexes regulate the native cellular and developmental processes [18,57].

Ligands of CnCC: DKeap1 can function as a sensor of oxidants and electrophiles, which react with its redox sensitive cysteine residues [58,59]. During oxidative stress or electrophilic xenobiotics, dKeap 1 undergoes destruction releasing CnCC that translocates to the nucleus [60,61]. CnCC is then stabilized and accumulates in the nucleus, where it binds to the Antioxidant Response Element (ARE) in the enhancers of its target genes [59,62].

Molecular Mechanisms of CNCC Functions in the Regulation of Cytochrome P450 genes

CNCC/DKEAP1 SIGNALLING PATHWAY

Antioxidant Response Element (ARE)-mediated response to oxidative stress is conserved from flies to humans. In unstressed conditions, Nrf2 (Nuclear factor erythroid-2 related factor- 2) in mammals, CncC (Cap 'n' collar isoform C) in *Drosophila*, is repressed by dKeap1 (*Drosophila* Kelch-like ECH-associated protein 1), which also functions as a sensor of oxidants and electrophilic compounds [59,63]. Under oxidative stress conditions, the inhibition of CncC by dkeap1 is abolished allowing this transcription factor to bind, with other proteins, to ARE sequences upregulating downstream genes such as P450s. The *Drosophila* dKeap1 contains Kelch repeats homologous to those that mediate Keap1 interaction with Nrf2 as well as a sequence motif that is required for mammalian Keap1 export from the nucleus [18]. Overexpression of CncC and depletion of dKeap1 in *Drosophila melanogaster* activates the transcription of many genes including *Cyp6g1* and *Cyp6a2* that protect cells from xenobiotic compounds, whereas dKeap1 overexpression represses their transcription, indicating that the functions of these protein families in the xenobiotic response are conserved between mammals and *Drosophila* (**FIGURE 2A-2C**) [18,25].

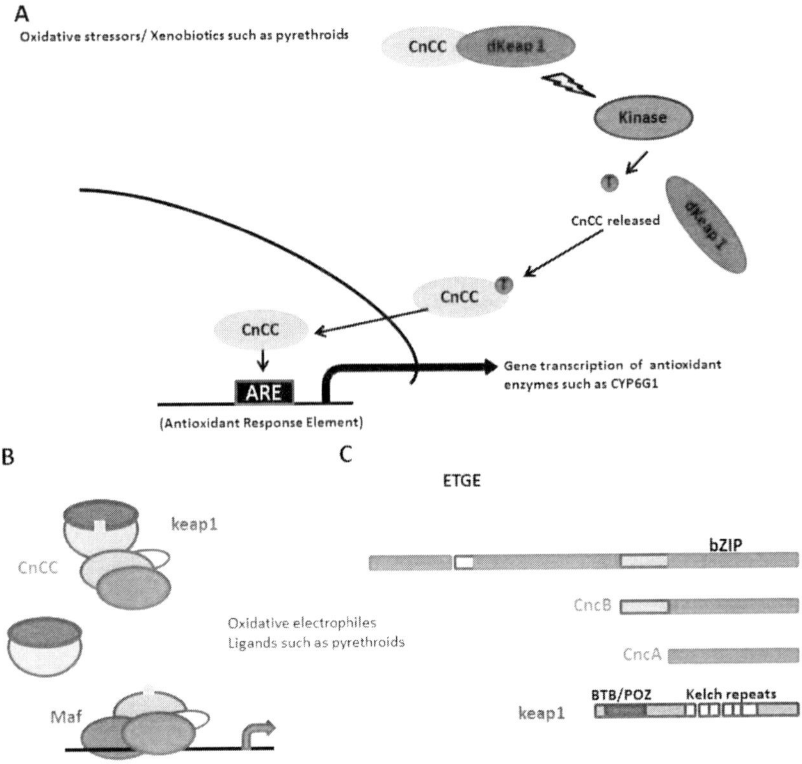

FIGURE 2: (A) General scheme for the induction of CnCC/dKeap 1-signaling pathway. The antioxidant response element (ARE) in the promoter region of select genes allows the coordinated upregulation of antioxidant and detoxifying enzymes in response to oxidative/electrophilic stress. This upregulation is mediated through Cap 'n' collar isoform C nuclear (CnCC) that may be activated by endogenous and exogenous molecules or stressful conditions. These agents disrupt the association between CnCC and dKeap1 with subsequent nuclear translocation of CnCC. In the cell nucleus CnCC interacts with small MAF protein, forming a heterodimer that binds to the ARE sequence in the promoter region and upregulates transcription of many genes encoding detoxifying enzymes such as *Cyp6g1*. It is therefore speculated that this signalling pathway is constitutively upregulated in long-lived individuals providing extension of longevity and health span. (Adopted and modified from [59,65]. (B) A detailed view at the promoter level of teh binding of CnCC to sMaf. (C) An illustration of the conservation of Nrf2 and Keap 1 in Drosophila (Adapted and Modified from [59,65].

Previous studies revealed that cytochrome P450 family members are modulated by Spineless/Tango and CnCC/ dKeap 1 pathways in *Drosophila melanogaster* (**TABLE 1**).

TABLE 1: Cytochrome P450s modulated by Spineless/Tango and CnCC/dKeap1 pathways.

S/No.	Cytochrome P450	Signalling pathway	Reference
1.	CYP6G1	ss/tango and CnCC/dKeap 1	[14,47]
2.	CYP6B1	ss/tango	[66]
3.	CYP6G2	CnCC/dKeap 1	[14]
4.	CYP12D1	CnCC/dKeap 1	[14]
5.	CYP6A2	CnCC/Keap 1	[25]
6.	CYP6A8	CnCC/Keap 1	[25]
7.	CYPA21	CnCC/Keap 1	[25]
8.	CYP6BQ9	CnCC/Keap 1	[25]
9.	CYP12A4	CnCC/Keap 1	[25]

Further studies have also shown that cytochrome P450 family genes contain elements responsive to the Spineless/Tango and CnCC transcription factors in *Drosophila melanogaster* (**TABLE 2**) [64-66].

TABLE 2: Response Elements for Spineless/Tango and CnCC/dKeap 1 transcription factors.

S/No.	P450 Family genes	Responsive elements	References
1	CYP6G1	XRE	[47]
3.	CYP6G2	ARE	[25]
4.	CYP12D1	ARE	[14]
5.	CYP6A2	ARE	[25]
6.	CYP6A8	ARE	[25]
7.	CYPA21	ARE	[25]
8.	CYP12A4	ARE	[25]

XRE: Xenobiotic Response Element; ARE: Antioxidant Response Element

CONCLUSIONS

Here we have reviewed the role of the spineless/tango and CnCC/ dKeap 1 signalling pathways for their mechanistic role in the regulation of Cytochrome P450s in the activation of xenobiotics in *Drosophila melanogaster*. Since *D. melanogaster* is a model insect system and the mechanisms of insecticide resistance in this species have been studied widely, this work has implications for the mechanistic understanding of the basis of insecticide resistance in insect disease vectors and hence the spread of Vector Borne Diseases.

ACKNOWLEDGEMENTS

This research was supported by Tertiary Education Trust Fund (TETFUND) and the University of Abuja-Nigeria.

REFERENCES

1. Scott JG, Wen Z (2001) Cytochromes P450 of insects: the tip of the iceberg Pest. Pest Manag Sci 57: 958-967.
2. Guo J, Thiess S, Johansson I, Mkrtchian S, Ingelman-Sundberg M (2016) Membrane topology and search for potential redox partners of colon cancer-specific cytochrome P450 2W1. FEBS lett 590: 330-339.
3. Ranson H, Claudianos C, Ortelli F, Abgrall C, Hemingway J, et al. (2002) Evolution of super gene families associated with insecticide resistance. Science 298: 179-181.
4. Fournier-Level A, Neumann-Mondlak A, Good RT, Green LM, Schmidt JM, et al. (2016) Behavioural response to combined insecticide and temperature stress in natural populations of *Drosophila melanogaster*. J Evol Biol 29: 1030-1044.
5. Nardini L, Christian RN, Coetzer N, Ranson H, Coetzee M, et al. (2012) Detoxification enzymes associated with insecticide resistance in laboratory strains of *Anopheles arabiensis* of different geographic origin. Parasit Vectors 5: 113.
6. Gong Y, Li T, Zhang L, Gao X, Liu N (2013) Permethrin Induction of Multiple Cytochrome P450 genes in insecticide resistant mosquitoes, *Culex quinquefasciatus*. Int J Biol Sci 9: 863-871.
7. Amichot M, Tares S, Brun-Barale A, Arthaud L, Bride JM, et al. (2004) Point mutations associated with insecticide resistance in the *Drosophila* cytochrome P450 *Cyp6a2* enable DDT metabolism. Eur J Biochem 271: 1250-1257.
8. Mitchell CL, Yeager RD, Johnson ZJ, D'Annunzio SE, Vogel KR et al. (2015) Long-Term Resistance of *Drosophila melanogaster* to the Mushroom Toxin Alpha-Amanitin. PloS One 10: e0127569.
9. Seong KM, Sun W, Clark JM, Pittendrigh BR (2016) Splice form variant and amino acid changes in MDR49 confers DDT resistance in transgenic *Drosophila*. Sci Rep 6: 23355.
10. Joussen N, Heckel DG, Haas M, Scuphan I, Schphan Schmidt B (2008) Metabolism of imidacloprid and DDT by P450 CYP6G1 expressed in cell cultures of *Nicotiana tabacum* suggests detoxification of these insecticides in *Cyp6g1*-

overexpressing strains of *Drosophilamelanogaster*, leading to resistance. Pest Manag Sci 64: 65-73.

11. Battlay P, Schmidt JM, Fournier-Level A, Robin C (2016) Genomic and Transcriptomic Associations Identify a New Insecticide Resistance Phenotype for the Selective Sweep at the *Cyp6g1* Locus of *Drosophila melanogaster*. G3: G3 GENES GENOM GENET 6: 2573-2581.
12. Chiu TL, Wen Z, Rupasinghe S, Schuler M (2008) Comparative molecular modeling of *Anopheles gambiae Cyp6z1*, a mosquito P450 capable of metabolizing DDT. Proc Natl Acad Sci, USA 105: 8855-8860.
13. Schrider DR, Hahn MW, Begun DJ (2016) Parallel evolution of copy number Variation across continents in *Drosophila melanogaster*. Mol Biol Evol 33: 1308-1316.
14. Loboda A, Damulewicz M, Pyza E, Jozkowicz A, Dulak J (2016) Role of Nrf2 /HO-1 system in development, oxidative stress response and diseases: an evolutionarily conserved mechanism. Cell Mol Life Sci 73: 3221-3247.
15. Feyereisen R (1999) Insect P450 enzymes. Annu Rev Entomol 44: 507-533.
16. Félix R, Silveira H (2012) The Role of *Anopheles gambiae* P450 Cytochrome in Insecticide Resistance and Infection, Insecticides-Pest Engineering. In: Dr. Perveen F (ed.).
17. Cederbaum AI (2013) *Nrf2* and Antioxidant Defence against *CYP2E1* Toxicity. Subcell Biochem 67: 105-130.
18. Deng H, Kerppola TK (2013) Regulation of *Drosophila* Metamorphosis by Xenobiotic Response Regulators. PloS Gene 9: e1003263.
19. Giantin M, Vascellari M, Lopparelli RM, Ariani P, Vercelli A, et al. (2013) Expression of the aryl hydrocarbon receptor pathway and cyclooxygenase-2 in dog tumors. Res Vet Sci94: 90-99.
20. Mohammed BR, Wilding CS, Collier PJ, Deeni YY (2014) Bioinformatic analysis of regulatory elements within the promoter region of the cytochrome P450 Gene, CYP6M2 in *Anopheles gambiae*. Eur J Biotechnol Biosci 2: 24 - 31.
21. Sakakibara Y, Katoh M, Kondo Y, Nadai M (2016) Effects of β-Naphthoflavone on Ugt1a6 and Ugt1a7 Expression in Rat Brain. Biol Pharm Bull 39: 78-83.
22. Pang G, Xie J, Chen Q, Hu Z (2012) How functional foods play critical roles in human health. Food Sci Hum Wellness 1: 26-60.
23. Bock KW (2014) Homeostatic Control of Xeno-and Endobiotics in the Drug-metabolizing enzyme system. Biochem Pharmacol90: 1-6.
24. Qin S, Hou DX (2016) Multiple regulations of Keap1/Nrf2 system by dietary phytochemicals. Mol Nutr Food Res 60: 1731-1755.
25. Misra JR, Lam G, Thummel CS (2013) Constitutive activation of the *Nrf2* / Keap 1 pathway in insecticide-resistant strains of *Drosophila*. Insect Biochem Molec Biol43: 1116-1124.
26. Reitzel AM, Passamaneck YJ, Karchner SI, Franks DG, Martindale MQ, et al. (2014) Aryl hydrocarbon receptor (AhR) in the cnidarian *Nematostella vectensis*: comparative expression, protein interactions and ligand binding. Dev Genes Evol 224: 13-24.
27. Mohammed BR (2014) Regulatory mechanisms involved in the control of CYP6M2 gene in insecticide resistant*Anopheles gambiae(Diptera: Culicidae)* Doctoral dissertation, Abertay University, Dundee, UK.
28. Peng T, Pan Y, Gao X, Xi J, Zhang L, et al. (2016) Cytochrome P450 CYP6DA2 Regulated by Cap 'n' collar isoform C (CncC) is Associated with Gossypol Tolerance in *Aphis gossypii*. Insect Mol Biol 25: 450-459
29. Bhattacharjee S, Renganaath K, Mehrotra R, Mehrotra S (2013) Combinatorial Control of Gene Expression. Bio Med Res Int, pp: 407263-407274.
30. Yamaguchi-Shinozaki K, Shinozaki K (2005) Organization of *cis*-acting regulatory elements in osmotic-and cold-stress-responsive promoters. Trends Parasitol 10: 88-94.

31. Symmons O, Spitz F (2013) From remote enhancers to gene regulation: charting the genome's regulatory landscapes. Phil Trans R Soc B 368: 20120358.
32. Stees JS, Varn F, Huang S, Strouboulis J, Bungert J (2012) Recruitment of Transcription Complexes to Enhancers and the Role of Enhancer Transcription. Biol 1: 778-793.
33. Atkinson TJ, Halfon MS (2014) Regulation of Gene Expression in the Genomic Context. Comput Struct Biotechnol J 9: 1-9.
34. Hernandez-Garcia CM, Finer JJ (2014) Identification and validation of promoters and cis-acting regulatory elements. Plant Sci 217: 109-119.
35. Huang W, Massouras A, Inoue Y, Peiffer J, Ràmia M, et al. (2014) Natural variation in genome architecture among 205 *Drosophila melanogaster* Genetic Reference Panel lines. Genome Res 24: 1193-1208.
36. Maston GA, Evans SK, Green MR (2006) Transcriptional regulatory elements in the human genome. Annu Rev Genomics Hum Genet 7: 29-59.
37. Gaston K, Jayaraman PS (2003) Transcriptional repression in eukaryotes: repressors and repression mechanisms. Cell Mol Life Sci 60: 721-741.
38. Vooght KM, Van Wijk R, Van Solinge WW (2009) Management of gene promoter mutations in molecular diagnostics. Clin Chem 55: 698-708.
39. Brun A, Cuany A, le Mouel T, Berge J, Amichot M (1996) Inducibility of the *Drosophila melanogaster*Cytochrome P450 Gene, *CYP6A2*, by Phenobarbital in Insecticide Susceptible or Resistant Strains. Insect Biochem Mol Biol 26: 697-703.
40. Maitra S, Dombrowski SM, Waters LC, Ganguly R (1996) Three second chromosome linked clustered Cyp6 genes show differential constitutive and barbital-induced expression in DDT-resistant and susceptible strains of *Drosophilamelanogaster*. Gene 180: 165-171.
41. Danielson PB, Maclntyre RJ, Fogleman JC (1997) Molecular Cloning of a Family of Xenobiotic-inducible Drosophilid cytochrome P450s: Evidence for Involvement in Host Plant Allo-chemical Resistance. Proc Natl Acad Sci USA 94: 10797-10802.
42. Dunkov BC, Rodriguez-ArnaizR, Pittendrough B, French-constant RH, Feyereisen R (1996) Cytochrome P450 gene clusters in *Drosophilia malangaster*. Mol Gen257: 290-297.
43. Dombrowski KE, Brewer KA, Kapp JA (1998) Ecto-ATPase: An activation marker necessary for effector cell function. Immunol Rev 161: 111-118.
44. McDonnell CM, Brown RP, Berenbaum MR, Schuler MA (2004) Conserved regulatory elements in the promoters of two allelochemical-inducible cytochrome P450 genes differentially regulate transcription. Insect Biochem Mol Biol 34: 1129-1139.
45. Maitra M, Menezes D, Pecchi S (2010) Biological characterization of NVP-BK M120, a novel inhibitor of phosphorinosotide 3-kinase in Phase I/II clinical trials. Presented at: 101st American Association for Cancer Research Congress. Washington DC, USA.
46. Duncan DM, Burgess EA, Duncan I (1998) Control of distal antennal identity and tarsal development in *Drosophila* by spineless-aristapedia, a homolog of the mammalian dioxin receptor. Genes Dev 12: 1290-1303.
47. Kuzin BA, Nikitina EA, Cherezov RO, Vorontsova JE, Slezinger MS, et al. (2014) Combination of Hypomorphic Mutations of the Drosophila Homologues of Aryl Hydrocarbon Receptor and Nucleosome Assembly Protein Family Genes Disrupts Morphogenesis, Memory and Detoxification. PloS ONE 9: e94975.
48. McMillan BJ, Bradfield CA (2007) The Aryl Hydrocarbon Receptor sans Xenobiotics: Endogenous Function in Genetic Model Systems. Mol Pharmacol 72: 487-498.
49. Emmons RB, Duncan D, Estes PA, Kiefel P, Mosher JT, et al. (1999) The Spineless-Aristapedia a Tango Bhlh-PAS proteins interact to control antennal and tarsal development in *Drosophila*. Development 126: 3937-3945.
50. Kuramoto N, Goto E, Masamune Y, Gion K, Yoneda Y (2002) Existence of xenobiotic response element binding in *Dictyostelium* spp. Biochimica et Biophysica Acta (BBA).Gene Struct Expr 1578: 1-11.

51. Zhang YL, Kulye M, Yang FS, Xiao L, Zhang YT, et al. (2011) Identification, characterization, and expression of a novel P450gene encoding CYP6AE25 from the Asian corn borer, *Ostrinia furnacalis*. J Insect Sci 11: 1-17.
52. Céspedes MA, Galindo MI, Couso JP (2010) Dioxin Toxicity *in vivo* Results from an Increase in the Dioxin-Independent Transcriptional Activity of the Aryl Hydrocarbon Receptor. PloS ONE5: e15382.
53. Misra JR, Horner MA, Lam G, Thummel CS (2011) Transcriptional regulation of xenobiotic detoxification in *Drosophila*. Genes Dev 25: 1796-1806.
54. Adachi J, Mori Y, Matsui S, Takigami H, Fujino J, et al. (2001) Indirubin and indigo are potent aryl hydrocarbon receptor ligands present in human urine. Biol Chem 276: 31475-31478.
55. Swanson HI, Chan WK, Bradfield CA (1995) DNA-Binding Specificities and Pairing Rules of the Ah Receptor, Arnt, and Sim Proteins. J Biol Chem 270: 26292-26302.
56. McMahon M,Thomas N, Itoh K, Yamamoto M, Hayes JD (2006) Dimerization of substrate adaptors can facilitate cullin-mediated ubiquitylation of proteins by a "tethering" mechanism: a two-site interaction model for the Nrf2-Keap1 complex. J Biol Chem 281: 24756-24768.
57. Grimberg KB, Beskow A, Lundin D, Davis MM, Young P (2011) Basic leucine zipper protein Cnc-C is a substrate and transcriptional regulator of the Drosophila 26S proteasome. Mol Cell Biol 31: 897-909.
58. Zhang DD (2006) Mechanistic studies of the Nrf2-Keap1 signaling pathway. Drug Metab Rev 38: 769-789.
59. Sykiotis GP, Bohmann D (2008) Keap1/Nrf2 signalling regulates oxidative stress tolerance and lifespan in *Drosophila*. Dev Cell 14: 76-85.
60. Itoh K, Mochizuki M, Ishii Y, Ishii T, Shibata T, et al. (2004) Transcription factor Nrf2 regulates inflammation by mediating the effect of 15-deoxy-Delta (12,14)-prostaglandin J2. Mol Cell Biol 24: 36-45.
61. Ma Q, He X (2012) Molecular basis of electrophilic and oxidative defence: promises and perils of Nrf2. Pharmacol Rev 64: 1055-1081.
62. Jaiswal AK (2004) Regulation of antioxidant response element-dependent induction of detoxifying enzyme synthesis. Method Enzymol 378: 221-238.
63. Nioi P, McMahon M, Itoh K, Yamamoto M, Hayes JD (2003) Identification of a novel Nrf2-regulated antioxidant response element (ARE) in the mouse NAD (P) H: quinone oxidoreductase 1 gene: reassessment of the ARE consensus sequence. Biochem J 374: 337-348.
64. Denison MS, Nagy SR (2003) Activation of the aryl hydrocarbon receptor by structurally diverse exogenous and endogenous chemicals. Annu Rev Pharmacol Toxicol 43: 309-334.
65. Guio L, Barron MG, Gonzalez J (2014) The Transposable Element Bari-Jheh Mediates oxidative stress response in *Drosophila*. Mol Ecol 23: 2020-2030.
66. Brown RP, McDonnell CM, Berenbaum MR, Schuler MA (2005) Regulation of an insect cytochrome P450 monooxygenase gene (CYP6B1) by aryl hydrocarbon and xanthotoxin response cascades. Gene 358: 39-52

CHAPTER 9

Bioefficacy of Organic Extracts of Ocimum basilicum against Sitophilus zeamais

Ouko RO*, Koech SC, Arika WM, Njagi SM, Oduor RO and Ngugi MP

Department of Biochemistry and Biotechnology, School of Pure and Applied Sciences, Kenyatta University, Nairobi, Kenya

ABSTRACT

Ocimum basilicum plant extracts in methanol, hexane and methanol: hexane solvent blend were evaluated for their fumigant toxicities, repellent and adult emergence inhibition activities against Sitophilus zeamais. Twenty grams of maize were used for each treatment and four different extract concentrations of 25, 50, 75 and 100% were tested for their activities. The highest mortality (100%) was observed after 72 h at 100% O. basilicum blend extract concentration. The methanolic extract recorded mortality rates of 74.36% at highest concentration while 78% mortality was true for hexane extract at 96 hours with 100% extract concentration. The three different solvents extracts of O. basilicum recorded high repellent and new adult emergence inhibition activities comparable to actellic (reference pesticide). In conclusion, blend extract of O. basilicum was the most effective in maize weevil control and therefore can be used in formulating eco-friendly bio-pesticide against S. zeamais.

Keywords: *Sitophilus zeamais*; *Ocimum basilica*; Adult emergence; Fumigant toxicity; Repellent activities

INTRODUCTION

Maize weevil (*Sitophilus zeamais*) (Coleoptera; Dryophthoridae), is one of the most important primary pest that attack stored-maize among others like larger grain borer (*Prostephanus truncatus*) [1]. *Sitophilus zeamais* are virtually found in every part of the world, especially Tropical regions where maize is grown [2]. The damage caused by weevils in maize grains reduces its nutritive value, weight and contaminates the stored grain rendering the grain unfit for consumption. The losses attributed to maize weevils are enormous and may lead to diverse effects such as hunger and food insecurity in affected areas [3]. There are a number of conventional methods of controlling weevils. Including, use synthetic pesticides. However, these methods have a number of short comings which include, residual toxicity, death of non-target organism, risk of user's contamination and pest resistance hence not preferable. Plants produce a diverse array of chemicals some of which have insecticidal and repellent activities [4]. This gives prospects of a source of safer, less expensive and easily processed insecticides against weevils after determination of their insecticidal and repellent efficacies [5]. The current study was designed to determine the fumigant toxicities, repellent activity and new adult emergence inhibition activities of methanolic, hexane and methanol: hexane blend extracts of *O. basilicum* to *S. zeamais* as a possible alternative pest control material.

MATERIALS AND METHODS

Plant Material Collection and Preparation of Crude Extracts

Fresh leaves of Ocimun basilicum (sweet basil) were collected from Siakago division, Mbeere North Sub County of Embu County, Kenya. They were then clean and transported to Kenyatta university biochemistry laboratory. The leaves were air dried indoors at room temperature for ten days until they were completely dry. The dried leaves were then crushed into powder using an electric mill to obtain fine powder. Seven hundred and fifty grams of the powder were soaked in a litre of each of the solvent for 72 h at room temperature. For blend extraction, hexane and methanol were mixed in the ratio 1: 1 followed by extraction. the extracts were then filtered and concentrated using rotary evaporator at 45ºC for eight hours and the final concentrates were kept as stock solution (100% concentrate) as suggested by Deshmukh and Borie [6]. Different extract proportions were then formulated from stock solution, i.e., 75% (3 stock solution: 1 solvent v/v), 50% (1:1) and 25% (3:1 v/v).

INSECT CULTURE

Maize weevils *Sitophilus zeamais* were reared at the Kenyatta university biochemistry laboratory growth chambers maintained at 27 ± 2ºC and 75% relative humidity. The weevils were fed on corn meal. The jars containing the weevils were covered with muslin cloth to ensure proper ventilation.

FUMIGANT TOXICITY ASSAY

Fumigant toxicities of the extracts on *S. zeamais* were carried out using plastic vials each containing twenty grams of whole maize grains divided into six different groups. Each group was separately treated with 1 ml of the different extract concentrates (25, 50, 75 and 100%) and four replicates made for each treatment. One control group was treated with solvent only while the other group with actellic (preference pesticide). The containers were shaken thoroughly and then left open for one hour so that traces of solvents can evaporate. Twenty adult *S. zeamais* (2 to 3 days old) were then placed into each container and ventilation made. The numbers of dead and alive weevils were counted after 6, 24, 48, 72 and 96 h. The insects were confirmed dead when they did not respond to probing with sharp object at the abdomen [7]. Corrected mortality was then computed following Abbort formula [8].

NEW ADULT EMERGENCE INHIBITION EFFECT

After 96 h, both dead and live insects were removed from the vials and the set up maintained for 49 days. The total number of *S. zeamais* that emerged from each vial was counted and percentage inhibition rates computed using the formula described by Tapondjou et al. [9].

REPELLENT ASSAY

The repellent activities of *O. basilicum* extracts were carried out following method described by Obeng-Ofori et al. [10]. The test areas were 9 cm diameter filter-paper circles cut into half. 1 ml of each of the extract was uniformly applied to one half paper-discs while the other half disc was treated with solvent only (control). The half-paper discs were then air dried and full discs re-made by attaching the half-paper discs with adhesive paper tape. The full discs were then placed into petri dish followed by release of twenty adult *S. zeamais* at the centre of the filter discs and covered. Four replicates were made for each treatment. The number of insects present on the control half (Nc) and treated half (Nt) were recorded hourly for five hours. Percentage Repellency (PR) was then calculated using the formula described by Thien et al. [11].

QUALITATIVE PHYTOCHEMICAL SCREENING

Freshly prepared organic extracts of *O. basilicum* were subjected to qualitative phytochemical screening to check for presence or absence of a number of secondary metabolites that have been have associated with fumigant toxicity and repellent activities of various plants. Particularly, terpenoids, flavonoids, phenolics, alkaloids, saponins, steroids and tannins were screened following the protocol described by Harborne [12] and Kotake [13].

DATA ANALYSIS

Data obtained from this work were subjected to descriptive statistics to obtain means ± SEM. One way analysis of variance (ANOVA) was then carried out followed by Tukey's post hoc test to compare and separate the means.

RESULTS

FUMIGANT TOXICITY OF METHANOLIC EXTRACT OF *O. basilicum* AGAINST *S. zeamais*

Low mortality rates of *S. zeamais* were recorded after exposure to methanolic extracts of *O. basilicum* in the sixth hour at all extract concentration used. Nonetheless, mortality rates increased in the subsequent time periods. Highest mortality rate of 74.36% was recorded by 100% extract concentration, followed by 65.38% mortality caused by 75% extract concentration within 96 hours (**TABLE 1**). The low extract concentrations of 25 and 50% had low fumigant toxicity effects of 30.77 and 33.38% within similar time period of 96 hours. Moreover, the fumigant toxicity effects of actellic (reference pesticide) was significantly higher compared to all extract treatments throughout the experimental period ($p<0.05$; **TABLE 1**).

TABLE 1: Fumigant toxicity effect of methanolic extract of *Ocimum basilicum* on adult *Sitophilus zeamais.*

Concentration	Mean % corrected mortality ± S.E with exposure period (hours)				
	6 h	24 h	48 h	72 h	96 h
Control	0.00 ± 0.00^{e}	0.00 ± 0.00^{c}	0.00 ± 0.00^{e}	0.00 ± 0.00^{d}	0.00 ± 0.00^{d}
25% extract	3.75 ± 2.39^{e}	20.51 ± 3.31^{b}	23.08 ± 3.63^{cd}	26.92 ± 4.38^{c}	30.77 ± 3.31^{c}
50% extract	5 ± 2.04^{b}	8.97 ± 2.45^{c}	20.51 ± 2.56^{d}	30.77 ± 4.44^{c}	33.33 ± 2.69^{c}
75% extract	8.75 ± 2.39^{c}	21.79 ± 3.23^{b}	34.62 ± 3.85^{c}	65.38 ± 4.38^{b}	65.38 ± 4.38^{b}
100% extract	12.5 ± 3.23^{b}	26.92 ± 1.28^{b}	48.72 ± 2.96^{b}	67.95 ± 4.38^{b}	74.36 ± 2.09^{b}
Actellic	91.25 ± 5.15^{a}	100 ± 0.00^{a}	100 ± 0.00^{a}	100 ± 0.00^{a}	100 ± 0.00^{a}

Values followed by the same superscript within the same column are not significantly different by one-way ANOVA ($P \leq 0.05$) followed by Tukey's test

FUMIGANT ACTIVITY OF HEXANE EXTRACTS OF *Ocimum basilicum* AGAINST *Sitophilus zeamais*

The hexane extract of *O. basilicum* showed varied fumigant toxicity effects against *S. zeamais* at all levels of extract concentration. In the 6th h, the acute fumigant toxicities effects of the four different extract concentrations were between 0-18.75% which was comparable to the control (**TABLE 2**). However, this was significantly lower than that of actellic treatment. In the 24th h, the actellic treatment still caused a higher acute fumigant toxicities compared to the different extract concentrations. The lowest mortality rate of 42.31% of *S. zeamais* was evoked by 25% extract concentration within the 96 hour period of exposure. The 75% and 100% extract concentrations recorded 66.67% and 78.21% mortality respectively. The fumigant toxicities effects of the extracts were significantly lower than actellic (reference pesticide) at 48, 72 and 96 hours of treatment period ($p<0.05$; **TABLE 2**).

TABLE 2: Fumigant toxicity effect of hexane extract of *Ocimum basilicum* on adult *Sitophilus zeamais.*

Concentration (% extract)	Mean % corrected mortality ± S.E with exposure period (hours)				
	6 h	24 h	48 h	72 h	96 h
Control	0.00 ± 0.00^{c}	0.00 ± 0.00^{e}	0.00 ± 0.00^{e}	0.00 ± 0.00^{e}	0.00 ± 0.00^{e}
25	0.00 ± 0.00^{c}	10.90 ± 4.84^{de}	28.21 ± 2.09^{d}	35.90 ± 4.44^{e}	42.31 ± 3.23^{d}
50	1.25 ± 0.25^{c}	25.64 ± 1.48^{cd}	44.87 ± 2.45^{c}	56.41 ± 1.48^{c}	62.82 ± 2.45^{c}
75	2.50 ± 1.44^{c}	26.92 ± 6.74^{c}	52.56 ± 1.28^{c}	61.54 ± 1.48^{bc}	66.67 ± 3.31^{c}
100	18.75 ± 4.27^{b}	48.72 ± 2.09^{b}	62.82 ± 2.45^{b}	70.51 ± 2.45^{b}	78.21 ± 1.28^{b}
Actellic	91.25 ± 5.15^{a}	100.00 ± 0.00^{a}	100.00 ± 0.00^{a}	100.00 ± 0.00^{a}	100.00 ± 0.00^{a}

Values followed by the same superscript within the same column are not significantly different one-way ANOVA ($P \leq 0.05$) followed by Tukey's test.

FUMIGANT TOXICITY OF HEXANE: METHANOL BLEND EXTRACT OF *Ocimum basilicum* AGAINST *Sitophilus zeamais*

When fumigant toxicity effects of hexane: methanolic extract of *O. basilicum* was examined against *S. zeamais*, extract concentration of 25-75% evoked 0-11.25% mortality rates within 6 hours of exposure period, which were not significantly different from effects of the control (P>0.05; **TABLE 3**). The 100% extract concentration recorded the highest percentage mortality at the 72nd h and this was comparable to actellic (reference pesticide) (p>0.05; **TABLE 3**). The effects of the lowest extract concentration (25%) caused the least mortality (32%) within 96 h of the study. When the extract concentration was increased to 50%, more than double mortality was recorded. The fumigant toxicity effects of the extract at 75% and 100% concentration were not significantly different from actellic (reference pesticide) 96 hours exposure period (p>0.05; **TABLE 3**). In general, the extract activity was dose dependent. Again, the hexane; methanolic blend extract of O. basiicum showed greater fumigant toxicity than individual solvent extracts. The difference in toxicities was even more marked with higher concentrations of the extracts (**FIGURES 1** and **2**).

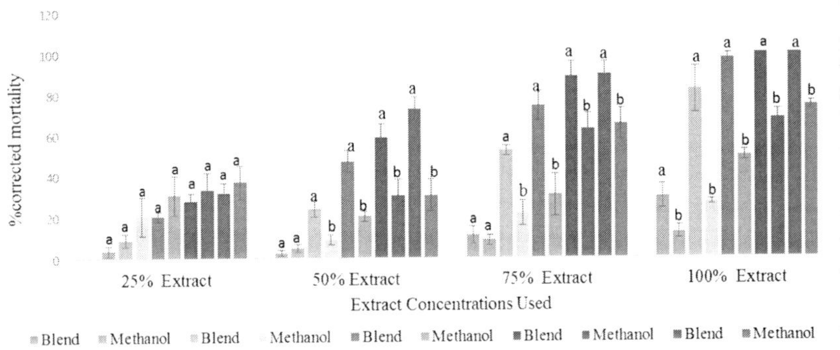

FIGURE 1: Comparison of fumigant toxicities of methanolic and hexane: methanolic blend extracts of *Ocimum basilicum* against *Sitophilus zeamais.*

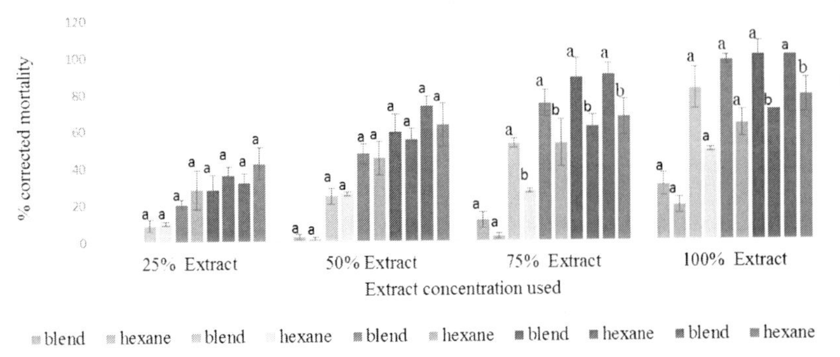

FIGURE 2: Comparison of fumigant toxicities of hexane and hexane: methanolic blend extracts of *Ocimum basilicum* against *Sitophilus zeamais.*

TABLE 3: Fumigant toxicity effect of mortality of hexane; methanolic blend extract of *Ocimum basilicum* on adult *Sitophilus zeamais.*

Concentration (% extract)	Mean % corrected mortality ± S.E with exposure period (hours)				
	6 h	24 h	48 h	72 h	96 h
Control	0.00 ± 0.00^{c}	0.00 ± 0.00^{e}	0.00 ± 0.00^{e}	0.00 ± 0.00^{d}	0.00 ± 0.00^{d}
25	0.00 ± 0.00^{c}	8.97 ± 3.23^{cd}	20.51 ± 2.56^{d}	28.21 ± 3.63^{c}	32.05 ± 4.85^{c}
50	2.50 ± 1.44^{c}	24.36 ± 4.38^{c}	47.44 ± 5.29^{c}	58.97 ± 6.62^{b}	73.08 ± 5.69^{b}
75	11.25 ± 4.27^{c}	52.56 ± 2.45^{b}	74.36 ± 7.25^{b}	88.46 ± 7.36^{a}	89.74 ± 6.28^{a}
100	30.00 ± 6.12^{b}	82.10 ± 11.40^{a}	97.44 ± 2.56^{a}	100.00 ± 0.00^{a}	100.00 ± 0.00^{a}
Actellic	91.25 ± 5.15^{a}	100.00 ± 0.00^{a}	100.00 ± 0.00^{a}	100.00 ± 0.00^{a}	100.00 ± 0.00^{a}

Values followed by the same superscript within the same column are not significantly different one-way ANOVA ($P \leq 0.05$) followed by Tukey's test.

REPELLENCE ACTIVITY OF METHANOLIC EXTRACTS OF *Ocimum basilicum* AGAINST *Sitophilus zeamais*

The methanolic extract of *O. basilicum* at all tested concentrations recorded repellence activities of 72.5% and above by the end of the fifth hour against *S. zeamais.* All the treatment tested showed activity above 57% during the first hour. The 25% extract concentration recorded upto 72.5% activity by the end of the fifth hour of exposure. A slightly higher repellence effect of 82.5% was realized when treatment concentration was increased to 50% extract. Both 75% and 100% extract concentration achieved greater repellence activities of 80% and 87.5% respectively. The mean Percentage Repellence (PR) for all extracts was above 68%. Among all the extract concentrations tested, the repellence activities were not significant difference in the first, third, fourth and fifth hours of exposure ($p>0.05$; **TABLE 4**). The repellent activities of actellic control was significantly higher than that of extracts in the first hour. However, in the following hours the repellent activities of actellic was comparable to that of the higher concentration of extracts (**TABLE 4**).

TABLE 4: Repellent activity of methanolic extract of *Ocimum basilicum* on *Sitophilus zeamais.*

Concentration (%extract)	PR (mean% ± S.E)^{m} with time					PR (mean%)^{n}
	1 h	2 h	3 h	4 h	5 h	
25	62.50 ± 4.79^{cb}	65.00 ± 2.89^{b}	65.00 ± 2.89^{b}	70.00 ± 4.08^{c}	72.50 ± 4.79^{b}	69.00
50	67.50 ± 4.79^{b}	57.50 ± 2.50^{b}	62.50 ± 4.79^{b}	72.50 ± 4.08^{c}	82.50 ± 2.89^{a}	68.5
75	57.50 ± 2.50^{c}	72.50 ± 2.50^{a}	75.00 ± 2.89^{a}	80.00 ± 4.08^{ab}	77.50 ± 4.79^{ab}	70.5
100	65.00 ± 6.45^{cb}	75.00 ± 8.66^{a}	75.00 ± 6.45^{a}	82.50 ± 4.79^{a}	87.50 ± 4.79^{a}	77
Actellic (control)	87.50 ± 4.79^{a}	87.50 ± 4.79^{a}	85.00 ± 6.45^{a}	97.50 ± 2.50^{a}	95.00 ± 5.00^{a}	90.50

Values followed by the same superscript within the same column are not significantly different ($P \leq 0.05$) determined by one way ANOVA following Tukey's test. Superscript m values were based on four extract concentrations, four replicates. Superscript n values were means obtained over the 5 hour duration

REPELLENT ACTIVITY OF HEXANE EXTRACTS OF *Ocimum basilicum* AGAINST *Sitophilus zeamais*

The hexane extract of *O. basilicum* demonstrated repellence activity on S. zeamias at different concentrations (**TABLE 5**). In the first hour of experimental period, all the extract concentrations tested caused repellent activities greater than 67%. The high repellent effect of 90% was caused by 100% concentration during the fifth hour which was not significant different from that caused by control pesticide. The 25% and 75% extract concentrations caused repellence activities of between 67- 80% during the entire experimental period, while 50% concentration caused repellence activities of between 65% and 77.5%. The repellence activities of all extract concentrations tested were not significantly different in the fourth hour of exposure ($p>0.05$; **TABLE 5**). Mean percentage repellence effects for the 5 hours experimental period were between 73 to 90.50%. The repellent activities of actellic were compared to that of the extract at higher doses (**TABLE 5**).

TABLE 5: Repellence activity of hexane extract of *Ocimum basilicum* on *Sitophilus zeamais.*

Concentration (%extract)	PR (mean% ± S.E)^m with time					PR (mean%)^n
	1 h	2 h	3 h	4 h	5 h	
25	80.00 ± 0.00^{ab}	70.00 ± 4.08^b	67.50 ± 2.50^c	77.50 ± 4.79^a	75.00 ± 2.89^b	74.00
50	70.00 ± 4.08^b	72.50 ± 4.79^{ab}	80.00 ± 4.08^{ab}	72.50 ± 2.50^b	77.50 ± 4.79^b	73.00
75	80.00 ± 4.08^{ac}	77.50 ± 4.79^{ab}	70.00 ± 4.08^b	67.50 ± 4.79^a	75.00 ± 2.89^b	74.00
100	85.00 ± 2.89^a	87.50 ± 2.50^a	82.50 ± 6.29^{ab}	75.00 ± 2.89^a	90.00 ± 0.0^a	84.00
Actellic (control)	87.50 ± 4.79^a	87.50 ± 4.79^a	85.00 ± 6.45^a	97.50 ± 2.50^a	95.00 ± 5.00^a	90.50

Values followed by the same superscript within the same column are not significantly different ($P \leq 0.05$) determined by one way ANOVA followed by Tukey's test. Superscript m values were based on four concentrations, four replicates. Superscript n values were means obtained over the 5 h duration.

REPELLENT ACTIVITY OF HEXANE: METHANOLIC BLEND EXTRACTS OF *Ocimum basilicum* ON *Sitophilus zeamais*

The blend extract concentrations of *O. basilicum* strongly repelled *S. zeamais* as indicated by results in **TABLE 6**. At the lowest extract concentration (25%), a repellence activity of 80% was achieved within the first hour, and then it decreased to 60% in the third hour but slightly increased to 75% and 72.5% by fourth and fifth hour respectively. When extract concentration was increased to 50%, the repellence effect increased to 95% in the end of fifth hour. Similarly, 75% extract concentration also had a strong repellence effect of 95.5% in the fourth hour. At the highest extract concentration (100%) repellence effect of 97.5% was recorded in the fourth and fifth hour. The repellent effects of 50, 75 and 100% extract concentration were not significantly different in the first and fourth hour of extract exposure ($p>0.05$). The mean percentage repellence of the extracts for five hour during was between 71.5 and 86.50 %. The repellent activities of actellic were significantly higher than extract in the first hour, but compared in the following hours (**TABLE 6**).

TABLE 6: Repellent activity of hexane; methanolic extract *Ocimum basilicum* on *Sitophilus zeamais.*

Concentration (%extract)	PR (mean% ± S.E)^{m} with time					PR (mean%)^{n}
	1 h	2 h	3 h	4 h	5 h	
25	80.00 ± 4.08^{ab}	70.00 ± 0.00^{b}	60.00 ± 4.08^{b}	75.00 ± 5.0^{b}	72.50 ± 4.79^{b}	71.50
50	70.00 ± 4.08^{bc}	85.00 ± 2.89^{a}	87.50 ± 4.79^{a}	92.50 ± 2.50^{a}	95.00 ± 2.89^{a}	86.00
75	75.00 ± 2.89^{ab}	82.5 ± 2.50^{a}	87.50 ± 4.79^{a}	95.00 ± 5.00^{a}	87.50 ± 2.50^{a}	84.00
100	67.50 ± 4.08^{c}	82.50 ± 2.89^{a}	87.50 ± 2.50^{a}	97.50 ± 2.50^{a}	97.50 ± 2.50^{a}	86.50
Actellic (control)	87.50 ± 4.79^{a}	87.50 ± 4.79^{a}	85.00 ± 6.45^{a}	97.50 ± 2.50^{a}	95.00 ± 5.00^{a}	90.50

Values followed by the same superscript within the same column are not significantly different ($P>0.05$) determined by one way ANOVA followed by Tukey's test. Superscript m values were based on four concentrations, four replicates. Superscript n values were means obtained over the 5 h duration.

QUALITATIVE PHYTOCHEMICAL SCREENING

Methanolic extract of *O. basilicum* contained phenolics, alkaloids, terpenoids, flavonoids, saponins and steroids. Hexane extract of *O. basilicum* contained phenolics, alkaloids, terpenoids, flavonoids, saponins and steroids.

DISCUSSION

When methanolic extracts of *O. basilicum* were tested against *S. zeamais*, the results obtained revealed that the extract had distinct effect on the mortality of the tested pest on treated maize. The methanolic extract was able to cause up 74% mortality at highest applied concentration and 37% mortality at lowest applied concentration. This corroborates with the work done by Mwangangi and Mutisya [14], who found out that *O. basilicum* leaf powder caused up to 90% mortality of maize weevils within two weeks when applied at 2 g powder per 100 g of maize grains. Other works by Boeke et al. [15] showed much greater effect of *O. basilicum* essential oils on closely related coleopteran, *C. maculatus* with a record of 49% mortality within 24 h of exposure.

The effectiveness of this extract in causing mortality could be due to the presence of active compounds such as terpenoids, alkaloids, steroids, glycosides and flavonoids that tested positive in the phytochemical analysis. Triterpenoids of various plant extract have be proven to have toxic effect against coleopterans [16,17]. Moreover, aromatic compounds such as glycosides, saponins, flavonoids and phenols have toxic effects on coleopterans too [18]. It is also possible that some of the extract came into contact with the insects' spiracles contributing to further mortality by suffocation. Since coating the grains with the extracts minimized contact between grains and weevils, some weevils might have died as a results of starvation [7]. Glycosides, terpenoids, tannins and certain monomeric flavonoids found among lamiaceae are excellent feeding deterrents against pest insect [18,19]. Therefore, could have enhanced more mortality again due to starvation.

Hexane extracts of *O. basilicum* had potent toxic activity against *S. zeamais.* The highest application rates caused 48% and 78% mortality at within 24 and 96 hours respectively, while the lowest application rates showed 10% and 42% mortality of the maize weevils within the same assessment period. The information available on the use of hexane extract of *O. basilicum* against insect pest is scanty. However, these results are in agreement with the findings of related studies done by Popović et al. [20] that reported 21.8% mortality of S. oryzae within 24 hours of exposure to *O. basilicum* essential oils at lower doses. Other studies done by Kerdchoechuen et al. [21] showed higher mortality rates of *S. zeamais* when subjected to *O. basilicum* essential oil treatment.

A range of phytochemical compounds such as glycosides, alkaloids, flavonoids and saponins were found to be present in this extract which could be responsible for its toxic activity against the tested pest. Saponins for example, have been shown to impair ecdysteroid synthesis [22]. Terpenoids especially monoterpenes have also been previously reported to possess insecticidal activity against coleopterans [23-25]. A combination of several mechanisms both physical and biological could have attributed to the efficacy the plant extract.

The hexane: methanolic blend extract of *O. basilicum* applied was found effective in controlling maize weevils. Higher efficacy rates of up to 100% mortality within 72 hour were recorded at the highest applied concentration. The extract also showed acute toxicity against the pest at higher dosage application. The findings of this present study were in agreement with earlier studies by Bekele and Hassanali, [26] that also recorded 100% mortality against *S. zeamais* and *Rhyzopertha* dominica when subjected to blend of various component of *Ocimum kilim* and scharica and *Ocimum kenyense*. Related studies by Ranger et al. [27] also revealed acute toxicities of blends of various plant components to coleopteran larvae. Toxicity of the blend extract could be attributed to presence of active constituents such as alkaloids, terpenoids, cadiac glycosides, saponins, steroids, tannins and phenols that are reported to be toxic to post harvest pest in several ways [18,28,29]. For example, alkaloids and tannins have been reported to impart toxicity by iron chelation and enzyme inhibition [30].

A complex mixture of these active compounds also contributes to synergism to greater extent which could have promoted much activity against the tested pest [29]. The greater toxicity of the blend extract *O. basilicum* could be in line with fact that higher concentration levels of the active constituents were obtained with blend maceration. Other factors such as suffocation and starvation might also have equally led to more mortality of *S. zeamais* within the treated grains [7].

In this study, hexane, methanolic and blend extracts of *O. basilicum* showed considerable potential as repellent against maize weevils. At low concentration of 25% extracts, above 68% mean repellency was true for all the tested combination of sweet basil. Similar observation has been reported on number of plant extract tested against a range of insect pest. Asawalam et al. [31] described moderate repellent activity of O. grattissium to *S. zeamais*. Essential oils of Eucalyptus globulus and *O. basilicum* have showed strong repellency against S. oryzae and T. casteneum at remarkably low concentration [32]. This supports the findings of this study.

Padin et al. [33] work on aqueous and methanolic extracts of *Jacaranda mimisifolia, matricaria chamomilla* and *T. minuta* also recorded high repellency index of these extracts against T. casteneum (Coleopteran). Many other related studies have also documented much on repellent potential of various plants against post-harvest pest [34- 39].

A number of secondary metabolites are linked to repellent activities of plants in which they occur. Hydrocarbons especially monoterpenes and oxygenated compounds like phenols and esters determine distinctive odor of plants [37]. Monoterpenes such as eugenol, limonene, camphor and thymol commonly found in basil have strong repellent activity against insects [40]. Therefore the activity of *O. basilicum* extracts could have been due presence of these monoterpenes. Odalo et al. [41] also found out alcohol components of basil (labiate) as effective repellents against A. gambiae (Diptera). Other studies have also indicated terpenoids as Arthropod-repellent compounds [37]. As much as repellent properties of these extracts could be attributed to specific compounds, a synergistic effect as a result of combination and interaction between various active phytochemicals could also have contributed to a greater activity of the extracts against the tested pest [42,43].

Adult emergence of *S. zeamais* was also affected by different *O. basilicum* extracts at all the tested extract of concentrations. High inhibition rate ranging between 86 and 100% were recorded in every study treatment. These results are in agreement with previous studies that showed various labiate plants as valuable source of potential grain protectant against development of all life stages of a number of postharvest grain pests. Keita et al. [44] observed zero emergence of *C. maculatus* F1 progeny in cowpea treated with *O. basilicum* extracts. A closely related study conducted by Vanmathi et al. [45] also reported that aqueous extracts of *O. tenuiflorum* greatly reduced F1 adult emergence of coleopterons. However, studies carried out by Vanmathi et al. [45] recorded much lower inhibition rates of *O. tenuiflorum* extracts against *C. maculatus*. A number of botanicals have also been shown to inhibit adult emergence of S. zeamias in treated grains [46].

The reduction in adult emergence could be as a result of high adult mortality of S. zeamias in treated grain. In addition, active phytochemicals such as alkaloids have been found to disrupt growth and reduce larval survival by hindering loss of exoskeleton during larval development [47]. Other active principles such as isoflavonoids, flavonoids and terpenoids have also been reported to inhibit reproduction and fertility among coleopterons [29,48].

Other studies have also shown phenolic compounds of basil to have ovicidal, larvicidal and nymphicidal properties [49]. Change in behavior and physiology of the insects following extract treatment due to chemical nature of the extract might have also led to low egg laying capacity of *S. zeamais* within the grain and subsequently few hatch adults [29]. These factors collectively may have led to high inhibition of rates F1 adult emergence within the treated grains.

CONCLUSION

This study recommends the use of hexane; methanolic blend extracts of *Ocimum basilicum* in control of *Sitophilus zeamais*. However further research should be undertaken to determine the mechanism of action of each active compound of *O. basilicum* extract.

REFERENCES

1. Vowotor KA, Meikle WG, Ayertey JN, Markham RH (2005) Distribution and association between the Larger Grain Borer *Prostephanus truncatus*(Horn) (Coleoptera: Bostrichidae) and the Maize Weevil *Sitophilus zeamais:* Motschulsky Coleoptera: Curculionidae) in Maize Stores. Stored Prod Res 41: 498-512.
2. Arthur FH, Throne JE (2003) Efficacy of diatomaceous earth to control internal infestations of rice weevil and maize weevil (Coleoptera: Curculionidae).J EconEntomol 96: 510-518.
3. Boxall RA (2002) Damage and loss caused by the larger grain borer *Prostephanus truncatus*. Integrated Pest Manag Rev 7: 105-121.
4. Koul O, Walia S, Dhaliwal GS (2008) Essential oils as green pesticides: potential and constraints. Biopestic Int 4: 63-84.
5. Viglianco AI, Novo RJ, Cragnolini CI, Nassetta M, Cavallo, A (2008) Antifeedant and repellent effects of extracts of three plants from Córdoba (Argentina) against *S. oryzae* L.) (Coleoptera: Curculionidae). Bio Assay 3: 1-6.
6. Deshmukh SD, Borle MN (1975) Studies on the insecticidal properties of indigenous plant products. Indian J Entomol 37: 11-18.

7. Kemabonta KA, Falodu BB (2013) Bioefficacy of three plant products as post-harvest grain protectants against *Sitophilus oryzae* Linnaeus (Coleoptera: Curculionidae) on stored wheat (*Triticum aestivum*). Int J Sci and Nat 4: 259-264.
8. Abbott WS (1925) A method of computing the effectiveness of insecticides. J Econ Entomol 18: 265-267.
9. Tapondjou LA, Adler C, Bouda H, Fontem DA (2002) Efficacy of powder and essential oil from *Chenopodium ambrosioides* leaves as postharvest grain protectants against six-stored product beetles. Stored Prod Res 38: 395-402.
10. Obeng-Ofori D, Reichmuth CH, Bekele AJ, Hassanali A (1998) Toxicity and protectant potential of camphor, a major component of essential oil of *Ocimum kilimandscharicum*, against four stored product beetles. Int J Pest Manage 44: 203-209.
11. Thein WM, Javier PA, Ceballo FA (2013) Insecticidal activity of crude plant extracts against *Sitophilus spp.* (Coleoptera: Curculionidae) and C*allosobruchus chinensis(L.)* (Coleoptera: Bruchidae). Philippine Agricultural Scientist96: 154-162.
12. Harbone JB (1998) Phytochemical Methods: A Guide to Modern Techniques of Plant Analysis. Chapman & Hal Publishersl, London, UK, pp 60-66.
13. Kotake CK (2000) Practical Pharmacognosy. Vallabh Prakashan, New Delhi, India 4: 107-111.
14. Mwangangi BM, Mutisya DL (2013) Performance of basil powder as insecticide against Maize Weevil, *Sitopillus zeamais* (Coleoptera: Curculionidae). Agric Food Sci 1: 196-201.
15. Boeke SJ, Barnaud C, van Loon JJ, Kossou DK, Dicke M (2004) Efficacy of plant extracts against the cowpea beetle, *Callosobruchus maculatus*. Int J Pest Manag 50: 251-258.
16. Mbailao M, Nanadoum M, Automne B (2006) Effect of six common seed oils on survival, egg lying and development of the cowpea weevil, *Callosobruchus maculatus*. Biol Sci 6: 420-425.
17. Ileke KD, Oni MO (2011) Toxicity of some plant powders to maize weevil, *Sitophilus zeamais* (Coleoptera: Curculionidae) on stored wheat grains. Afr JAgric Resour 6: 3043-3048.
18. Ekeh FN, Onah IE, Atama CI, Ivoke N, Eyo JE (2013) Effectiveness of botanical powders against *Callosobruchus maculatus* (Coleoptera: Bruchidae) in some stored leguminous grains under laboratory conditions. Afr J Biotechnol 12: 1384-1391.
19. Shadia E, El-Aziz A, Omer EA, Sabra AS (2007) Chemical composition of *Ocimum americanum* essential oil and its biological effects against *Agrotis ipsilon* (Lepidoptera: Noctuidae). Res J Agric Biol Sci 3: 740-747.
20. Popović Z, Kostić M, Popović S, Skorić S (2006) Bioactivities of essential oils from basil and sage to *Sitophilus oryzae* L. Biotechnol Biotechnol Equip 20: 36-40.
21. Kerdchoechuen O, Laohakunjit N, Singkornard S, Matta FB (2010) Essential oils from six herbal plants for biocontrol of the maize weevil. Horticulture Sci 45: 592-598.
22. Al-Rajhy DH, Alahmed AM, Hussein HI, Kheir SM (2003) Acaricidal effects of cardiac glycosides, azadirachtin and neem oil against the camel tick, *Hyalomma dromedarii*(Acari: Ixodidae). Pest Manage Sci 59: 1250-1254.
23. Waliwitiya R, Isman MB, Vernon RS, Riseman A (2005) Insecticidal activity of selected monoterpenoids and rosemary oil to *A. obscurus* (Coleoptera: Elateridae). J Econ Entomol 98: 1560-1565.
24. Kouninki H, Hance T, Noudjou FA (2007) Toxicity of some terpenoids of essential oils of *Xylopia aethiopica* from Cameroon against *Sitophilus zeamais* Motschulsky. J Appl Entomol 131: 269-274.
25. Ngamo TSL, Ngatanko I, Ngassoum MB, Mapongmestsem PM, Hance T (2007) Persistence of insecticidal activities of crude essential oils of three aromatic plants towards four major stored product insect pests. Afr J Agric Res 2: 1730-177.

26. Bekele AJ, Hassanali A (2001) Blend effects in the toxicity of the essential oil constituents of *Ocimumkilmandscharicum* and *Ocimum kenyense* (Labiateae) on two Post-harvest insect pests. Phytochemistry 57: 385-391.
27. Ranger CM, Reding ME, Oliver JB, Moyseenko JJ, Youssef N, et al. (2013) Acute toxicity of plant essential oils to scarab larvae (Coleoptera: Scarabaeidae) and their analysis by gas chromatography-mass spectrometry. J Econ Entomol 106: 159-167.
28. Ali WK, Mohammed HH (2014) Toxic effect of some plant extracts on the mortality of flour beetle *Tribolium confusum* (Duval) (Coleoptera: Tenebrionidae). Entomol Ornithol Herpetol 2: 1-3.
29. Adesina JM, Jose AR, Rajashaker Y, Afolabi LA (2015) Entomo Toxicity of *Xylopia aethiopica* and *Aframomum melegueta* in suppressing oviposition and adult emergence of *Callasobruchus maculatus* (Fabricus) (Coleoptera: Chrysomelidae) infesting stored cowpea seeds. Jordan J Biol Sci 8: 263-268.
30. Karamanoli K, Bouligaraki P, Constantinidou HI, Lindow SE (2011) Polyphenolic compounds on leaves limit iron availability and affect growth of epiphytic bacteria. Appl Biol 159: 99-108.
31. Asawalam EF, Emosairue SO, Hassanali A (2008) Essential oil of *Ocimum grattissimum* (Labiatae) as *Sitophilus zeamais* (Coleoptera: Curculionidae) protectant. Afr J Biotechnol 7: 3771-3776.
32. Mishra BB, Tripathi SP, Tripathi CPM (2012) Repellent effect of leaves essential oils from (*Eucalyptus globulus*) (Mirtaceae) and (*Ocimum basilicum*) (Lamiaceae) against two major stored grain insect pests of Coleopterons. Nature Sci 10: 50-54.
33. Padín SB, Fusé C, Urrutia MI, Dal Bello GM (2013) Toxicity and repellency of nine medicinal plants against *Tribolium castaneum* in stored wheat. Bull Insectol 66: 45-49.
34. Juan HV, Saad J, Giordano O, García C, Martín T, et al. (2008) Insect growth regulatory effects of linear diterpenoids and derivatives from *Baccharis thymifolia*. Natural Products 71: 190-194.
35. Ogendo JO, Kostyukovsky M, Ravid U, Matasyoh JC, Deng AL, et al. (2008) Bioactivity of *Ocimum gratissimum* oil and two constituents against five insect pests attacking stored food products. Stored Prod Res 44: 328-334.
36. Benzi V, Stefanazzi N, Ferrero AA (2009) Biological activity of essential oils from leaves and fruits of pepper tree (*Schinus molle* L.) to control rice weevil (*Sitophilus oryzae* L.). Chilean J Agric Res 69: 154-159.
37. Nerio LS, Olivero-Verbel J, Stashenko E (2010) Repellent activity of essential oils. Bioresour Technol 101: 372-378.
38. Auamcharoen W, Chandrapatya A, Kijjoa A, Kainoh Y (2012) Toxicity and repellency activities of the crude methanol extract of *Duabanga grandiflora* (Lythraceae) against *Sitophilus oryzae* (Coleoptera: Curculionidae). Pak J Zool 44: 227-232.
39. Niroumand MC, Farzaei MH, Razkenari EK, Amin G, Khanavi M, et al. (2016) An Evidence-Based Review on Medicinal Plants Used as Insecticide and Insect Repellent in Traditional Iranian Medicine. Iran Red Crescent Med J 18: 1-13.
40. Yang FL, Zhu F, Lei CL (2012) Insecticidal activities of garlic substances against adults of grain moth, *Sitotroga cerealella* (Lepidoptera: Gelechiidae). Insect Sci 19: 205-212.
41. Odalo JO, Omolo MO, Malebo H, Angira J, Njeru PM, et al. (2005) Repellency of essential oils of some plants from the Kenyan coast against *Anopheles gambiae*. Acta Trop 95: 210-218.
42. Omolo MO, Okinyo D, Ndiege IO, Lwande W, Hassanali A (2004) Repellency of essential oils of some Kenyan plants against *Anopheles gambiae*. Phytochemistry 65: 2797-2802.
43. Liu CH, Mishra AK, Tan RX, Tang C, Yang H, et al. (2006) Repellent and insecticidal activities of essential oils from *Artemisia princeps* and *Cinnamomum camphora*and their effect on seed germination of wheat and broad bean. Bioresour Technol 97: 1969-1973.

44. Kéïta SM, Vincent C, Schmit JP, Ramaswamy S, Bélanger A (2000) Effect of various essential oils on *Callosobruchus maculatus* (F.) (Coleoptera: Bruchidae). Stored Prod Res 36: 355-364.
45. Vanmathi JS, Padmalatha C, Sing AJA, Suthakar S (2010) Efficacy of selected plant extracts on the oviposition deterrent and adult emergence activity of *Callosobruchus maculatus*. F (Bruchidae; Coeleoptera). Glob J Sci Front Res 10: 2-6.
46. Akinbuluma MD, Adepetun MT, Yeye EO (2015) Insecticidal Effects of Ethanol Extracts of *Capsicumfrutescens* and *Dennettia tripetala* against *Sitophilus zeamais* Motschulsky on Stored Maize. Int Res Agric For 2: 1-7.
47. Ileke KD, Ogungbite OC (2014) Entomocidal activity of powders and extracts of four medicinal plants against *Sitophilus oryzae* (L), *Oryzaephilus mercator* (Faur) and *Ryzopertha dominica* (Fabr.). Biol Sci 7: 57-62.
48. Chebet F, Deng AL, Ogendo JO, Kamau AW, Bett PK (2013) Bioactivity of Selected Plant Powders against *Prostephanus truncatus*. Plant Prot Sci 49: 34-43.
49. Belong P, Akono E, Bakwo E, Foko G, Tamesse J (2013) Chemical composition and residue activities of *Ocimum canun* Sims and *Ocimum basilicum* L. essential oils on adult female *Anopheles funestus* ss. Anim Plant Sci 19: 2854-2863.

CHAPTER 10

Evaluating Alpha and Beta Taxonomy in Ant-Nest Beetles (Coleoptera, Carabidae, Paussini)

Simone Fattorini,[1,2] Emanuela Maurizi,[3] and Andrea Di Giulio[3]

[1]*Water Ecology Team, Department of Biotechnology and Biosciences, University of Milano Bicocca, Piazza della Scienza 2, 20126 Milan, Italy*
[2]*Azorean Biodiversity Group (ABG, CITA-A) and Portuguese Platform for Enhancing Ecological Research & Sustainability (PEERS), University of the Azores, Rua Capitão João de Àvila, 9700-042 Angra do Heroísmo, Portugal*
[3]*Department of Sciences, University of Roma Tre, Viale G. Marconi 446, 00146 Rome, Italy*

ABSTRACT

We evaluated completeness, accuracy, and historical trend of the taxonomic knowledge on the myrmecophilous ground beetle tribe Paussini (Coleoptera, Carabidae, Paussinae). Accumulation curves for valid names and synonyms of species, subgenera, and genera were modelled using logistic functions. Analyses of trends in synonymies suggest that few currently accepted taxa will be recognized to be synonymous in the future. This may indicate that Paussini are a taxonomically relatively stable tribe of carabid beetles. However, this result might also be due to the lack of recent taxonomic work in some biogeographical regions.

1. INTRODUCTION

Arthropods are the most diversified animal group [1, 2]. Although it is widely acknowledged that only a small fraction of the extant arthropod species has been described, the magnitude of the so-called Linnean shortfall (i.e., the discrepancy between the number of described species and the number of living species) is a matter of discussion [2]. Also for relatively well-investigated arthropod groups, there is few information about the quality of the taxonomic knowledge [3, 4]. The most basic question is to establish how complete and accurate the taxonomic status of a given group is. With the word completeness we refer here to the problem whether the species list of a given group can be considered fairly complete or if there are still many species to describe. A completely known group is one for which there is no longer a need of an alpha taxonomic work (the discovering and naming of new species [5]). With accuracy we refer to taxonomic stability. An accurately known group is one for which there is no more need of a beta taxonomic work (the study of the relationships between the already described taxa, through systematic revisional work of higher taxa [5]). Because it is not rare that species are redundantly described under different names (i.e., synonyms), a group is known with accuracy when no relevant taxonomic change is expected.

Although the two aspects tend to be interrelated, they are not necessarily redundant, because revisional works are much rarer than descriptions of new taxa.

In this paper, we evaluated the completeness and accuracy of the taxonomic knowledge about a group of myrmecophilous beetles, the tribe Paussini ("ant-nest beetles") of the family Carabidae (Coleoptera, Adephaga, Paussinae), at a global level.

All Paussini are highly specialized social parasites, depending on ants (mainly associated with Myrmicinae and Formicinae) during any stage of their development [6–9]. Adults prey on ants and their broods without any obvious benefit for the ant colonies [10–15]. Because of their specialised behavioural and morphological adaptations, Paussini have long attracted the interest of entomologists working on myrmecophilous insects [13], and they have been recently into focus because of strong incertitude about their relationships with other Paussinae lineages [6, 7]. These studies have prompted our knowledge of Paussini biology, with emphasis on their immature stages and microscopic morphology, but taxonomical work seems to receive little attention.

In general, for assessing the status of the taxonomical process in a given group, the study should be addressed to describe (1) the growth through time of the cumulative number of valid names to estimate the number of species that remain to be discovered in a given taxonomic group, globally or regionally [4, 16–19], (2) the progression of the cumulative number of invalid names (synonyms), and (3) the temporal trends in the proportion of synonyms [20, 21]. Presence of a plateau is considered evidence that no, or few, species remain to be described, but it can be also due to a stop in taxonomic research [22]. In this paper we present an extension and continuation of a recently published study [22] where we have presented a comprehensive treatment of point 1. In the present paper we will treat the additional aspects of points 2 and 3 taking advantage of the statistical methodologies developed in the former paper.

2. MATERIAL AND METHODS

2.1. Data Collection

We used a computerized database including 572 species and 17 subspecies of the tribe Paussini.

The following information was recorded for each species and subspecies: generic assignment, subgeneric assignment, author, year of description, synonyms, and the biogeographical region of species distribution. We also recorded authorship and year of description of genera (see [22] for details).

2.2. Historical Accumulation Curves of Valid Names and Synonyms

We extracted the year of description of all valid species and subspecies names, as well as the year of description of the names that are currently considered synonyms and grouped years into decades. We plotted the raw number of described taxa, and the raw number of valid taxa, the raw number of synonyms, as well as their cumulative number, against the decade of description.

To model species accumulation curves we used the logistic function , where , , and are estimated parameters, because it gave excellent fits and the first parameter is the upper asymptote, thus providing an immediate estimate of the expected number of taxa. Similar analyses were conducted for genera and subgenera. The use of subgenera in the tribe Paussini is very controversial. For this reason, as in our companion study [22], we used subgenera as currently accepted by most authors [23].

2.3. Trends in Synonymies

Both the historical accumulation of species names and the relationship between valid species and synonyms may provide information about the status of the taxonomical knowledge in a given group [24].

Thus, we modelled synonym accumulation curves and measured the temporal variation in the taxonomical efficiency through time in three ways: (1) as the relationship between the number of synonyms versus the number of total described taxa in each decade, (2) as the proportion of names that are now regarded as synonyms over the total number of taxa described in that decade, (3) as the cumulative proportion of synonyms through decades.

Relationship between the number of synonyms versus the number of total described taxa was substantially linear, and we used an ordinary least squares (OLS) regression to model it. We forced the regression to pass through the origin, because when no taxon is described, the number of synonyms must be zero. We used the coefficient of the regression line as a measure of the number of synonyms introduced—on average—for each species in each decade. We used the 95% confidence limits to identify decades with exceptional number of synonyms.

Proportion of synonyms was used as a measure of the relationship between descriptive (alpha) and revisional (beta) taxonomy. We calculated the proportion of synonymous taxa described in each decade to identify a possible temporal trend in synonym proliferation.

The cumulative proportion of synonyms through decades was used as a rough measure of the quality of currently valid names. Following Baselga et al. [24] we assumed that the more taxonomical revisions are carried out, the higher is the probability for a given species name to be synonymized. Given that the synonyms are assigned to the date when the name was introduced, rather than the date when it was recognized as a synonym, the percentage of synonyms will show a diminishing trend with time, as newly described species will have had less time to be reviewed and eventually synonymized [24]. Irrespective of that, the steepness of the decay of this percentage through time can help us to measure the quality of currently valid names.

3. RESULTS

The rate of species description per decade, when the absolute numbers are considered, is very irregular (Figure 1(a)). Between 1775 (when the first species of Paussini was described by Linnaeus) and 1840 only 31 species were described, and no species was described in the decade 1820–1829. In the latter half of the 19th century species were described at an increasing rate, with two peaks, respectively, in the decades 1870–1879 and 1890–1899, in which a total of 38 Paussini taxa were described. However, the description of species peaked between 1920 and 1929, during which period 131 taxa were described, covering almost 17% of the available names. The low level of species descriptions in the decades 1910–1919 and 1940–1949 may be explained by the effects of the First and Second World Wars. Since the 1950s descriptions decreased progressively. When the cumulative numbers are considered, the increase per decade was low until the 1840s. The cumulative numbers of species/subspecies descriptions have reached a plateau, the estimated asymptotic value for the fitted curve being 768 taxa (Figure 1(b)).

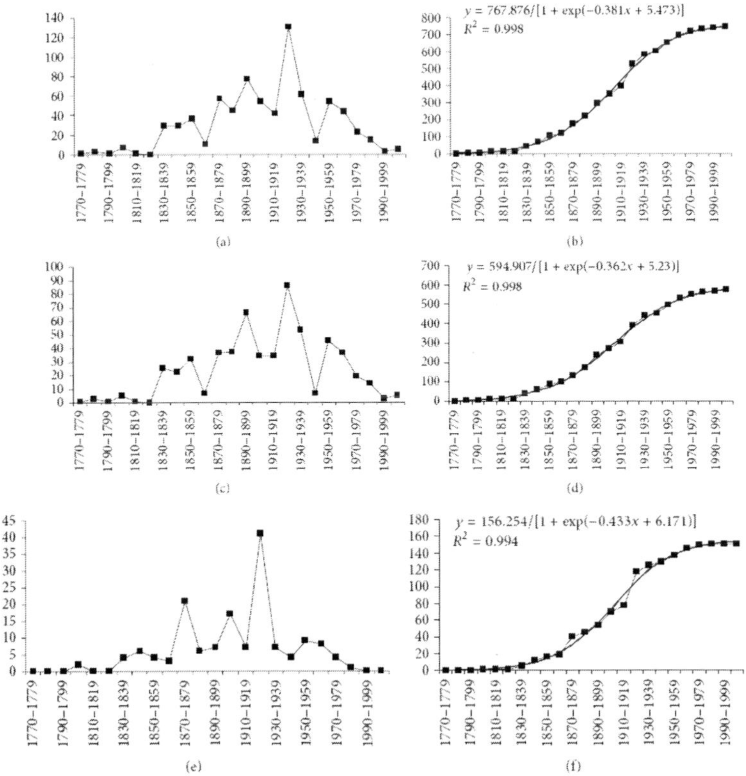

Figure 1: Numbers of total described taxa (a, b), valid species (c, d), and synonyms (e, f) of Paussini by decade. Figures (a), (c), and (e) report the absolute numbers, and Figures (b), (d), and (f) the cumulative numbers along with the equations of the fitted curves.

Analyses omitting synonyms and subspecies revealed similar patterns (Figure 1(c)), with an estimated asymptotic value of species number at 595 (Figure 1(d)). Because the number of currently recognized valid species is 572, the model predicts the existence of 23 undescribed species, with about 96 per cent of the world fauna described.

Patterns in synonyms were also similar to the general trend (Figures 1(e) and 1(f)). The asymptotic value for the number of synonyms is 156 names, very close to the current number of recognized synonyms.

Number of synonyms per decade was directly proportional to the number of described taxa, with a mean rate of one synonym per four taxa each decade (Figure 2). However, the decades 1870–1879, 1900–1909, and 1920–1929 were characterized by an exceptional high number of synonyms. An in-depth analysis of the percentage of synonyms per decade shows a roughly humped trend, with proportion of synonymous taxa increasing from 1830–1849 to 1870–1879, and then decreasing to very low values (Figure 3(a)), which determines a plateau in the accumulation curve of synonymies (Figure 3(b)).

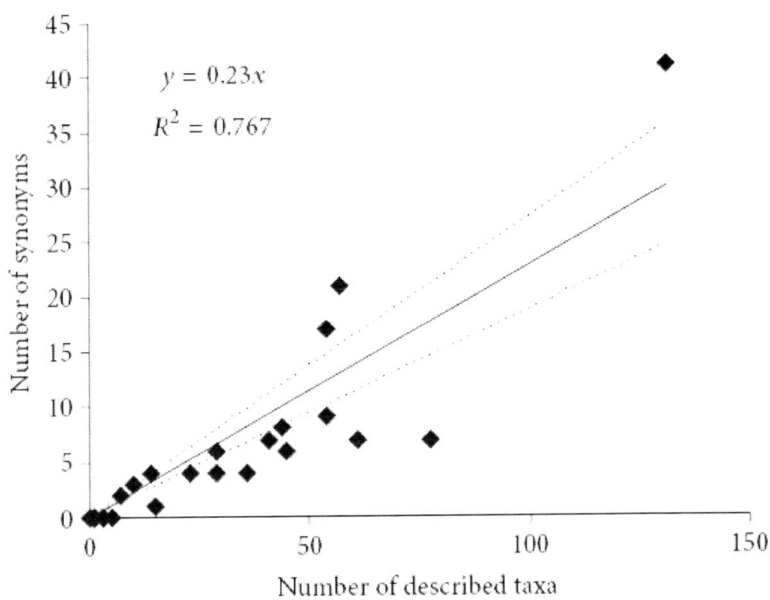

Figure 2: Relationship between number of synonyms and total number of described taxa per decade. Ordinary least square (OLS) regression forced to pass through the origin.

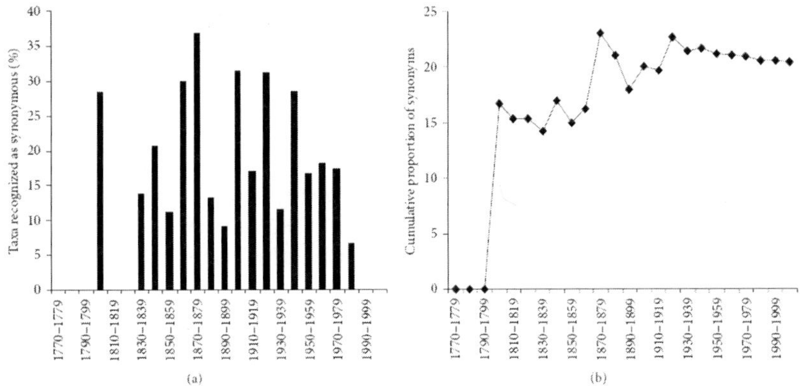

Figure 3: Percentages of synonymous taxa described in each decade (a) and their historical process of accumulation (b).

The historical process of variation in proportion of synonymized names defines the following time spans that correspond to periods of roughly homogenous taxonomical work (Figure 3(a)): (1) the very early stage was obviously characterized by few descriptions (cf. Figure 1) which are still valid species; (2) the relative rate of redescriptions was nearly constant between 1800 and 1870; (3) between 1870 and 1930 we found that at increasing description of species there was also an increasing number of species subsequently found to be synonymous; and (4) finally, from 1930 to present time, the relative rate of descriptions subsequently synonymized diminishes drastically, as less than 20% of the species described during this period have been synonymized (Figure 3(b)).

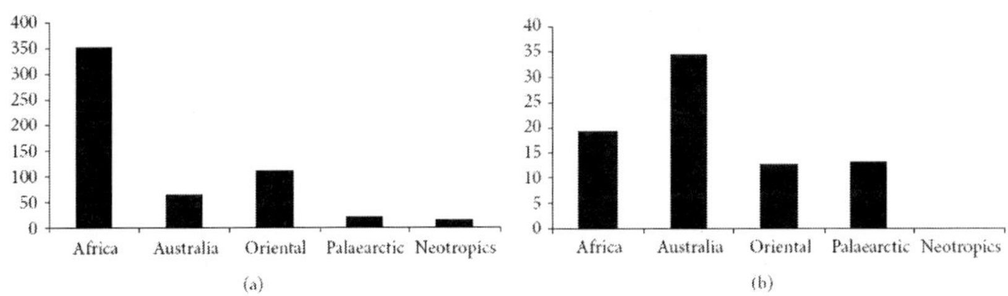

Figure 4: Number of valid species of Paussini per biogeographical region (a) and percentage of synonymous taxa of Paussini per biogeographical region (b).

The largest numbers of described species occur in Africa, followed by the Oriental and Australian regions (Figure 4(a)). This pattern is not paralleled by proportion of synonymies, with the Australian fauna being that with the highest percentage of synonymized taxa (Figure 4(b)).

The distribution of the numbers of authors that have described Paussini taxa is strongly right-skewed (Figure 5). Over 52% of authors have described only one species. The most productive author, Reichensperger, described 88 species, followed by Westwood (79), Wasmann (73), and Luna de Carvalho (71). Altogether, these four authors described more than 40% of known species.

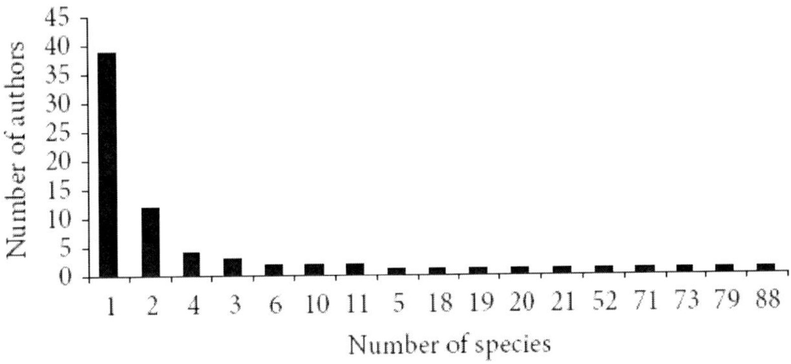

Figure 5: Number of authors in relation to the numbers of species of Paussini that they described.

Reichensperger published his descriptions between 1913 and 1958 (with an average of ca. 2 species per year), covering all biogeographical regions except the Australian. Most of his species (ca. 94%) were described from Africa. Westwood made his descriptions between 1833 and 1874 (with an average of more than 6 species per year) covering all biogeographical regions with a high proportion (ca. 41%) of Oriental taxa. Wasmann also covered all biogeographical regions between 1892 and 1930, with similar proportion of African (49%) and Oriental (42%) taxa and a mean rate of ca. 2 species per year. Finally, Luna de Carvalho described most of his species from Africa (ca. 85%), with a few species from the Oriental and the Palaearctic regions.

Paussini species are currently allocated in 23 genera. The total number of described genera is 51, with 28 synonyms (55%). Most of the genera were described in the decade 1920–1929 (Figure 6(a)). Although some decades were characterized by a high number of descriptions of genera, many were recognized as synonyms (especially among those described in the decade 1920–1929) (Figure 6(e)), so the decade with the highest number of valid genera (4 genera) was 1830–1839 (Figure 6(c)): 75% of the genera described in the decade 1920–1929, and 50% of those described in 1980–1989, were subsequently recognized as synonyms.

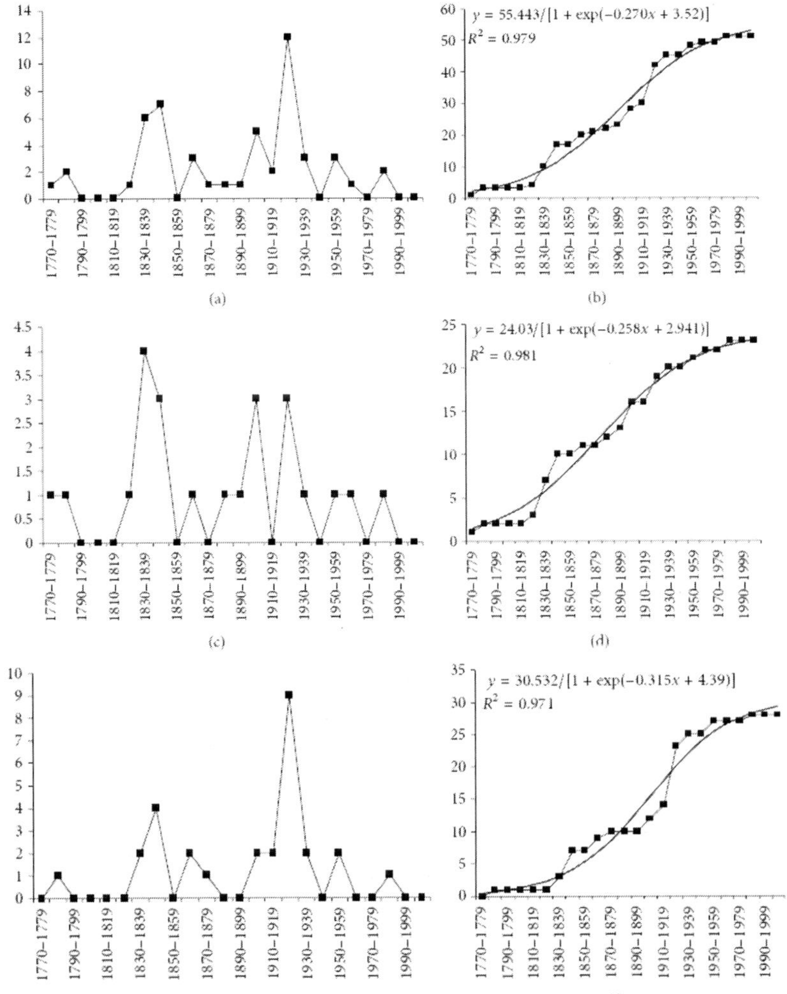

Figure 6: Numbers of total described genera (a, b), valid genera (c, d), and synonyms (e, f) of Paussini by decade. Figures (a), (c), and (e) report the absolute numbers, and Figures (b), (d), and (f) the cumulative numbers along with the equations of the fitted curves.

Patterns of genera accumulation through time indicate a good sigmoid shape for the total number of species (Figure 6(b)), valid genera (Figure 6(d)), and synonyms (Figure 6(f)). In all cases, a plateau has been reached, so virtually no new genus is expected for the future. The historical process of variation in proportion of synonymized genera indicates that after the 1930s there is a substantial stabilization (Figure 7).

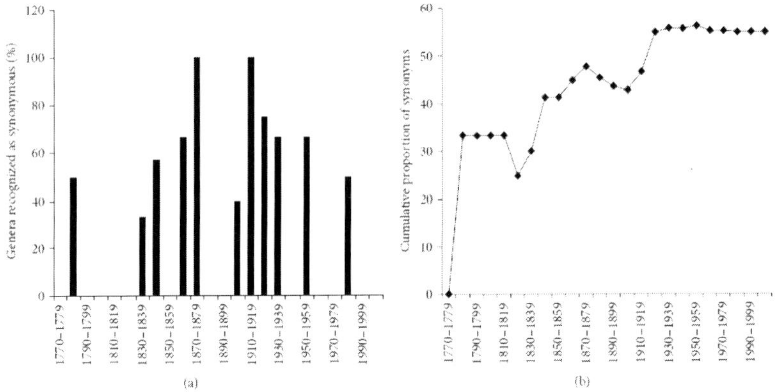

Figure 7: Percentage of synonymous genera described in each decade (a) and their historical process of accumulation (b).

The study of subgenera indicates a proliferation of names in the periods 1920–1929 and 1980–1989 as for the genera (Figure 8). Although these were the two decades which mostly contributed to the current accepted subgenera, these were also the decades in which a large number of synonymous subgenera were described, with proportions of synonyms of more than 54% and 64%, respectively. Accumulation curves showed a stair shape pattern, with apparent plateaus, and were therefore not modelled with fitting curves (Figure 8). Moreover, the historical process of variation in proportion of synonymized genera indicates that there is no substantial stabilization (Figure 9). This was mostly due to the large number of subgenera proposed in a recent time (1980–1989) and subsequently synonymized (Figure 9). These patterns suggest that subdivision into subgenera is not reaching a definitive solution.

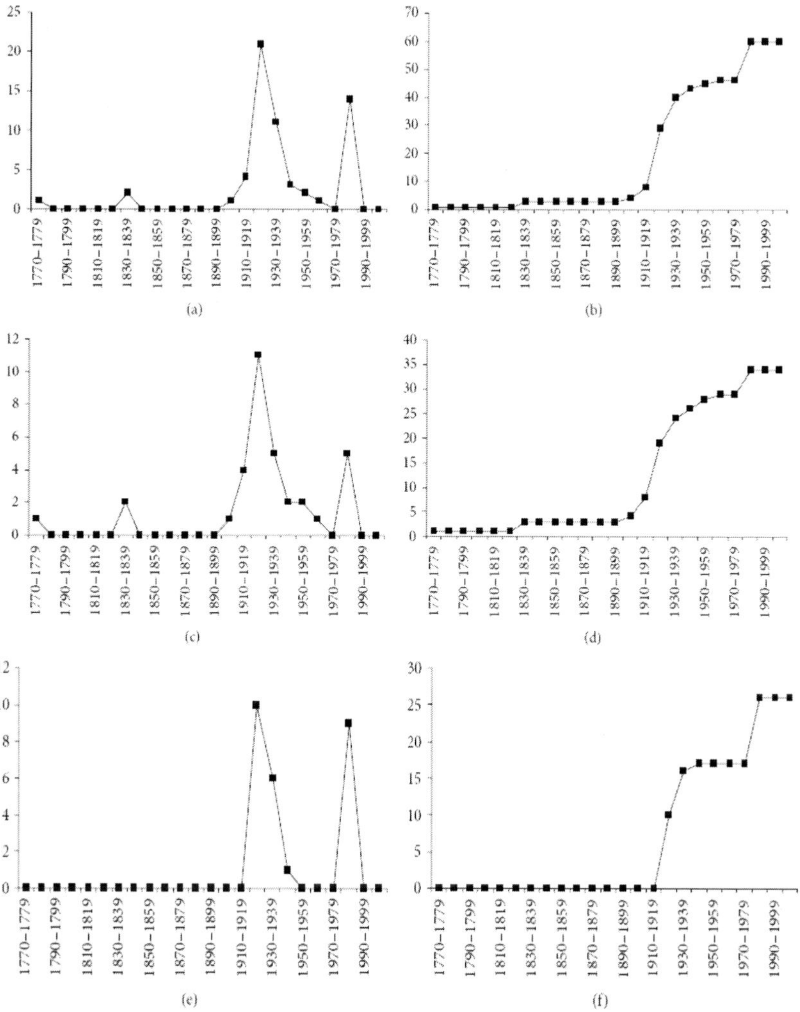

Figure 8: Numbers of total described subgenera (a, b), valid subgenera (c, d), and synonyms (e, f) of Paussini by decades. Figures (a), (c), and (e) report the absolute numbers, and Figures (b), (d), and (f) the cumulative numbers.

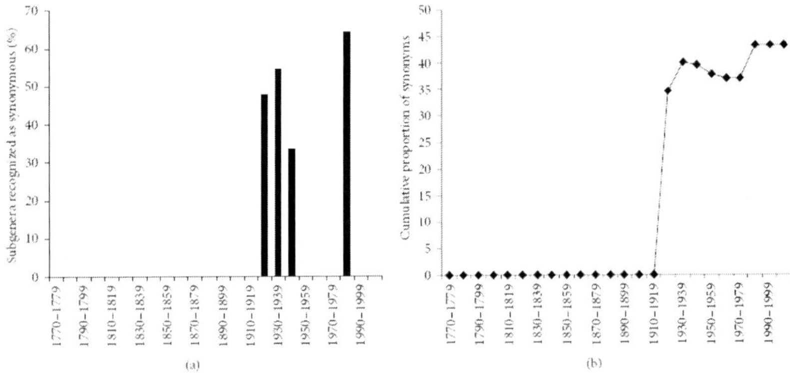

Figure 9: Percentage of synonymous subgenera described in each decade (a) and historical process of accumulation of percentage of synonyms over the total number of names in the Paussini, according to the date of their description (b).

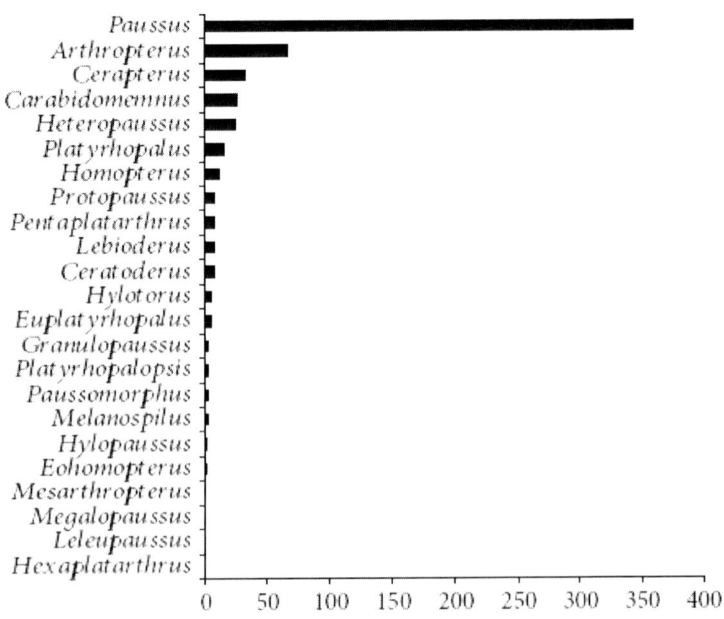

Figure 10: Number of species per genus in the tribe Paussini.

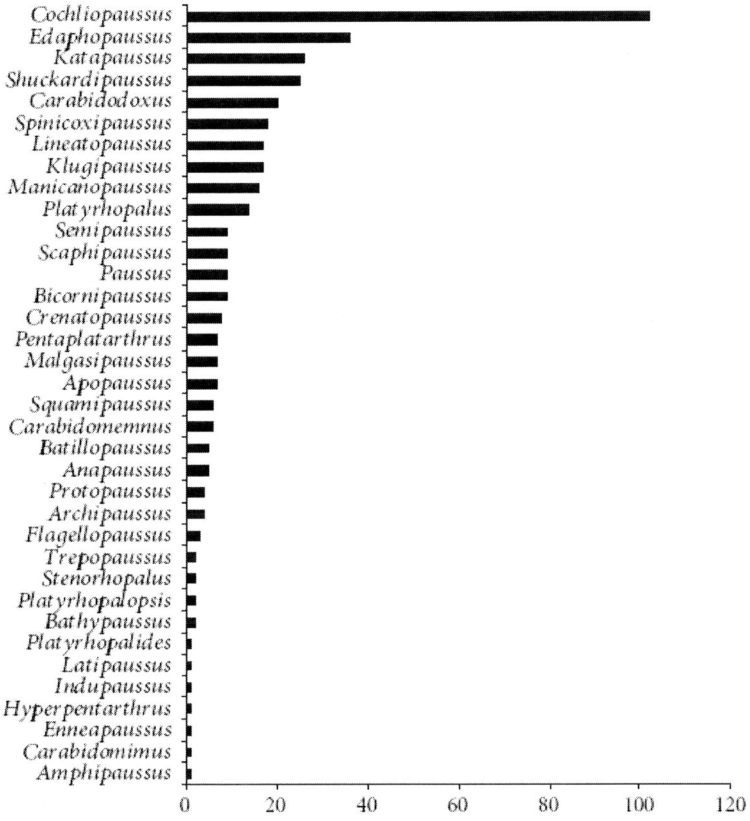

Figure 11: Number of species per subgenus in the tribe Paussini.

Species allocation among genera is strongly dominated by the richest genus (Paussus), with 342 ascribed species (Figure 10) and 25 subgenera. Species distribution among subgenera is also very uneven: the subgenus with the highest number of species is Cochliopaussus (Figure 11).

4. DISCUSSION

Species accumulation curves of the world Paussini fauna indicate that this tribe of carabid beetle is taxonomically stable but do not prove that knowledge is exhaustive. According to the trends analyzed in the present paper, relatively few species are expected to be described in the future on morphological basis and few currently accepted taxa will be recognized to be synonymous. However, if this situation may reflect a true state of affairs in the best explored regions, it may be an artefact when stabilization is merely due to prolonged taxonomic inactivity.

In general, temporal trends in species descriptions mirror dramatic events in human history. The first peak in African species description occurred in the decade 1880–1899, which can be considered an indirect reflection of the first phase of African explorations that occurred between 1840 and 1870 and especially a direct effect of the German expansion in Africa in the 1880s. The second peak occurred in the decade 1920–1939, which coincides with the third phase of the African colonialism, during which the most influential European states organized and stabilised their territories. The overall trend in species descriptions shows two falls in correspondence with the First and Second World Wars. If taxonomic research was frozen at those dates, we would have a completely false signal of stability. For example, taxonomic knowledge in Australia rested at the 1930s [22]. The lack of recent taxonomic activity, coupled with the low number of described species and the high percentage of synonyms, indicates that the fauna of this region is still poorly known.

Most of taxonomic work on Paussini has been produced by few but very prolific authors. Moreover, the authors that described most species during the 20th century were the same that realised the most comprehensive revisions. This has created a self-referenced system, with an almost complete lack of plurality of views. Therefore, taxonomic stability is largely an effect of the "monopolistic" position of certain taxonomists (e.g., Reichensperger, Westwood, Wasmann, and Luna de Carvalho) for long times. Moreover, each of the most active taxonomists was mostly interested in a different biogeographical region, thus with limited taxonomic overlap.

At global level, the asymptotic value calculated for the synonym curve is very close to the current number of synonyms (151); thus we expect that virtually no taxa will be recognized as synonymous in the next future. This indicates that new species are still being described (alpha taxonomy), albeit at decreasing rate in the best explored regions, whereas virtually no synonyms are currently being described, implying either a lack of beta taxonomy (i.e., redundant descriptions are still considered valid because of the reduced revisional work) or an excellent efficiency of alpha taxonomy (i.e., all new species are valid and none is redundantly described) [24]. We think that failure to recognize synonymies is likely high in the less studied faunas, for which most species have been described from sparse individuals, but this is balanced by the presence of still undescribed species. This may be the case of the Oriental region, which seems to have few species and a moderate percentage of synonymies, but from which so many species are being discovered and no further synonymies established.

Stability in species beta taxonomy indicates that Paussini species are recognized as discrete entities by most researchers. Paussini species were described and are currently recognized on the basis of morphological traits, that is, as groups of phenetically similar individuals that can be separated from other analogous groups by means of phenetic gaps, thus corresponding to a morphological concept of species [25]. Stabilization in synonymies suggests that most taxonomists agree in considering the diagnostic characters presented in species descriptions as gaps sufficiently strong to mark discontinuities among populations. Morphologically defined species do not necessarily correspond to "biological" species (defined as reproductively isolated populations [5]). However, the application of a morphological approach for discriminating species was the practical methodology most frequently used by taxonomists in the past, and the same approach still dominates (and likely will dominate) daily work of the majority of taxonomists. Stability in beta taxonomy of morphological species makes Paussini an ideal candidate for future works using molecular approach to investigate how morphological discontinuities are paralleled by molecular divergences. This would be particularly important to clarify relationships among species. Current taxonomic patterns suggest that most species were allocated into the genus Paussus probably reflecting a real phylogenetic proximity. However, subgeneric divisions appear instable and based on subtle and controversial morphological characters. This suggests that morphological characters are not fully adequate to resolve infrageneric relationships.

ACKNOWLEDGMENTS

The authors are grateful to Wendy Moore (AZ, USA), Augusto Vigna Taglianti (Roma, Italy), and Peter Nagel (Basel, Switzerland) for the continued stimulating discussions about the taxonomy of Paussini and two anonymous referees for their useful comments. Wendy Moore also provided them with an electronic species level database of Paussini and revised the language of the paper. The authors thank Valerio Viglioglia for his help in the data entry and Peter Hlaváč for his suggestions.

REFERENCES

1. K. J. Gaston, "The magnitude of global insect species richness," Conservation Biology, vol. 5, no. 3, pp. 283–296, 1991.
2. R. J. Ladle and R. J. Whittaker, Conservation Biogeography, Springer, West Sussex, UK, 2011.

3. A. Baselga, J. Hortal, A. Jiménez-Valverde, J. F. Gómez, and J. M. Lobo, "Which leaf beetles have not yet been described? Determinants of the description of Western Palaearctic Aphthona species (Coleoptera: Chrysomelidae)," Biodiversity and Conservation, vol. 16, no. 5, pp. 1409–1421, 2007.
4. J. Hortal, A. Jiménez-Valverde, J. F. Gómez, J. M. Lobo, and A. Baselga, "Historical bias in biodiversity inventories affects the observed environmental niche of the species," Oikos, vol. 117, no. 6, pp. 847–858, 2008.
5. E. Mayr, Principles of Systematic Zoology, McGraw-Hill, New York, NY, USA, 1969.
6. A. Di Giulio, S. Fattorini, A. Kaupp, A. V. Taglianti, and P. Nagel, "Review of competing hypotheses of phylogenetic relationships of Paussinae (Coleoptera: Carabidae) based on larval characters," Systematic Entomology, vol. 28, no. 4, pp. 509–537, 2003.
7. A. Di Giulio and W. Moore, "The first-instar larva of the genus Arthropterus (Coleoptera:Carabidae: Paussinae): implications for evolution of myrmecophily and phylogenetic relationships within the subfamily," Invertebrate Systematics, vol. 18, no. 2, pp. 101–115, 2004.
8. A. Di Giulio, "Fine morphology of the myrmecophilous larva of Paussus kannegieteri (Coleoptera: Carabidae: Paussinae: Paussini)," Zootaxa, no. 1741, pp. 37–50, 2008.
9. W. Moore, X. B. Song, and A. di Giulio, "The larva of Eustra (Coleoptera, Paussinae, Ozaenini): a facultative associate of ants," ZooKeys, vol. 90, pp. 63–82, 2011.
10. C. Escherich, "Zur naturgeschichte von Paussus favieri Fairm," Verhandlungen der Zoologisch-Botanischen Gesellschaft, vol. 49, pp. 278–283, 1899.
11. G. Le Masne, "Recherches sur la biologie des animaux myrmécophiles IV: observations sur le comportement de Paussus favieri Fairm, hôte de la fourmi Pheidole pallidula Nyl.," Annales de la Faculté des Sciences de Marseille, vol. 31, pp. 111–130, 1961.
12. G. Le Masne, "Recherches sur la biologie des animaux myrmécophiles I: L'adoption des Paussus favieri Fairm, par une nouvelle societe de Pheidole pallidula Nyl.," Comptes Rendus Hebdomadaires des Scéances de L'Academie des Scéances, vol. 253, pp. 1621–1623, 1961.
13. S. F. Geiselhardt, K. Peschke, and P. Nagel, "A review of myrmecophily in ant nest beetles (Coleoptera: Carabidae: Paussinae): linking early observations with recent findings," Naturwissenschaften, vol. 94, no. 11, pp. 871–894, 2007.
14. A. di Giulio, E. Maurizi, P. Hlaváč, and W. Moore, "The long-awaited first instar larva of Paussus favieri (Coleoptera: Carabidae: Paussini)," European Journal of Entomology, vol. 108, no. 1, pp. 127–138, 2011.
15. E. Maurizi, S. Fattorini, and A. Di Giulio, "Behavior of Paussus favieri (Coleoptera, Carabidae, Paussini), a myrmecophilous beetle associated with Pheidole pallidula (Hymenoptera, Formicidae)," Psyche, vol. 2012, Article ID 940315, 9 pages, 2012.
16. G. C. Steyskal, "Trend curves of the rate of species description in zoology," Science, vol. 149, no. 3686, pp. 880–882, 1965.
17. R. A. Medellín and J. Soberón, "Predictions of mammal diversity on four land masses," Conservation Biology, vol. 13, no. 1, pp. 143–149, 1999.
18. F. J. Cabrero-Sañudo and J. M. Lobo, "Estimating the number of species not yet described and their characteristics: the case of Western Palaearctic dung beetle species (Coleoptera, Scarabaeoidea)," Biodiversity and Conservation, vol. 12, no. 1, pp. 147–166, 2003.
19. A. Baselga and F. Novoa, "Diversity of Chrysomelidae (Coleoptera) in Galicia, Northwest Spain: estimating the completeness of the regional inventory," Biodiversity and Conservation, vol. 15, no. 1, pp. 205–230, 2006.
20. A. R. Solow and W. K. Smith, "On estimating the number of species from the discovery record," Proceedings of the Royal Society B, vol. 272, no. 1560, pp. 285–287, 2005.
21. J. Alroy, "How many named species are valid?" Proceedings of the National Academy of Sciences of the United States of America, vol. 99, no. 6, pp. 3706–3711, 2002. •View at Scopus

22. S. Fattorini, E. Maurizi, and A. Di Giulio, "Tackling the taxonomic impediment: a global assessment for ant-nest beetle diversity (Coleoptera: Carabidae: Paussini)," Biological Journal of the Linnean Society, vol. 105, pp. 330–339, 2012.
23. W. Lorenz, A Systematic List of Extant Ground Beetles of the World (Coleoptera "Geadephaga": Trachypachidae and Carabidae Incl. Paussinae, Cicindelinae, Rhysodinae), W. Lorenz, Tutzing, Germany, 2nd edition, 2005.
24. A. Baselga, J. M. Lobo, J. Hortal, A. Jiménez-Valverde, and J. F. Gómez, "Assessing alpha and beta taxonomy in eupelmid wasps: determinants of the probability of describing good species and synonyms," Journal of Zoological Systematics and Evolutionary Research, vol. 48, no. 1, pp. 40–49, 2010.
25. R. R. Sokal and J. Crovello, "The biological species concept: a critical evaluation," The American Naturalist, vol. 104, pp. 127–153, 1970.

CHAPTER 11

Propolis Counteracts Some Threats to Honey Bee Health

Michael Simone-Finstrom [1,*], Renata S. Borba [2,3], Michael Wilson [4] and Marla Spivak [5]

[1]*USDA-ARS Honey Bee Breeding, Genetics, and Physiology Laboratory, Baton Rouge, LA 70820, USA*

[2]*Centre for High-Throughput Biology, University of British Columbia, Vancouver, V6T 1Z4, Canada*

[3]*Beaverlodge Research Farm, Agriculture and Agri-Food Canada, Beaverlodge, AB T0H 0C0, Canada*

[4]*Center for Drug Design, University of Minnesota, Minneapolis, MN 55018, USA*

[5]*Department of Entomology, University of Minnesota, St. Paul, MN 55108, USA*

ABSTRACT

Honey bees (*Apis mellifera*) are constantly dealing with threats from pathogens, pests, pesticides and poor nutrition. It is critically important to understand how honey bees' natural immune responses (individual immunity) and collective behavioral defenses (social immunity) can improve bee health and productivity. One form of social immunity in honey bee colonies is the collection of antimicrobial plant resins and their use in the nest architecture as propolis. We review research on the constitutive benefits of propolis on the honey bee immune system, and its known therapeutic, colony-level effects against the pathogens *Paenibacillus larvae* and *Ascosphaera apis*. We also review the limited research on the effects of propolis against other pathogens, parasites and pests (*Nosema*, viruses, *Varroa destructor*, and hive beetles) and how propolis may enhance bee products such as royal jelly and honey. Although propolis may be a source of pesticide contamination, it also has the potential to be a detoxifying agent or primer of detoxification pathways, as well as increasing bee longevity via antioxidant-related pathways. Throughout this paper, we discuss opportunities for future research goals and present ways in which the beekeeping community can promote propolis use in standard colonies, as one way to improve and maintain colony health and resiliency.

Keywords: *Apis mellifera*; social immunity; plant resin; plant insect interactions

1. INTRODUCTION

Honey bee (*Apis mellifera*) populations in North America and Europe are currently experiencing high annual losses throughout the year due to various, often interacting factors including pathogens, parasites, pesticides, poor nutrition and management [1,2,3,4]. It is critically important to understand the impact of individual stressors and the interactions among stressors in order to develop solutions to increase colony health and survival. It is equally important to understand how honey bees' natural immune responses (individual immunity) and collective behavioural defenses (social immunity) can improve and maintain bee health and counteract stressors without human intervention. One form of social immunity in honey bee colonies is the formation of a propolis envelope within the nest that acts as an important antimicrobial layer. While a propolis envelope cannot mitigate all colony stressors, we review research to date on its known benefit to individual immunity and effect on reducing colony pathogen loads. We also suggest research avenues that could reveal additional ways propolis may improve colony health and resiliency.

1.1. The Role of Propolis in the Hive

Propolis is the apicultural term for the plant resins that honey bees collect, bring back into the hive, and then deposit throughout the nest [5]. Resin-handling bees within the nest mix varying amounts of wax, which they produce themselves, with the resins during deposition [5,6]. The primary chemical components of propolis are derived from the plant-produced resins, though there is some evidence that honey bee glandular secretions (i.e., β-glucosidase) may also be added, potentially as an artifact of handling [7]. Various species across Hymenoptera (e.g., ants [8], stingless bees [9], and other *Apis* species [10,11]) collect and use resins for various purposes, including nest construction, and as a defense against predators, microbes and pathogens.

Honey bee colonies typically nest in tree hollows. Once a swarm (thousands of non-reproductive workers and one queen) finds a suitable cavity, they line the interior walls and often the nest entrance with a layer of propolis 0.3–0.5mm thick [10]. This propolis envelope surrounds the colony and likely serves many non-mutually exclusive purposes, including waterproofing and preventing fungal decay of the hive walls [5,12], reducing cracks and hive entrances, helping promote stable temperature and humidity through condensation [13], and reducing hive microbial loads [5,14]. In addition, there have now been several, mainly recent, studies examining the indirect and direct effects that propolis has on bees' immune systems, pathogens and parasites [5,14,15,16,17,18,19,20,21,22] (Figure 1).

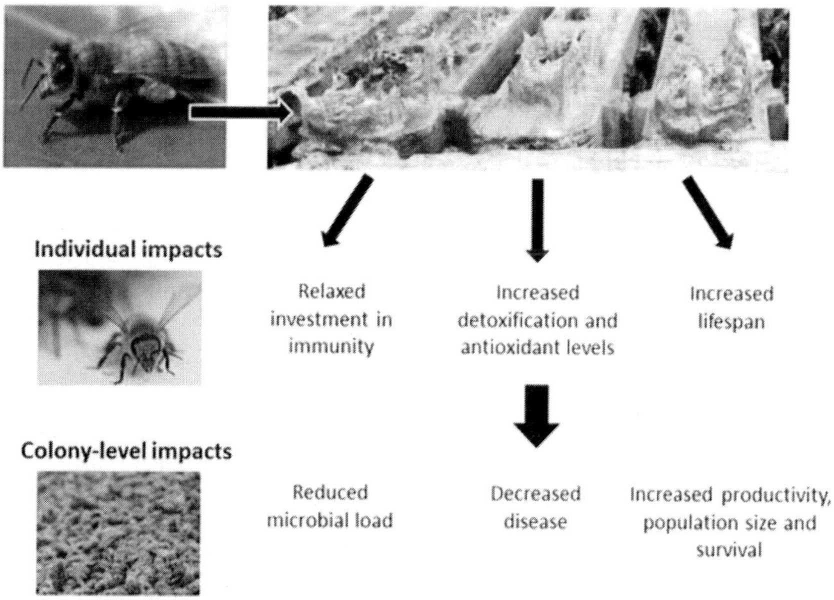

Figure 1. Overview of the impacts of honey bee-harvested plant resins on individual and colony health (photos by Michael Simone-Finstrom).

1.2. Propolis as Social Immunity

Social immunity describes cooperative behavioral defenses among members of a social group that result in the avoidance, control or elimination of parasitic infections [23]. A social immune system supplements and enhances the individual immune system, benefitting the health of a colony by decreasing the risk of microbe exposure and disease transmission among group members [24].

The nests of densely populated honey bee and other social insect colonies provide a favourable habitat for a wide range of parasites and pathogens [25] that have evolved to overwhelm or suppress their hosts' immune defenses. Honey bees were perhaps not subject to the multitude of pathogens and other stressors that confront them today within the context of modern beekeeping practices (e.g., transportation of hives, density of hives in close proximity, and exchange of comb across beekeeping operations [3]), but, as cavity nesters within trees, they likely encountered saprophytic, symbiotic and parasitic microbes and fungi. Some of these endemic microbes can induce an immune response in individual bees [26,27]. The evolution of resin use by honey bees could have been to combat fungal growth and potential fungal pathogens within the nest. Although both bacteria and fungi are common bee immune stressors, the honey bee immune system appears to be more attuned against bacterial pathogens [28]. One hypothesis is that social immune behaviors, such as the collection of plant resins and their deposition in the nest as an antimicrobial propolis envelope, evolved to compensate for deficiencies in innate or physiological immunity. Supporting this notion, there is some evidence that propolis may reduce impacts of mycotoxins produced by fungi [29]. In this case, feeding propolis extracts, rich in flavonoids and phenolic compounds, to adult honey bees reduced adverse effects of exposure to toxins produced by *Aspergillus*, a common hive fungus.

The immune system is the most costly physiological system in insects [30]. An elevated immune response can lead to reduced colony productivity in honey bees [31] and decreased individual survival in bumble bees [32]. Thus, it is not surprising that honey bees have evolved behavioral mechanisms of social immunity to reduce activation of the individual immune system against these microbes to ultimately benefit colony health [33,34,35].

2. CONSTITUTIVE EFFECTS OF PROPOLIS ON THE HONEY BEE IMMUNE SYSTEM

Our experiments on the effects of resin on honey bee health followed findings of other researchers on the use of resin by social wood ants (*Formica paralugubris*) in Switzerland. This ant species constitutively collects spruce tree resins and places globules of resin near the brood, which results in reduced growth of microorganisms in the nest mound [8,36], lower immune system activity of adult worker ants [37], and increased survival during pathogen challenge [38]. Based on these findings, we hypothesized that, for a honey bee colony, the antimicrobial properties of propolis would reduce the general microbial load within the nest cavity, thereby reducing the production of antimicrobial peptides by the innate immune system of individual bees. The first study to test if propolis exposure influenced honey bee immune function confirmed our prediction [14]. Bees were hived in small man-made bee boxes, enriched with an experimentally applied propolis envelope (ethanol extract of propolis painted inside the hive box; Figure 2A). Extracts were made from propolis previously collected from colonies at the University of Minnesota (resins collected mostly from plants in the genus *Populus* [17]), or from Brazilian green propolis (from the tropical shrub *Baccharis dracunculifolia* [39]). General bacterial loads (16S rRNA expression) and transcription levels of the gene encoding for the antimicrobial peptide, *hymenoptaecin*, and a gene involved in cellular immunity, *AmEater*, were found in significantly reduced quantities in seven-day-old bees from colonies with the propolis-extract envelope compared to bees in colonies with no propolis envelope. In summary, bees in colonies with the propolis-extract envelope had an overall reduction in general bacterial loads, and a corresponding lower investment in immune gene expression compared to bees in boxes without the propolis envelope [14].

A second study [20] also confirmed the prediction that bees in colonies with a propolis envelope had reduced investment in individual bee immune function, but, in contrast to Simone et al. [14], no reduction in overall bacterial load was observed. The methods of the second study differed: rather than painting a propolis-extract envelope within the bee boxes, the colonies were allowed to deposit their own propolis envelope from resins collected within a foraging range of the University of Minnesota. Twenty-four colonies were hived from "packages" (10,000 bees and a mated queen) in new Langstroth bee boxes. Twelve of the colonies were hived in boxes fitted with commercial propolis traps

Figure 2. Examples of propolis envelopes in managed hives. (**A**) a propolis extract was used to "varnish" the interior hive body walls. The left was treated with 70% ethanol, the solvent for the extract, while the right was treated with poplar-derived propolis extract (photos by Michael Simone-Finstrom). (**B**) commercially purchased "propolis traps" were cut and stapled to the interior walls (**left**). Once the bees deposit sufficient propolis in the trap, the trap can be removed leaving the propolis attached to the wall (**right**; photos by Renata S. Borba).

(plastic grids) along all inner walls of the brood boxes, which stimulated the bees to form a propolis envelope (Figure 2B). The effect of the propolis envelope on colony strength and health, and on individual bee immune function was measured over a full year. As a validation of our methods, the entire experiment was repeated a second year using 24 new colonies. The expression of immune-related genes, particularly *hymenoptaecin* and *abaecin*, of individual seven-day-old bees was significantly lower over the entire foraging season in colonies with a propolis envelope in both trials, indicating the long-term effect of the propolis on baseline expression of immune-related genes [20]. By the following spring, there were no significant differences in gene expression levels for most immune genes between bees from the two treatment groups. The effect on the immune system was diminished [20], likely due to the decrease in biological activity of the propolis during the winter months when bees were not foraging or seasonal variation in availability of resin sources in early spring, which can vary in active compounds [20,40,41]. Presumably, when the growing season commenced, bees would add new propolis on top of the old and restore the antimicrobial properties of the envelope. Bees in colonies with a propolis envelope did have significantly higher levels of the blood storage protein vitellogenin (Vg) compared to control colonies in spring of both years. Higher levels of Vg are an indicator of well-nourished bees [42,43,44]. A decrease in energetic costs associated with the maintenance of an efficient immune system during the foraging season might help bees maintain higher storage protein levels (e.g., vitellogenin) required for overwintering success [45] and allocate energy to perform vital tasks the next spring (e.g., foraging, rearing brood).

The results of these two experiments demonstrated that costly immune gene expression in bees was lowered by the presence of the propolis envelope within the nest, but clear effects of propolis on colony fitness were not observed. We did not measure the effects of propolis envelope on individual bee fitness, which, in retrospect, may have been more revealing. Nicodemo et al. [18] found positive effects of propolis on individual bee fitness by measuring brood viability and adult bee longevity of Africanized honey bee colonies, an *A. mellifera* subspecies notable for depositing high amounts of propolis in some naturally selected colonies and those bred for increased propolis use due to industry export demands [18,46,47,48]. Colonies selected for high propolis collection for one generation produced 34 times more propolis compared to colonies selected for low propolis collection, consistent with the high heritability estimates for propolis collection ranging from 0.66 [49] to 0.87 [46]. Brood survival rates (survival from egg to adult stage) and individual adult bee longevity were both significantly higher in the high-propolis collecting colonies. The effects of propolis on colony-level measures have been less clear [46,47,48,49]. High propolis collecting colonies had more pollen and honey stores within the nest in some experiments [47,48,49] (but see [46]); it is unclear, however, if high propolis collection was correlated with foraging rates in general in these experiments. Bees within the high propolis collecting colonies also exhibited increased hygienic behaviour [46,48,49], assayed by the rate of removal of pin-killed brood (a proxy for removal of diseased, parasitized and dead brood), although there were no differences in levels of *Varroa* mites between sets of colonies [49].

Future studies might continue the work of Borba et al. [20] to examine the effect of the propolis envelope on better metrics of colony-level fitness such as queen longevity or colony reproduction (swarming) and survivorship, particularly for European honey bees. Furthermore, the potential benefits of propolis enrichment for colonies in a commercial, migratory or honey production operation need to be explored more fully. For the effects of propolis on the fitness of individual bees, experiments could investigate whether the reduced investment in immune function by individual bees after exposure to propolis corresponds with increased bee weight at emergence and longevity.

3. DOES PROPOLIS SUPPRESS IMMUNE FUNCTION?

The reduction in immune function observed in bees surrounded by a propolis envelope [14,20] does not imply the bees' immune system was suppressed or inhibited. A recent study using honey bees maintained in "bee cups" coated with propolis extracts collected from North Dakota or left propolis-poor by coating with an equal amount of 70% Ethanol (the solvent of the extracts), confirmed that propolis does not cause immune suppression [50]. Individual, newly emerged bees were fed either 2 μL of sucrose solution (unchallenged controls) or 2 μL of sucrose containing 40 mg of lipopolysaccharides (LPS) from *Escherichia coli* (challenged). LPS is the major outer surface membrane component of Gram-negative bacteria and a known immune-stimulant for honey bees (and other insects [51,52]), and, as such, acts to activate the immune system without creating an active infection. Analysis of immune gene expression of individual bees from the cages showed bees exposed to propolis from North Dakota had an enhanced immune response in challenged individuals relative to unchallenged controls for the genes encoding the antimicrobial peptides *defensin1* and *abaecin* (Figure 3; [50]). Unchallenged control or challenged LPS-fed bees were also monitored for survival. Immune activation via LPS significantly reduced lifespan of all bees, but there was no effect of propolis exposure (Figure 4; [50]). While, in this controlled cage study, a general reduction in immune gene expression was not observed as in colony-level field exposures [14,20], this could be due to the sterile environment of the incubator cages. Earlier work suggested that the reduction in immune gene expression could be, at least in part, due to the reduction of hive microbes [14] (but, see [20]), which would not be present in the incubator cages. In summary, the challenged bees in a propolis-enriched environment had an increased immune response compared to challenged bees in the cage environment with no propolis, indicating that the presence of propolis does not appear to directly suppress the immune system. Recent evidence suggests that at least some components of propolis may be able to influence lifespan in the cage setting, as bees fed with *p*-coumaric acid and quercetin, common phytochemicals in hive products like propolis, had longer lifespans compared to control bees under specific conditions. Though exposure to propolis volatiles in the cage study did not affect lifespan (Figure 4), possibly due to smaller samples sizes than those in the recent cage study with oral treatments [53], it remains to be verified if propolis exposure or hive supplementation in general increases lifespan of individual bees under more realistic conditions, as observed by Nicodemo et. al. [18] in Africanized honey bees bred for increased propolis production.

Bees' exposure to propolis may also "prime" the immune system in the way that it appears to prime detoxification pathways, at least for experiments involving oral doses of propolis extracts or chemical components [29,55,56]. In another field-based experiment, colonies with and without a natural propolis envelope were either challenged or not with spores of a honey bee bacterial pathogen, *Paenibacillus larvae* (see section below for details, [19]). In the presence of clinical signs of the disease, adult bees showed significantly higher expression of two antimicrobial peptides, *hymenoptaecin* and *apidaecin*, compared to bees in colonies without a propolis envelope. This further provides support that the immune system is not supressed when bees are exposed to propolis, but simply can allow for a reduced investment in immune expression when unchallenged [14,20]. In fact, bees in an environment with propolis invested more in the production of antimicrobial peptides when the colony was exposed to pathogenic microbes compared to same-aged bees in colonies without propolis [19], similar to what was found in the cage study (Figure 3). Individual investment in inducible immune defences (e.g., antimicrobial peptide production) therefore may be at least somewhat context dependent on the colony investment in constitutive defenses (i.e., propolis deposition; [34]), given that reliance on both types of broad immune defenses may also be regulated by type of pathogen exposure [57]. Thus, the propolis envelope can be viewed as an external and vital component of the bees' immune defense and the colony's social immunity.

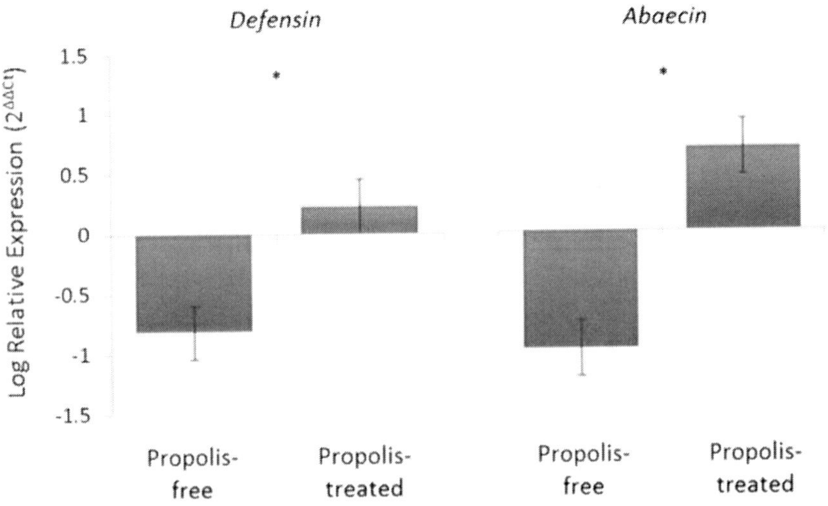

Figure 3. Propolis exposure may boost individual immunity when challenged. The log relative expression was calculated using the ΔΔCt method [54] with expression for LPS-fed bees is shown relative to expression of unchallenged control bees for the propolis-free and propolis-treated cages. Expression of the antimicrobial peptides is relative to the reference *β-actin*. The propolis-treated cages resulted in a significant increase in immune gene expression in LPS-fed, immune-challenged bees for both *defensin* ($F_{1,31}$ = 12.46, p = 0.001) and *abaecin* ($F_{1,31}$ = 9.06, p = 0.005), as indicated by the "*" between treatment groups. n = 15 bees per treatment group.

4. THE EFFECT OF PROPOLIS AGAINST PATHOGENS

The majority of propolis-related research has focused on identifying chemical components, and understanding the bioactivity of whole extracts, fractions and specific compounds. Much of this is in relation to identifying novel compounds that could be used for human health applications [58,59,60,61,62]. Several reviews have examined this subject over the years, consistently updating the numbers of new chemical compounds being discovered in propolis samples collected across the globe and often including measures of biological activity. Over 300 compounds with variable biological activities have been identified in propolis samples [63] (see Box 1).

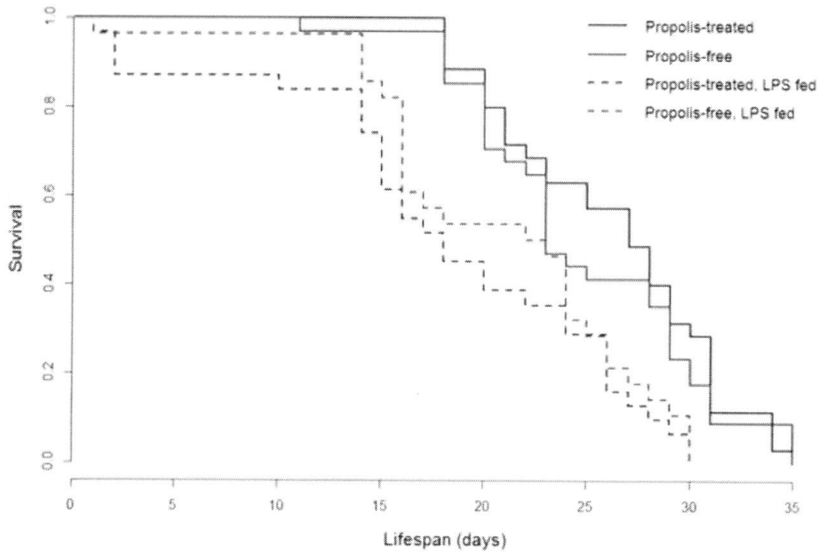

Figure 4. Immune challenge reduces honey bee lifespan in caged bees (Wald $X^2 = 14.18, p = 0.002$). In the incubator setting, which is likely more stressful than a hive, there was no effect of the propolis enriched environment on lifespan ($p = 0.83$). $n = 35$ bees per treatment group.

Box 1. Highlight of recent work to identify the major bioactive components of propolis

Propolis is a rich and diverse source of biologically active compounds due to the diversity of resin-producing plants throughout the world [64]. Propolis has been widely investigated for compounds useful to human health and is regularly screened for cytotoxicity, antibacterial, and antifungal activity. In addition, propolis has been studied for its effects on the immune system and anti-protozoal activity. Several flavonoids and their esters isolated from Sonoran propolis induce apoptosis in a β-cell lymphoma cell line [65], while the polyisoprenylated benzophenone nemorosone from *Clusia*-derived Cuban propolis is cytotoxic against four different cancer cell lines [66]. Prenylated cinnamic acids isolated from Brazilian propolis [67] and terpenes isolated from both Brazilian propolis and Cretan propolis [68] have reported antimicrobial activity against bacterial and/or fungal pathogens. These compounds from Brazilian and Cretan propolis originated from *Baccharis dracunculifolia* and likely *Cupressaceae* conifers, respectively [69]. Caffeic acid phenethyl ester (CAPE) from poplar-derived propolis has been a target of much interest, as it appears to be a major factor modulating anti-inflammatory and immunomodulatory properties of some propolis varieties [61]. Lastly, diterpenes from Libyan propolis were found to have antiprotozoal activity against *Trypanosoma brucei* and *Leishmania donovani* [70].

Recent studies have also focused on propolis compounds important for bee health, with an emphasis on brood diseases. Several known compounds with activity against *P. larvae* have been isolated from Bulgarian propolis including pinocembrin, pinobanksin-3-acetate, and a mixture of caffeic acid esters [71]. However, these compounds may not be responsible for the majority of anti-*P. larvae* activity observed in U.S. propolis samples [21]. Using bioassay-guided fractionation, Wilson et al. [72] isolated pinobanksin-3-octanoate and five other 3-acyl-dihydroflavonols that showed inhibitory activity against both *P. larvae* and *A. apis*. These 3-acyl-dihydroflavonols originated from *Populus fremontii* resin, but varying amounts were also found in other North American *Populus* resins. The amount of pinobanksin-3-octanoate and pinobanksin-3-hexanoate in regional samples of U.S. propolis was strongly correlated with anti-*P. larvae* activity, suggesting that these compounds could be markers of anti-*P. larvae* activity. Voigt and Rademacher [73] found that the common propolis phenolics cinnamic acid and pinocembrin had a fungistatic effect on *A. apis*. We do not understand how propolis compounds interact with bee pathogens in the hive, so this question should be a priority in future studies.

Propolis is most well-known for its antimicrobial activity against a number of bacterial, fungal and viral pathogens, particularly human pathogens [74,75,76,77]. A honey bee colony collects resin from multiple plant sources, though there is likely source fidelity for individual bees [5,17]. The antimicrobial activity of the propolis depends on the antimicrobial properties of its source plant resins [21,72], and the fact that colonies mix sources of resins creates an extremely complex mixture against which susceptible parasites and pathogens would have difficulty developing resistance. A previous study investigating the biological activity of single and multiple-resin source against stingless bee parasites and pathogens showed that, although some single-resin sources may be more effective against a single parasite or pathogen, mixed sources are considerably more effective overall due to increased variety of chemical compounds [78].

The therapeutic value of propolis for honey bees has been understudied. Most studies have concentrated on the high in vitro activity of propolis from various regions of the world against two pathogens: *Paenibacillus larvae*, the causative agent of American foulbrood (AFB) [15,19,21,79,80,81]; and *Ascosphaera apis*, the fungal agent of chalkbrood disease [16,21,73]. Early studies on the therapeutic effects of propolis on individual bees and colonies yielded positive results when propolis was fed to the bees in sugar solution (e.g., [80,81]). In these studies, clinical signs of AFB in field colonies, and number of *P. larvae* spores in honey stores were reduced when bees were fed propolis in sugar syrup. Feeding bees propolis would be similar to administering oral antibiotics, but as the antimicrobial properties of propolis vary widely across plant sources and regions, oral application risks under- or over-dosing the bees and potentially harming the beneficial microbiota in bees' guts [82]. To our knowledge, honey bees do not naturally consume propolis. Therefore, the mode of action of a therapeutic effect of propolis on colony pathogens is probably via volatile compounds [83] or direct contact [84] either on the hive walls, a barrier at the nest entrance, or along the rims of comb cells.

Our in vivo studies at the colony level first examined the effectiveness of the propolis envelope within the nest to reduce chalkbrood infection [16]. Chalkbrood is one of the major fungal diseases affecting honey bees, which only infects the larval stage [85]. Adult bees are carriers of the fungal spores and feed young larvae contaminated brood food, presumably by adding nectar from the bee crop contaminated with spores into brood food. Alternatively, the food fed to young larvae may become contaminated with fungal spores left in the wax comb from a previous infection in the colony. As the larva develops, the spores germinate in the gut and eventually penetrate into the hemocoel, overwhelming the bee with mycelia and turning it into a "chalkbrood mummy." In one field test of potential effects of propolis against chalkbrood, colonies were either enriched with an experimentally applied propolis envelope (ethanol extract of Minnesota-derived propolis painted inside the hive box) or not propolis-enriched (as in Figure 2A). Colonies were then challenged by feeding them pollen patties containing *A. apis* spores, prepared by homogenizing chalkbrood mummies in the pollen [86]. After three weeks, the colonies with a propolis envelope had significantly fewer chalkbrood mummies in the combs (14.7 + 7.5 mummies; n = 6 colonies), compared to colonies without the propolis envelope (108.2 + 40.0 mummies, n = 6 colonies) [16]. One known mechanism of resistance to chalkbrood disease is through hygienic behaviour, whereby the adult bees detect and quickly remove chalkbrood-infected brood from the nest before the pathogen becomes infectious [86]. In this experiment, none of the colonies were considered hygienic, as determined by freeze-killed brood assay (reviewed in [33]); thus, the difference in disease levels between the two sets of colonies was not due to differences in hygienic behaviour. While in vitro work has shown that propolis extracts do directly inhibit growth of *A. apis* in liquid media [21], the mode of action for how the propolis extract painted on the inside of the bee box reduced fungal infection in the larvae, maintained mainly in the center of the nest, remains unknown.

In subsequent tests, the effectiveness of the propolis envelope to reduce severity of AFB, which is similar in etiology to chalkbrood disease, was examined [19]. Adult bees are only the carriers of *P. larvae* spores and are not susceptible to infection. As with chalkbrood, young larvae may become infected via contaminated larval food fed by adult bees, or the larval food may become contaminated with spores already in the wax comb. Bees resist AFB infection in a number of ways: in addition to hygienic behaviour of adult bees toward infected larvae [87,88], some larvae demonstrate genetic resistance against *P. larvae* [89,90], and young adult bees from some genetic lines secrete antimicrobial compounds into larval food, which protect the larvae from infection [91,92]. Our experiment consisted of challenging non-hygienic colonies in the field by spraying combs with sugar solution that contained spores of *P. larvae* (following [93,94]). Five challenged colonies were fitted with commercial propolis traps along all inner walls of the brood boxes, stimulating the bees to form a natural propolis envelope as in Borba et al. [20] (Figure 2B), and five other challenged colonies were not provided with propolis traps and did not construct a propolis envelope. We investigated the effects of the propolis envelope on the overall reduction of clinical signs of AFB, and on the antimicrobial activity of larval food fed to 1–2 day old larvae. The presence of the propolis envelope did not completely clear AFB infection; all colonies had clinical signs two months later at the end of the experiment, similar to early studies (e.g., [80]). However, the severity of AFB, in colonies with a propolis envelope was relatively mild (severity score just over 1, equating to 1–5 infected larvae per comb, following [87]) compared to the colonies without the propolis envelope (severity score just over 2, or 6–25 infected larvae per comb). Moreover, the ability of the larval food from challenged colonies with a propolis envelope to inhibit the growth of *P. larvae,* in vitro, was significantly higher compared to the activity of larval food from challenged colonies without a propolis envelope [19]. It is unclear if the increased antimicrobial activity of the larval food was due to the presence of antimicrobial peptides produced by adult bees and incorporated into larval food, or to the presence of compounds from the propolis in the food. Since antimicrobial peptides and other antimicrobials (e.g., glucose oxidase) are secreted by nurse bees into brood food, the same mechanism that allows adult bees to alter investment in innate immunity may allow nurse bees to invest more in these compounds as a social immune defense [95]. These studies emphasize the critical importance of the propolis envelope to honey bees' health and demonstrate its role in larval defense against bacterial infections.

Field experiments have also provided evidence of self-medication in honey bees as the rate of resin foraging increased in colonies after challenge with the fungal pathogen, *A. apis*. Colonies have been shown to consistently increase the number of resin foragers after a colony-level infection with *A. apis* over several years of study [16,19]. This case of self-medication is particularly interesting because it occurs at the colony level and does not simply involve individuals ingesting compounds to self-medicate themselves [96]. Because only larvae are infected by *A. apis*, the increased resin collection is a great example of a social immune response increasing collection of antimicrobial products by adult bees to protect younger nestmates. Subsequent studies have determined that this self-medication behavior is pathogen-specific and does not occur for colony-level infections with the bacterial brood disease AFB [16,19]. This finding makes the response to chalkbrood all the more interesting and raises questions related to the mechanisms regulating the behavioural change and how colonies may differentially invest in social and physiological immune defences against fungal versus bacterial pathogens.

5. EFFECT OF PROPOLIS ON OTHER BEE PATHOGENS, PARASITES AND PESTS

Honey bee colonies are susceptible to various pathogens, parasites and pests [85,97]. The main health threats contributing to colony losses include the microsporidian gut parasite *Nosema* spp., the ectoparasitic mite *Varroa destructor*, and some of the 20 viruses known to infect bees [97]. Other secondary pests can kill weakened hives and destroy hive products, such as the greater wax moth, small hive beetle, and small rodents.

5.1 Nosema Ceranae

Nosema ceranae is a gut parasite of honey bees. This parasite was originally found in Asian honey bees, *Apis cerana*, and recently described in European honey bees, *A. mellifera* [98]. In the last 15 years, *N.*

ceranae has spread throughout Europe and the U.S and largely replaced *N. apis* in these two regions [98,99]. To our knowledge, no studies have investigated the effects of propolis on *Nosema* spp. infection in honey bees (*A. mellifera*). However, two studies have investigated the effect of stingless bee (*Trigona apicalis*) propolis on *N. ceranae* infections in *A. cerana* [100] and *A. florea* [101]. While the methods used in these experiments are not ideal to fully answer the question because of the origin of propolis and the propolis extraction methods, they raise interesting questions that should be explored further. In both studies, individual bees were inoculated per os with sucrose solution containing spores of *N. ceranae*. After inoculation, bees were fed propolis extract in water. Similar results were found in both experiments: all challenged bees treated with propolis lived longer than untreated challenged bees (but not longer than unchallenged bees), and those fed with a high concentration of propolis (a 50% solution of propolis extract—60 g propolis/100 mL ethanol—in sugar syrup) had lower infection rate [100]. These studies provide a starting point to investigate the biological relevance of these findings within an *A. mellifera* colony and how propolis in the nest environment may also be impacting gut parasites. The results indicate that propolis may have toxicity to this gut pathogen, but also raises concerns about toxicity to beneficial microbiota in bees' guts, and whether the natural mode of action is per os. This issue is of particular relevance since the bee microbiome consists of a large proportion of gram-positive bacteria [102], and propolis is generally more effective against Gram-positve bacteria verus Gram-negative bacteria [41].

5.2 Varroa Destructor

Of considerable interest is whether propolis has biological activity against *Varroa destructor*. A series of laboratory assays have shown that directly exposing mites held in Petri dishes to relatively low concentrations of ethanolic propolis extracts causes high mite mortality [22,84,103,104,105]. Ethanolic extracts from German propolis caused 100% mortality due to contact with 10% propolis extract [84]. Furthermore, mite exposure to extracts at concentrations as low as 0.5% caused narcotic effects leading to reduced heat production and metabolic rates [84,103]. Sublethal effects of low concentrations of propolis may debilitate a *Varroa* mite's mobility, as well as make them less capable of dealing with environmental stressors (e.g., high temperatures [103]). Moreover, ethanolic extracts from Argentina also caused narcotic effects and significantly higher mite mortality (60.5% to 90%) compared to control treatments [105]. However, a recent study using raw propolis, not extracts, found no effect of propolis exposure on mite survival in the laboratory [22]. Given that ethanolic extractions of propolis are essentially chemically enriched fractions because they contain no wax, it is unclear when using raw propolis how much of the material is bioactive. Furthermore, the propolis used in the study was a year old and so likely lost a portion of its bioactive compounds. With all of this in mind, effects of more natural applications of propolis versus *Varroa* need to be explored. Furthermore, differences in the acaricidal effects of German and Argentinian propolis in these studies may be due to the differences in the concentration of compounds, the botanical origin and, more importantly, the chemical properties of the propolis from these two regions. With respect to colony-level differences in chemical composition of propolis, Popova et al. [106] showed that propolis from colonies able to maintain very low mite infestation levels without acaracide treatment were distinct in its chemical composition compared to propolis from colonies with high level of mites. Propolis from mite-tolerant colonies had significantly higher concentration of four compounds (i.e., caffeic acids and pentenyl caffeates). Borba et al. [20] found no differences in mite levels between large field colonies with a propolis envelope and those without an envelope over two years of study.

5.3 Viruses

Related to *Varroa* are the numerous honey bee viruses that can be transmitted through mite infestation, but also are transmitted horizontally from worker to worker or vertically via the queen. Since propolis has known activity against several viruses of human interest (reviewed in [62]), there has been considerable interest in potential effects of propolis against bee viruses, despite little study on the subject. Current evidence suggests that the presence of a propolis envelope within the honey bee nest cavity may not influence viral loads. Borba et al. [20] found no difference in the viral loads of deformed wing virus (DWV), black queen cell virus or Israeli acute bee paralysis virus in bees from propolis-rich or propolis-poor colonies in fall or spring over two years of study. However, a recent study suggests that increased propolis in colonies may impact the levels of DWV, but not sacbrood virus [22]. When raw propolis was added to the tops of colonies, DWV levels did not increase with mite infestation as was expected and observed in the propolis-deprived colonies [22]. Furthermore, they show data that indicates that colonies increase the rate of resin foraging in relation to DWV infection. Given the low mite levels in these colonies overall and the fact that they were previously treated with miticides, these intriguing findings warrant further investigation. Particularly, future work should aim to disentangle the combined effects of DWV-infection and *Varroa* infestation on resin foraging behaviour.

5.4 Other Hive Pests

In regard to other hive pests, two studies have examined the effectiveness of propolis extracts against the greater wax moth (*Galleria mellonella*), an opportunistic pest that mainly affects weakened hives [107,108]. Propolis extract concentrations of 4%–10% (g propolis/100 mL ethanol) showed high toxicity to *G. mellonella* when early stage larval instars had direct contact with the propolis solution for 30 s [108]. Lower concentrations of propolis extract showed sub-lethal effects by reducing pupal metamorphosis duration. However, when fractionated propolis (80% methanol extraction of propolis followed by physical and chemical fractionation) was added to artificial diet fed to *G. mellonella*, it did not influence larval growth, and no toxic effect was observed [107]. The lack of propolis fractions toxicity to *G. mellonella* when mixed into food indicates that this species may tolerate low concentrations of phenolics in their diet. This is not unusual in lepidopterans as other species in this family have also shown tolerance to some flavonoids (quercetin and rutin [109]) and, in some cases, the presence of flavonoids in the diet improves the performance of silkworms (*Bombyx mori* [110]).

Cape honey bees, *A.m. capensis*, have been observed encapsulating the parasitic small hive beetle, *Aethina tumida*, in "propolis prisons", which serves to prevent the beetles from successfully reproducing [111,112]. The extent of this behavior in other subspecies has not been fully investigated, but it is not a trait unique to the Cape honey bee and has been noted to be variably expressed in European honey bee colonies [113,114]. The European honey bee, *A. mellifera*, will also embalm other intruders that are presumably too large to remove from the nest after being killed [5]. Hoyt [115] observed a mouse encased in propolis and suggested that the bees covered it in propolis to prevent odor and decay from affecting the rest of the hive.

6. PROPOLIS AND PESTICIDES

Another area of special interest to the beekeeping community is the presence of contaminants in-hive products, like propolis (i.e., [116]). While commercial hives are often given a variety of chemical treatments to control various hive diseases and parasites, investigations into the residues that these may leave behind are relatively new. There is limited evidence that acaricides can occasionally be found in propolis collected from a hive [22,117,118], as well as the antibiotic (tylosin) used to treat the bacterial diseases American foulbrood and European foulbrood (two of 30 samples from China had detectable amounts [119]). Similarly low levels of pesticide residues likely from treatments on the plant sources or contaminated pollen have been detected in some propolis samples [120,121], but not in others [122,123]. Further study on the frequency and abundance of these chemicals in propolis samples needs to be conducted as well as the possible antagonistic effect that these compounds could have on the chemical constituents of propolis or possibly the synergistic effects that the residues have with those chemicals found in wax, honey and pollen stores (i.e., [124,125]).

Recent evidence also indicates that honey bees may "entomb" fungicide-contaminated pollen in cells with propolis, but the frequency of this behavior and subsequent effect on colony health is currently unclear [126]. Older comb tends to have more "entombed" cells and a regular rotation of new comb into beekeeping operations would reduce the presence of these cells in colonies. It also raises the question of how bees manipulate contaminated propolis samples and they simply layer more propolis over contaminated areas. One other example seen during mite treatments with thymol products is that colonies often surround the thymol tray in a thick layer of propolis, entirely encasing it on occasion (Figure 5). It is unclear if they reuse and redistribute this propolis once the treatment is gone and if it could extend the length of the treatment or cause other negative effects at the colony level. One recent study suggests that propolis could be a "reservoir" of miticides. *Varroa* exposed in a laboratory setting to propolis harvested from a thymol-treated colony had higher mortality than mites exposed to propolis from an untreated colony, even though the propolis was tested a year after collection [22]. This result and the potential impacts warrant further study.

7. PROPOLIS AND OTHER ASPECTS OF BEE HEALTH

It is possible that the antimicrobial properties of materials used and stored in combs (e.g., royal jelly, honey) are enhanced by the addition of propolis [8,98]. Current work showing the effect of a propolis-enriched environment on the antimicrobial activity of larval food continues to

Figure 5. Thymol treatment on the top of a hive encased by propolis. Bees can be seen surrounding the tray with the miticide with a wall of propolis. (Photo by Michael Simone-Finstrom).

raise this possibility [19]. Additionally, it has been suggested that some of the phenolic compounds present in honey may be derived from propolis itself [55]. Depending on how much propolis is incorporated into the comb or along the rims of cells, which is seen in feral colonies and in some managed colonies [5,127], the potential of propolis-related compounds and volatiles "leeching" into other hive products remains a possibility and needs to be further examined. Alternatively, honey bees could potentially actively "enhance" the antimicrobial activity of propolis or these other hive products by combining bee-produced compounds with the plant-produced resins, as has recently been documented in wood ants [128]. This potentially adds another dynamic to the fact that glucose oxidase (a compound produced in the mandibular glands) has been found in propolis [7]. The assumption was that it was found in propolis simply as an artefact of handling, but perhaps this needs to be investigated more fully.

In particular, a series of studies has documented the role of one particular compound, *p*-coumaric acid [56]. *P*-coumaric acid is found in pollen and nectar but is also a common component of propolis worldwide [60,75]. The ingestion of *p*-coumaric acid by honey bees increases expression of detoxification genes and increases metabolism of common beekeeper-applied miticides [56]. In addition, bees fed a diet of *p*-coumaric and another phytochemical, quercetin, also exhibited increased lifespans in some cage scenarios and after exposure to pyrethroid insecticides, though effects were complex based on the addition of protein supplements [53]. It may be that the presence of propolis in the hive and compounds like *p*-coumaric acid actually primes the detoxification pathways so that, upon exposure to certain pesticides, the honey bee is more quickly and effectively able to induce a detoxification response [55]. Ingestion of propolis extracts also decreases mortality of bees exposed to aflatoxins produced by fungi like *Aspergillus* that naturally grow on stored pollen in combs within colonies [29]. These toxic compounds are detoxified by cytochrome P450 activation, which is induced after ingestion of propolis and propolis-related compounds. The big question remains of whether or not simply exposure to propolis in the nest environment has these effects, since honey bees are not thought to typically ingest it.

Other than the potential of propolis as a detoxifying agent or primer of detoxification pathways, another hypothesized mechanism that could influence honey bee longevity via antioxidant-related pathways regards increased resistance to oxidative stress. The free radical theory of aging [129,130] bases the aging process on the production of reactive oxygen species (ROS) via typical cellular metabolism. These ROS result in the oxidation of lipids, proteins and even DNA, disrupting cellular membrane stability and eventually causing apoptosis. Aging, in some ways, is thought to be an accumulation of all of this damage due to oxidative stress. Propolis extracts, through various studies in mainly vertebrate models [131,132], have been shown to widely inhibit the formation of reactive oxygen species, which reduces oxidative damage to proteins, lipids and DNA (reviewed in [133]). Reducing damage from oxidative stress accumulation through diet, by consuming antioxidant rich foods or supplements has been sought as a way to reduce these effects of aging with limited success in humans [130]. However, antioxidants need to get into cells to quench ROS, so studies examining appropriate delivery mechanisms are important. Propolis, as a natural mixture rich in antioxidants, is certainly understudied in regard to its potential effects against oxidative stress. Impacts of oxidative stress are natural in the progression of aging even in honey bees [134,135,136] and are known to be influenced by the environmental conditions in which colonies are maintained [137]. Work on potential effects that propolis may have on colonies in a commercial beekeeping operation to help reduce possible impacts of some of the stressors, including oxidative stress, are important moving forward.

8. APPLICATIONS FOR THE BEEKEEPING COMMUNITY

Several studies have now clearly documented the benefits of a propolis envelope, particularly an envelope naturally constructed by the bees, to bee health and immune system functioning. However, the effects of propolis enrichment in colonies managed in a commercial setting need to be explored more fully. The collection of resins to construct a natural propolis envelope is performed by a rare subset of foraging bees, so increased resin use should not negatively affect honey production, and, in fact, the opposite effect has actually been observed [47,48]. It is estimated that the number of resin foragers is less than 1% of the total number of foragers in the hive, but this may be influenced by the bees' genetics [138,139]. Resin collection is partly a genetic tendency and partly a demand-driven process [16,140,141]. How and what resin collectors detect inside the nest to determine the need for resin is not clear. When resin foragers encounter rough surfaces and gaps inside the hive, they respond by collecting more resin to seal these cracks in the nest architecture and resin foragers are more sensitive to tactile information [142]. Therefore, a colony of bees can be encouraged to build a natural propolis envelope within standard beekeeping equipment by modifying the inner walls of bee boxes. Commercial propolis traps can be cut to fit the four inside walls of the hive boxes and stapled with the smooth side of the trap facing the wood and the rough side facing the colony ([20]; Figure 2B). It is recommended to manage colonies using nine frames instead of ten when using this method due to the space required for the traps. If the inside of the bee box is built with unfinished, rough lumber, scraped briskly with a wire brush, or if small grooves are cut in the interior walls of the box, the bees will apply a layer of propolis in the grooves of rough surfaces, forming a natural propolis envelope.

With this in mind, however, we must consider how altering the nest structure can also potentially adversely affect colony dynamics. The initial experimental design for the studies on the long-term effects of the propolis envelope consisted of three treatments: colonies with no extra propolis (control), colonies with a propolis envelope (Figure 2B), and colonies fitted with a propolis trap on top of the frames of the top box, as is done to collect propolis commercially. Bees from colonies with the propolis traps on top of the frames showed inconsistent, and sometimes higher immune-related gene expression, compared to bees in the propolis envelope and control colonies [19]. Moreover, bees from colonies with a propolis trap on top of the frames had significantly higher levels of virus (DWV) compared to bees in control and propolis envelope colonies in September 2012, May 2013 and May 2014 (but, see [22], which shows reduction in DWV relative to the controls). The hypothesis is that the presence of the water-resistant propolis trap throughout the year on top of the colony could have altered the microenvironment of the colony (e.g., increasing humidity levels or affecting air circulation within the nest), leading to favorable conditions for the growth of pathogens and maybe viruses. Thus, it appears that leaving a propolis trap on top of a colony for a long period of time, and, especially over the winter, is not beneficial to bee health and is not recommended. Finally, there is no evidence that bees consume resins or propolis. We do not recommend that beekeepers feed propolis solution to bees until studies adequately address the long-term effects of such a treatment. Because of the highly antibacterial and antifungal properties of propolis, it could risk poisoning bees and killing the beneficial microbiome in bees' guts that is also so critical to their health and survival [27,82].

9. CONCLUSIONS

Honey bees collect plant-produced antimicrobial compounds and incorporate them into the nest environment. Honey bees, other social and solitary insects, and even some vertebrates co-opt these defensive compounds as their own form of defense. The specific role that this behavior plays in behavioral immunity is ripe for study. Here, we have described recent work that has been done on the interactions between propolis, the honey bee immune system, and honey bee pathogens, parasites and pests. Managed honey bee colonies are currently experiencing high rates of annual mortality, largely due to pathogens, parasites, pesticides and poor nutrition. Perhaps propolis can, at least in part, help mitigate effects from these threats. Understanding the role that propolis plays as a social immune defense directly against parasites and pathogens and through subtle, indirect effects on individual immunity and detoxification enzymes could be a key part of the puzzle to improve bee health. More research needs to be conducted on the long-term effects on the role of propolis on colony health and productivity in order to garner support from the beekeeping community to start selecting for propolis collection in the U.S., something that has been historically, and likely passively, selected against because of its sticky nature. In this way, propolis can be one part of an effective strategy to improve selection in U.S. stocks for resistance traits.

ACKNOWLEDGMENTS

This research was funded by the National Science Foundation IOS 0717530 and IOS 1256992 to Marla Spivak, and the National Honey Board, Project Apis m, and the California State Beekeeping Association to Renata S. Borba and Marla Spivak. Funding was received to cover costs of publishing in open access.

CONFLICTS OF INTEREST

The authors declare no conflict of interest.

REFERENCES

1. Evans, J.D.; Schwarz, R.S. Bees brought to their knees: Microbes affecting honey bee health. Trends Microbiol. 2011, 19, 614–620.
2. Sánchez-Bayo, F.; Goulson, D.; Pennacchio, F.; Nazzi, F.; Goka, K.; Desneux, N. Are bee diseases linked to pesticides?—A brief review. Environ. Int. 2016, 89–90, 7–11.
3. vanEngelsdorp, D.; Meixner, M.D. A historical review of managed honey bee populations in europe and the united states and the factors that may affect them. J. Invertebr. Pathol. 2010, 103, S80–S95.
4. Seitz, N.; Traynor, K.S.; Steinhauer, N.; Rennich, K.; Wilson, M.E.; Ellis, J.D.; Rose, R.; Tarpy, D.R.; Sagili, R.R.; Caron, D.M.; et al. A national survey of managed honey bee 2014–2015 annual colony losses in the USA. J. Apic. Res. 2015, 54, 292–304.
5. Simone-Finstrom, M.; Spivak, M. Propolis and bee health: The natural history and significance of resin use by honey bees. Apidologie 2010, 41, 295–311.
6. Meyer, W. The propolis bees and their activities. Z. Bienenforsch. 1954, 2, 185–200.
7. Zhang, C.-P.; Zheng, H.-Q.; Hu, F.-L. Extraction, partial characterization, and storage stability of β-glucosidase from propolis. J. Food Sci. 2011, 76, C75–C79.
8. Castella, G.; Chapuisat, M.; Christe, P. Prophylaxis with resin in wood ants. Anim. Behav. 2008, 75, 1591–1596.
9. Leonhardt, S.D.; Bluthgen, N. A sticky affair: Resin collection by bornean stingless bees. Biotropica 2009, 41, 730–736.
10. Seeley, T.D.; Morse, R.A. Nest of the honey bee (Apis mellifera L.). Insectes Soc. 1976, 23, 495–512.
11. Seeley, T.D.; Seeley, R.H.; Akratanakul, P. Colony defense strategies of the honeybees in thailand. Ecol. Monogr. 1982, 52, 43–63.
12. Visscher, P.K. Adaptations of honey bees (Apis mellifera) to problems of nest hygiene. Sociobiology 1980, 5, 249–260.

13. Clark, E.H. Constructive Beekeeping; Ulsaker Printing Company: Fargo, ND, USA, 1918; p. 52.
14. Simone, M.; Evans, J.D.; Spivak, M. Resin collection and social immunity in honey bees. Evolution 2009, 63, 3016–3022.
15. Bastos, E.M.A.F.; Simone, M.; Macedo Jorge, D.; Egea Soares, A.E.; Spivak, M. In vitro study of the antimicrobial activity of brazilian propolis against Paenibacillus larvae. J. Invertebr. Pathol. 2008, 97, 273–281.
16. Simone-Finstrom, M.D.; Spivak, M. Increased resin collection after parasite challenge: A case of self-medication in honey bees? PLoS ONE 2012, 7, e34061.
17. Wilson, M.B.; Spivak, M.; Hegeman, A.D.; Rendahl, A.; Cohen, J.D. Metabolomics reveals the origins of antimicrobial plant resins collected by honey bees. PLoS ONE 2013, 8, e77512.
18. Nicodemo, D.; Malheiros, E.B.; De Jong, D.; Couto, R.H.N. Increased brood viability and longer lifespan of honeybees selected for propolis production. Apidologie 2014, 45, 269–275.
19. Borba, R.S. Constitutive and Therapeutic Benefits of Plant Resins and a Propolis Envelope to Honey Bee, Apis mellifera L., Immunity and Health. Ph.D. Thesis, University of Minnesota, Ann Arbor, MA, USA, 2015.
20. Borba, R.S.; Klyczek, K.K.; Mogen, K.L.; Spivak, M. Seasonal benefits of a natural propolis envelope to honey bee immunity and colony health. J. Exp. Biol. 2015, 218, 3689–3699.
21. Wilson, M.B.; Brinkman, D.; Spivak, M.; Gardner, G.; Cohen, J.D. Regional variation in composition and antimicrobial activity of us propolis against Paenibacillus larvae and Ascosphaera apis. J. Invertebr. Pathol. 2015, 124, 44–50.
22. Drescher, N.; Klein, A.-M.; Neumann, P.; Yañez, O.; Leonhardt, S. Inside honeybee hives: Impact of natural propolis on the ectoparasitic mite Varroa destructor and viruses. Insects 2017, 8, 15.
23. Cremer, S.; Armitage, S.A.O.; Schmid-Hempel, P. Social immunity. Curr. Biol. 2007, 17, R693–R702.
24. Meunier, J. Social immunity and the evolution of group living in insects. Philos. Trans. R. Soc. B-Biol. Sci. 2015, 370, 10.
25. Schmid-Hempel, P. Parasites in Social Insects; Princeton University Press: Princeton, NJ, USA, 1998.
26. Evans, J.D.; Lopez, D.L. Bacterial probiotics induce an immune response in the honey bee (hymenoptera: Apidae). J. Econ. Entomol. 2004, 97, 752–756.
27. Anderson, K.E.; Sheehan, T.H.; Eckholm, B.J.; Mott, B.M.; DeGrandi-Hoffman, G. An emerging paradigm of colony health: Microbial balance of the honey bee and hive (Apis mellifera). Insectes Soc. 2011, 58, 431–444.
28. Evans, J.D.; Aronstein, K.; Chen, Y.P.; Hetru, C.; Imler, J.L.; Jiang, H.; Kanost, M.; Thompson, G.J.; Zou, Z.; Hultmark, D. Immune pathways and defence mechanisms in honey bees Apis mellifera. Insect Mol. Biol. 2006, 15, 645–656.
29. Niu, G.; Johnson, R.M.; Berenbaum, M.R. Toxicity of mycotoxins to honeybees and its amelioration by propolis. Apidologie 2011, 42, 79–87.
30. Schmid-Hempel, P. Evolutionary ecology of insect immune defenses. Annu. Rev. Entomol. 2005, 50, 529–551.
31. Evans, J.D.; Pettis, J.S. Colony-level impacts of immune responsiveness in honey bees, Apis mellifera. Evolution 2005, 59, 2270–2274.
32. Moret, Y.; Schmid-Hempel, P. Survival for immunity: The price of immune system activation for bumblebee workers. Science 2000, 290, 1166–1168.
33. Wilson-Rich, N.; Spivak, M.; Fefferman, N.H.; Starks, P.T. Genetic, individual, and group facilitation of disease resistance in insect societies. Annu. Rev. Entomol. 2009, 54, 405–423.
34. Simone-Finstrom, M. Social immunity and the superorganism: Behavioral defenses protecting honey bee colonies from pathogens and parasites. Bee World 2017, 94, 21–29.

35. Erler, S.; Moritz, R.F. Pharmacophagy and pharmacophory: Mechanisms of self-medication and disease prevention in the honeybee colony (Apis mellifera). Apidologie 2016, 47, 389–411.
36. Christe, P.; Oppliger, A.; Bancala, F.; Castella, G.; Chapuisat, M. Evidence for collective medication in ants. Ecol. Lett.2003, 6, 19–22.
37. Castella, G.; Chapuisat, M.; Moret, Y.; Christe, P. The presence of conifer resin decreases the use of the immune system in wood ants. Ecol. Entomol. 2008, 33, 408–412.
38. Chapuisat, M.; Oppliger, A.; Magliano, P.; Christe, P. Wood ants use resin to protect themselves against pathogens. Proc. R. Soc. B-Biol. Sci. 2007, 274, 2013–2017.
39. Park, Y.K.; Paredes-Guzman, J.F.; Aguiar, C.L.; Alencar, S.M.; Fujiwara, F.Y. Chemical constituents in Baccharis dracunculifolia as the main botanical origin of southeastern brazilian propolis. J. Agric. Food Chem. 2004, 52, 1100–1103.
40. Bankova, V.; Boudourova-Krasteva, G.; Popov, S.; Sforcin, J.M.; Funari, S.R.C. Seasonal variations of the chemical composition of brazilian propolis. Apidologie 1998, 29, 361–367.
41. Sforcin, J.; Fernandes, A.; Lopes, C.; Bankova, V.; Funari, S. Seasonal effect on brazilian propolis antibacterial activity. J. Ethnopharmacol. 2000, 73, 243–249.
42. Engels, W.; Kaatz, H.; Zillikens, A.; Simões, Z.P.; Trube, A.; Braun, R.; Dittrich, F. Honey bee reproduction: Vitellogenin and caste-specific regulation of fertility. In Honey Bee Reproduction: Vitellogenin and Caste-Specific Regulation of Fertility; Elsevier Science Publishers B.V.: Amsterdam, The Netherlands, 1990; pp. 495–502.
43. Amdam, G.V.; Norberg, K.; Hagen, A.; Omholt, S.W. Social exploitation of vitellogenin. Proc. Natl. Acad. Sci. USA2003, 100, 1799–1802.
44. Amdam, G.V.; Simões, Z.L.; Hagen, A.; Norberg, K.; Schrøder, K.; Mikkelsen, Ø.; Kirkwood, T.B.; Omholt, S.W. Hormonal control of the yolk precursor vitellogenin regulates immune function and longevity in honeybees. Exp. Gerontol. 2004, 39, 767–773.
45. Smart, M.; Pettis, J.; Rice, N.; Browning, Z.; Spivak, M. Linking measures of colony and individual honey bee health to survival among apiaries exposed to varying agricultural land use. PLoS ONE 2016, 11, e0152685.
46. Garcia, C.R.; Oliveira, N.T.E.; Camargo, S.C.; Pires, B.G.; Oliveira, C.A.L.; Teixeira, R.A.; Pickler, M.A. Honey and propolis production, hygiene and defense behaviors of two generations of africanized honey bees. Sci. Agric. 2013, 70, 74–81.
47. Manrique, A.J.; Soares, A.E.E. Start of africanized honey bee selection program for increased propolis production and its effect on honey production. Interciencia 2002, 27, 312–316.
48. Nicodemo, D.; De Jong, D.; Couto, R.H.N.; Malheiros, E.B. Honey bee lines selected for high propolis production also have superior hygienic behavior and increased honey and pollen stores. Genet. Mol. Res. 2013, 12, 6931–6938.
49. Padilha, A.H.; Sattler, A.; Cobuci, J.A.; McManus, C.M. Genetic parameters for five traits in africanized honeybees using bayesian inference. Genet. Mol. Biol. 2013, 36, 207–213.
50. Simone-Finstrom, M.; Hayworth, M. USDA-ARS Honey Bee Breeding, Genetics, and Physiology Laboratory, Baton Rouge, LA 70820, USA. Unpublished data. 2012.
51. Mallon, E.B.; Brockmann, A.; Schmid-Hempel, P. Immune response inhibits associative learning in insects. Proc. R. Soc. B-Biol. Sci. 2003, 270, 2471–2473.
52. Richard, F.J.; Aubert, A.; Grozinger, C.M. Modulation of social interactions by immune stimulation in honey bee, Apis mellifera, workers. BMC Biol. 2008, 6.
53. Liao, L.-H.; Wu, W.-Y.; Berenbaum, M. Impacts of dietary phytochemicals in the presence and absence of pesticides on longevity of honey bees (Apis mellifera). Insects 2017, 8, 22.
54. Schmittgen, T.D.; Livak, K.J. Analyzing real-time pcr data by the comparative ct method. Nat. Protoc. 2008, 3, 1101–1108.

55. Johnson, R.M.; Mao, W.; Pollock, H.S.; Niu, G.; Schuler, M.A.; Berenbaum, M.R. Ecologically appropriate xenobiotics induce cytochrome p450s in Apis mellifera. PLoS ONE 2012, 7, e31051.
56. Mao, W.; Schuler, M.A.; Berenbaum, M.R. Honey constituents up-regulate detoxification and immunity genes in the western honey bee Apis mellifera. Proc. Natl. Acad. Sci. USA 2013, 110, 8842–8846.
57. Westra, E.R.; van Houte, S.; Oyesiku-Blakemore, S.; Makin, B.; Broniewski, J.M.; Best, A.; Bondy-Denomy, J.; Davidson, A.; Boots, M.; Buckling, A. Parasite exposure drives selective evolution of constitutive versus inducible defense. Curr. Biol. 2015, 25, 1043–1049.
58. Burdock, G.A. Review of the biological properties and toxicity of bee propolis. Food Chem. Toxicol. 1998, 36, 347–363.
59. Bankova, V.; Marcucci, M.C. Standardization of propolis: Present status and perspectives. Bee World 2000, 81, 182–188.
60. Bankova, V.S.; de Castro, S.L.; Marcucci, M.C. Propolis: Recent advances in chemistry and plant origin. Apidologie2000, 31, 3–15.
61. Sforcin, J.M.; Bankova, V. Propolis: Is there a potential for the development of new drugs? J. Ethnopharmacol. 2011, 133, 253–260.
62. Silva-Carvalho, R.; Baltazar, F.; Almeida-Aguiar, C. Propolis: A complex natural product with a plethora of biological activities that can be explored for drug development. Evid.-Based Complement. Alternat. Med. 2015, 2015, 206439.
63. Huang, S.; Zhang, C.-P.; Wang, K.; Li, G.Q.; Hu, F.-L. Recent advances in the chemical composition of propolis. Molecules 2014, 19, 19610–19632.
64. Langenheim, J.H. Plant Resins: Chemistry, Evolution, Ecology and Ethnobiology; Timber Press: Portland, OR, USA, 2003; p. 586.
65. Alday, E.; Valencia, D.; Carreño, A.L.; Picerno, P.; Piccinelli, A.L.; Rastrelli, L.; Robles-Zepeda, R.; Hernandez, J.; Velazquez, C. Apoptotic induction by pinobanksin and some of its ester derivatives from sonoran propolis in a b-cell lymphoma cell line. Chem.-Biol. Interact. 2015, 242, 35–44.
66. Cuesta-Rubio, O.; Frontana-Uribe, B.A.; Ramírez-Apan, T.; Cárdenas, J. Polyisoprenylated benzophenones in cuban propolis: Biological activity of nemorosone. Z. Naturforsch. C 2002, 57, 372–378.
67. Aga, H.; Shibuya, T.; Sugimoto, T.; Kurimoto, M.; Nakajima, S. Isolation and identification of antimicrobial compounds in brazilian propolis. Biosci. Biotechnol. Biochem. 1994, 58, 945–946.
68. Popova, M.P.; Chinou, I.B.; Marekov, I.N.; Bankova, V.S. Terpenes with antimicrobial activity from cretan propolis. Phytochemistry 2009, 70, 1262–1271.
69. Salatino, A.; Weinstein Teixeira, W.; Negri, G.; Message, D. Origin and chemical variation of brazilian propolis. Evid. Based Complement. Alternat. Med. 2005, 2, 33–38.
70. Siheri, W.; Igoli, J.O.; Gray, A.I.; Nasciemento, T.G.; Zhang, T.; Fearnley, J.; Clements, C.J.; Carter, K.C.; Carruthers, J.; Edrada-Ebel, R. The isolation of antiprotozoal compounds from libyan propolis. Phytother. Res. 2014, 28, 1756–1760.
71. Bilikova, K.; Popova, M.; Trusheva, B.; Bankova, V. New anti-paenibacillus larvae substances purified from propolis. Apidologie 2013, 44, 278–285.
72. Wilson, M.B.; Pawlus, A.D.; Brinkman, D.; Gardner, G.; Hegeman, A.D.; Spivak, M.; Cohen, J.D. 3-acyl dihydroflavonols from poplar resins collected by honey bees are active against the bee pathogens Paenibacillus larvae and Ascosphaera apis. Phytochemistry 2017, 138, 83–92.
73. Voigt, K.; Rademacher, E. Effect of the propolis components, cinnamic acid and pinocembrin, on Apis mellifera and Ascosphaera apis. J. Apic. Sci. 2015, 59, 89–95.
74. Ghisalberti, E.L. Propolis: A review. Bee World 1979, 60, 263–282.
75. Marcucci, M.C. Propolis: Chemical composition, biological properties and therapeutic activity. Apidologie 1995, 26, 83–99.

76. Gekker, G.; Hu, S.X.; Spivak, M.; Lokensgard, J.R.; Peterson, P.K. Anti-HIV-1 activity of propolis in CD4(+) lymphocyte and microglial cell cultures. J. Ethnopharmacol. 2005, 102, 158–163.
77. Sforcin, J.M. Biological properties and therapeutic applications of propolis. Phytother. Res. 2016, 30, 894–905.
78. Drescher, N.; Wallace, H.M.; Katouli, M.; Massaro, C.F.; Leonhardt, S.D. Diversity matters: How bees benefit from different resin sources. Oecologia 2014, 176, 943–953.
79. Lindenfelser, L.A. Antimicrobial activity of propolis. Am. Bee J. 1967, 107, 90–92.
80. Lindenfelser, L.A. In vivo activity of propolis against Bacillus larvae. J. Invertebr. Pathol. 1968, 12, 129–131.
81. Antunez, K.; Harriet, J.; Gende, L.; Maggi, M.; Eguaras, M.; Zunino, P. Efficacy of natural propolis extract in the control of american foulbrood. Vet. Microbiol. 2008, 131, 324–331.
82. Tian, B.; Fadhil, N.H.; Powell, J.E.; Kwong, W.K.; Moran, N.A. Long-term exposure to antibiotics has caused accumulation of resistance determinants in the gut microbiota of honeybees. MBio 2012, 3.
83. Messer, A.C. Fresh dipterocarp resins gathered by megachilid bees inhibit growth of pollen-associated fungi. Biotropica 1985, 17, 175–176.
84. Garedew, A.; Lamprecht, I.; Schmolz, E.; Schricker, B. The varroacidal action of propolis: A laboratory assay. Apidologie 2002, 33, 41–50.
85. Morse, R.; Flottum, K. Honey Bee Pests Predators and Diseases; AI Root Company: Medina, OH, USA, 1997.
86. Gilliam, M.; Taber, S.; Lorenz, B.J.; Prest, D.B. Factors affecting development of chalkbrood disease in colonies of honey bees, Apis mellifera, fed pollen contaminated with Ascosphaera apis. J. Invertebr. Pathol. 1988, 52, 314–325.
87. Spivak, M.; Reuter, G.S. Resistance to american foulbrood disease by honey bee colonies (Apis mellifera) bred for hygienic behavior. Apidologie 2001, 32, 555–565.
88. Rothenbuhler, W.C. Behaviour genetics of nest cleaning in honey bees. I. Responses of four inbred lines to disease-killed brood. Anim. Behav. 1964, 12, 578–583.
89. Rothenbuhler, W.C.; Thompson, V.C. Resistance to american foulbrood in honey bees. I. Differential survival of larvae of different genetic lines. J. Econ. Entomol. 1956, 49, 470–475.
90. Bambrick, J.F.; Rothenbuhler, W.C. Resistance to american foulbrood in honey bees. 4. Relationship between larval age at innoculation and mortality in a resistant and in a susceptible line. J. Invertebr. Pathol. 1961, 3, 381–390.
91. Thompson, V.C.; Rothenbuhler, W.C. Resistance to american foulbrood in honey bees. Ii. Differential protection of larvae by adults of different genetic lines. J. Econ. Entomol. 1957, 50, 731–737.
92. Rose, R.I.; Briggs, J.D. Resistance to american foulbrood in honey bees ix. Effects of honey-bee larval food on the growth and viability of Bacillus larvae. J. Invertebr. Pathol. 1969, 13, 74–80.
93. Seeley, T.D.; Tarpy, D.R. Queen promiscuity lowers disease within honeybee colonies. Proc. R. Soc. B-Biol. Sci. 2007, 274, 67–72.
94. de Graaf, D.C.; Alippi, A.M.; Antunez, K.; Aronstein, K.A.; Budge, G.; De Koker, D.; De Smet, L.; Dingman, D.W.; Evans, J.D.; Foster, L.J.; et al. Standard methods for american foulbrood research. J. Apic. Res. 2013, 52, 1–28.
95. López-Uribe, M.M.; Fitzgerald, A.; Simone-Finstrom, M. Inducible versus constitutive social immunity: Examining effects of colony infection on glucose oxidase and Defensin-1 production in honey bees. R. Soc. Open Sci. 2017. accepted.
96. Gherman, B.I.; Denner, A.; Bobis, O.; Dezmirean, D.S.; Marghitas, L.A.; Schluens, H.; Moritz, R.F.A.; Erler, S. Pathogen-associated self-medication behavior in the honeybee Apis mellifera. Behav. Ecol. Sociobiol. 2014, 68, 1777–1784.
97. Cox-Foster, D.L.; Conlan, S.; Holmes, E.C.; Palacios, G.; Evans, J.D.; Moran, N.A.; Quan, P.-L.; Briese, T.; Hornig, M.; Geiser, D.M.; et al. A metagenomic survey of microbes in honey bee colony collapse disorder. Science 2007, 318, 283–287.

98. Chen, Y.; Evans, J.D.; Smith, I.B.; Pettis, J.S. Nosema ceranae is a long-present and wide-spread microsporidian infection of the european honey bee (apis mellifera) in the united states. J. Invertebr. Pathol. 2008, 97, 186–188.
99. Klee, J.; Besana, A.M.; Genersch, E.; Gisder, S.; Nanetti, A.; Tam, D.Q.; Chinh, T.X.; Puerta, F.; Ruz, J.M.; Kryger, P. Widespread dispersal of the microsporidian Nosema ceranae, an emergent pathogen of the western honey bee, Apis mellifera. J. Invertebr. Pathol. 2007, 96, 1–10.
100. Yemor, T.; Phiancharoen, M.; Eric Benbow, M.; Suwannapong, G. Effects of stingless bee propolis on nosema ceranae infected asian honey bees, apis cerana. J. Apic. Res. 2015, 54, 468–473.
101. Suwannapong, G.; Maksong, S.; Benbow, M. Stingless bee propolis effects on experimental infection of apis florea with nosema ceranae. J. Agric. Sci. Technol. A 2011, 1, 818–825.
102. Kwong, W.K.; Moran, N.A. Gut microbial communities of social bees. Nat. Rev. Microbiol. 2016, 14, 374–384.
103. Garedew, A.; Schmolz, E.; Lamprecht, I. Microcalorimetric and respirometric investigation of the effect of temperature on the antivarroa action of the natural bee product-propolis. Thermochim. Acta 2003, 399, 171–180.
104. Damiani, N.; Daniel Maggi, M.; Brenda Gende, L.; Faverin, C.; Javier Eguaras, M.; Augusto Marcangeli, J. Evaluation of the toxicity of a propolis extract on Varroa destructor (acari: Varroidae) and Apis mellifera (hymenoptera: Apidae). J. Apic. Res. 2010, 49, 257–264.
105. Damiani, N.; Fernandez, N.J.; Maldonado, L.M.; Alvarez, A.R.; Eguaras, M.J.; Marcangeli, J.A. Bioactivity of propolis from different geographical origins on Varroa destructor (acari: Varroidae). Parasitol. Res. 2010, 107, 31–37.
106. Popova, M.; Reyes, M.; Le Conte, Y.; Bankova, V. Propolis chemical composition and honeybee resistance against Varroa destructor. Nat. Prod. Res. 2014, 28, 788–794.
107. Johnson, K.; Eischen, F.; Giannasi, D. Chemical composition of north american bee propolis and biological activity towards larvae of greater wax moth (lepidoptera: Pyralidae). J. Chem. Ecol. 1994, 20, 1783–1791.
108. Garedew, A.; Schmolz, E.; Lamprecht, I. Effect of the bee glue (propolis) on the calorimetrically measured metabolic rate and metamorphosis of the greater wax moth Galleria mellonella. Thermochim. Acta 2004, 413, 63–72.
109. Shaver, T.; Lukefahr, M. Effect of flavonoid pigments and gossypol on growth and development of the bollworm, tobacco budworm, and pink bollworm. J. Econ. Entomol. 1969, 62, 643–646.
110. Kato, M. Phenols as indispensible components of the synthetic diet of the silkworm, Bombyx mori. Entomol. Exp. Appl.1978, 24, 485–490.
111. Neumann, P.; Pirk, C.; Hepburn, H.; Solbrig, A.; Ratnieks, F.; Elzen, P.; Baxter, J. Social encapsulation of beetle parasites by cape honeybee colonies (Apis mellifera capensis esch.). Naturwissenschaften 2001, 88, 214–216. [PubMed]
112. Ellis, J.D.; Hepburn, H.R. A note on mapping propolis deposits in cape honey bee (Apis mellifera capensis) colonies. Afr. Entomol. 2003, 11, 122–124.
113. Ellis, J. Life behind bars: Why honey bees feed small hive beetles. Am. Bee J. 2002, 142, 267–269.
114. Ellis, J.D.; Hepburn, H.R.; Ellis, A.M.; Elzen, P.J. Social encapsulation of the small hive beetle (Aethina tumida murray) by european honeybees (Apis mellifera L.). Insectes Soc. 2003, 50, 286–291.
115. Hoyt, M. The World of Bees; Coward McCann Inc.: New York, NY, USA, 1965.
116. Bogdanov, S. Contaminants of bee products. Apidologie 2006, 37, 1–18.
117. Bogdanov, S.; Kilchenmann, V.; Imdorf, A. Acaricide residues in some bee products. J. Apic. Res. 1998, 37, 57–67.
118. Wallner, K. Varroacides and their residues in bee products. Apidologie 1999, 30, 235–248.
119. Zhou, J.; Xue, X.; Li, Y.; Zhang, J.; Chen, F.; Wu, L.; Chen, L.; Zhao, J. Multiresidue determination of tetracycline antibiotics in propolis by using hplc-uv detection with

ultrasonic-assisted extraction and two-step solid phase extraction. Food Chem. 2009, 115, 1074–1080.

120. Chen, F.; Chen, L.; Wang, Q.; Zhou, J.; Xue, X.; Zhao, J. Determination of organochlorine pesticides in propolis by gas chromatography–electron capture detection using double column series solid-phase extraction. Anal. Bioanal. Chem.2009, 393, 1073–1079.

121. Pérez-Parada, A.; Colazzo, M.; Besil, N.; Geis-Asteggiante, L.; Rey, F.; Heinzen, H. Determination of coumaphos, chlorpyrifos and ethion residues in propolis tinctures by matrix solid-phase dispersion and gas chromatography coupled to flame photometric and mass spectrometric detection. J. Chromatogr. A 2011, 1218, 5852–5857.

122. Acosta-Tejada, G.M.; Medina-Peralta, S.; Moguel-Ordóñez, Y.B.; Muñoz-Rodríguez, D. Matrix solid-phase dispersion extraction of organophosphorus pesticides from propolis extracts and recovery evaluation by gc/ms. Anal. Bioanal. Chem. 2011, 400, 885–891.

123. dos Santos, T.F.S.; Aquino, A.; Dórea, H.S.; Navickiene, S. Mspd procedure for determining buprofezin, tetradifon, vinclozolin, and bifenthrin residues in propolis by gas chromatography–mass spectrometry. Anal. Bioanal. Chem.2008, 390, 1425–1430.

124. Johnson, R.M. Honey bee toxicology. Annu. Rev. Entomol. 2015, 60, 415–434.

125. Frazier, M.; Mullin, C.; Frazier, J.; Ashcraft, S. What have pesticides got to do with it? Am. Bee J. 2008, 148, 521–524.

126. vanEngelsdorp, D.; Evans, J.D.; Donovall, L.; Mullin, C.; Frazier, M.; Frazier, J.; Tarpy, D.R.; Hayes, J., Jr.; Pettis, J.S. "Entombed pollen": A new condition in honey bee colonies associated with increased risk of colony mortality. J. Invertebr. Pathol. 2009, 101, 147–149.

127. Tautz, J. The Buzz about Bees: Biology of a Superorganism; Springer: Berlin, Germany, 2008.

128. Brütsch, T.; Jaffuel, G.; Vallat, A.; Turlings, T.C.J.; Chapuisat, M. Wood ants produce a potent antimicrobial agent by applying formic acid on tree-collected resin. Ecol. Evol. 2017, 7, 2249–2254.

129. Harman, D. Aging: A theory based on free radical and radiation chemistry. J. Gerontol. 1956, 11, 298–300.

130. Camici, G.G.; Shi, Y.; Cosentino, F.; Francia, P.; Luscher, T.F. Anti-aging medicine: Molecular basis for endothelial cell-targeted strategies—A mini-review. Gerontology 2011, 57, 101–108.

131. Yonar, S.M.; Ural, M.S.; Silici, S.; Yonar, M.E. Malathion-induced changes in the haematological profile, the immune response, and the oxidative/antioxidant status of cyprinus carpio carpio: Protective role of propolis. Ecotoxicol. Environ. Saf. 2014, 102, 202–209.

132. Ichikawa, H.; Satoh, K.; Tobe, T.; Yasuda, I.; Ushio, F.; Matsumoto, K.; Endo, K.; Ookubo, C. Free radical scavenging activity of propolis. Redox Rep. 2002, 7, 347–350.

133. Farooqui, T.; Farooqui, A.A. Beneficial effects of propolis on human health and neurological diseases. Front. Biosci.2012, 4, 779–793.

134. Seehuus, S.C.; Norberg, K.; Gimsa, U.; Krekling, T.; Amdam, G.V. Reproductive protein protects functionally sterile honey bee workers from oxidative stress. Proc. Natl. Acad. Sci. USA 2006, 103, 962–967.

135. Li-Byarlay, H.; Huang, M.H.; Simone-Finstrom, M.; Strand, M.K.; Tarpy, D.R.; Rueppell, O. Honey bee (Apis mellifera) drones survive oxidative stress due to increased tolerance instead of avoidance or repair of oxidative damage. Exp. Gerontol. 2016, 83, 15–21.

136. Williams, J.B.; Roberts, S.P.; Elekonich, M.M. Age and natural metabolically-intensive behavior affect oxidative stress and antioxidant mechanisms. Exp. Gerontol. 2008, 43, 538–549.

137. Simone-Finstrom, M.; Li-Byarlay, H.; Huang, M.H.; Strand, M.K.; Rueppell, O.; Tarpy, D.R. Migratory management and environmental conditions affect lifespan and oxidative stress in honey bees. Sci. Rep. 2016, 6, 32023.
138. Butler, C. The Honeybee: An Introduction to Her Sense-Physiology and Behaviour; Oxford University Press: London, UK, 1949.
139. Page, R.E.; Fondrk, M.K. The effects of colony-level selection on the social organization of honey bee (Apis mellifera L.) colonies: Colony-level components of pollen hoarding. Behav. Ecol. Sociobiol. 1995, 36, 135–144.
140. Nakamura, J.; Seeley, T.D. The functional organization of resin work in honeybee colonies. Behav. Ecol. Sociobiol. 2006, 60, 339–349.
141. Martinez, O.A.; Soares, A.E.E. Melhoramento genético na apicultura comercial para produção da própolis. Rev. Bras. Saúde Prod. Anim. 2012, 13, 982–990.
142. Simone-Finstrom, M.; Gardner, J.; Spivak, M. Tactile learning in resin foraging honeybees. Behav. Ecol. Sociobiol. 2010, 64, 1609–1617.

CHAPTER 12

A Butterfly Picks Its Poison: Cycads (Cycadaceae), Integrated Pest Management and Eumaeus atala Poey (Lepidoptera: Lycaenidae)

Koi S*

McGuire Center for Lepidoptera Associate, University of Florida, Gainesville, FL, USA

ABSTRACT

The imperiled Atala hairstreak butterfly, Eumaeus atala Poey 1832 (Lepidoptera: Lycaenidae) is a specialist species historically living in southeast Florida's endangered pine rockland ecosystems. Until relatively recently, the butterfly used North America's only native cycad Zamia integrifolia L. (Zamiaceae: Cycadales), commonly called "coontie," as the host plant for its offspring. With the introduction of many non-native and valuable cycads into botanical and domestic gardens, the butterfly has expanded its choices to include these exotic species. Conservation of both the plant and insect is complex as herbivory can damage the plant, but control of the larval damage may be detrimental to the fragile populations of the butterfly. Larval and adult host plant choice tests were implemented to compare larval survival, development rates and subsequent adult choice between Z. integrifolia and non-native Zamia vazquezii L. (Zamiaceae:

Cycadales), a popular garden cycad that is critically endangered in its native Mexico. Results indicate that both adults and larvae chose native more often than non-native; larval survival decreased, but development time increased, as did adult lifespan when utilizing Z. vazquezii. Adult survival to successful mating and fecundity occurred with either host choice. Integrated pest management techniques are discussed for avoiding pesticide use to control larval herbivory on valuable cycads.

Keywords: *Sitophilus zeamais*; *Ocimum basilica*; Adult emergence; Fumigant toxicity; Repellent activities

INTRODUCTION

Botanists have recently established global networks to monitor the highly endangered populations of cycads worldwide [1-4], now recognized as the most highly endangered plant species on earth [5]. These severe global declines mirror the losses of many other species recognized as part of the "sixth great extinction" on our planet [6-8]. Causes of cycad decline include legal and illegal collection, removal of seed heads and root calyxes for food and/or "bush medicines," religious ceremonies, and overexploitation, including the two biggest on-going threats: unsustainable trade and habitat loss [1,9]. Another possible reason for declining colonies of these ancient plants is the possibility that the specialist pollinators associated with an individual cycad species have been extirpated, or is extinct [1,2,10]. Notwithstanding legislation and laws designed to protect them, both legal and illegal trade negatively impact global cycad colonies, many of which are isolated and highly vulnerable [1-3]. In Colombia, González [10] indicated that the native cycad, *Zamia encephalartoides*, its pollinator and the herbivores are all in danger.

Found historically from southern Georgia to the Florida Keys, North America's only native cycad, *Zamia integrifolia* L.f. (Zamiaceae: Cycadales), commonly known as "coontie," is still listed as "Threatened" in wild natural areas [11]. Although different cycad species contain various mixes of neurotoxins, the roots of the cycads plants can be harvested and washed to remove the water-soluble toxins in order to make mildew-resistant flour [11,12]. In south Florida, the original resident Calusa Indians processed coontie root for flour; ensuing inhabitants, from the Seminole and Miccosukee Indians to the European settlers, followed suit. The first industries in south Florida were the starch industries of the last century, which depleted wild populations of the cycad, almost to the point of state-wide extirpation. In many undeveloped and underdeveloped nations, cycad-root flour is still a staple [1,2,4] and often, the reproductive cones are removed as well [4,13], hindering seed recruitment for new plants.

Nevertheless, during the past twenty years, native North American coontie has been recovering due to the Florida landscaping industry, as nurseries and homeowners discovered how well-adapted the luxurious leafy plant is to Southeast Florida's diverse ecosystems and stochastic weather cycles [14,15]. More exotic non-native cycads have been added to the repertoire available to landscape planners, increasing urban use and world nursery trade. The coontie plants are not increasing their populations in wild colonies, however, partially because the slow-growing plants are located in habitats with very thin nutrient-poor soils resting on limestone outcrops (**FIGURE 1**).

FIGURE 1: Native *Zamia integrifolia* (coontie) growth habit varies greatly. A) In the oolitic limestone substrate of south Miami-Dade County coontie struggles to survive. B) A domestic garden provides high nutrients. C) A ruderal area provides slightly more organic material in the substrate, but it is still compact rock. D) Coontie is found in scattered pockets in its native pine rockland ecosystem (Photo credit: Koi S).

Two species of native lepidopteran herbivores attack cycads in North America: the Echo Moth (*Seirarctia echo* Smith 1797) and the Atala butterfly (*Eumaeus atala* Poey 1832). The Echo moth range covers Georgia south to the Florida Keys, and west to Mississippi, which is wider than the range of the cycad. The moth is widely polyphagous, utilizing Sabal Palmetto (*Sabal palmetto*), and various species of crotons, lupines, oaks and persimmons, as well as coontie and other plants.

Called "destructive" in the early part of this century, the Echo moth was lamented as being a possible competitor of the other lepidopteran cycad-herbivore, the Atala butterfly, at a time when the butterfly was thought to be extinct or nearly so [16-18].

Historically, the Atala hairstreak butterfly was only found in southeast Florida, in the now endangered pine rockland ecosystemsand tropical hardwood hammocks. Unlike the Echo moth's eclectic choice of foods, the Atala is a specialist butterfly that used our only native cycad, Z. *integrifolia* as its sole host plant. The decline of native cycads caused the radical decline in the butterfly's population and by 1951 it was thought to be extinct [16]. Unbeknownst to almost everyone, the butterfly had survived in isolated refugia (safe havens) far from the increased habitat destruction as the cities in southeast Florida grew.

Chemoecology of cycads and butterfly associates

Each species of cycad contains unique mixes of neurotoxins, such as cycasins and macrozamins in different concentrations and amounts. Butterflies recognize and respond to chemical volatiles released by their host plants, and by larval herbivory, to help them locate potential oviposition sites [1,2,15,19-21].

The Atala has utilized as many as 38 additional cycad species in southeast Florida [22-28], all of which contain similar chemoecology [1,2,15,21].

INTEGRATED PEST MANAGEMENT

Integrated Pest Management (IPM) is a multifaceted methodology developed to prevent or control pest insect damage to economically valuable plants, including ornamentals that may be expensive, rare and/or aesthetically significant to botanical gardens and homeowners. Various methods for detecting, identifying, controlling and managing insect pests are integrated by combining biological, cultural, physical and chemical means to keep insect damage at an acceptable level. IPM controls damages caused by insects in the most economically balanced way, preventing potential health hazards to humans and animals [29-31]. It is one of the safest methods for controlling infestations of unwanted insect pests, while conserving beneficial insect species, as well as other imperiled or endangered biota sharing the same habitat.

The four systems of IPM are: Cultural control, physical control, biological control and chemical control. The ultimate goal of IPM is to use the first three to prevent using the fourth. Cultural control is the foundation for everything else; it includes a thorough understanding of the biology of the plant and the pest insects' life, sanitation practices (such as removing sick plants promptly) and both planting and harvesting times.

Physical control includes proper pruning methods, mulching appropriately for the plant species, and may include mixed plantings that encourage beneficial insects and discourage pests. Biological controls may include releasing commercially propagated beneficial insects into a garden and/or providing the plants and cultural practices that promote naturally occurring beneficial insects. Chemical means are used as a last resort for severe infestations, to target specific pest species, maximizing effectiveness while minimizing hazards to humans and other animals [29-31].

One of the valuable non-native cycads being used by the Atala butterfly as an alternative host plant is *Zamia vazquezii* D. W. Stev, Sabato and DeLuca (Zamiaceae: Cycadales) [23-27], a critically endangered cycad in its native Mexico [27,32] and a highly employed cycad in upscale urban developments, and botanical and domestic gardens in southeast Florida.

Herbivory of *Z. vazquezii* is a pest management concern of economic importance not only because it is considered endangered in its native Mexico (The IUCN 2013), but also because it is one of the exotic cycad species grown in highly specialized south Miami-Dade nurseries; it is considered a valuable addition to a prestigious landscape (i.e., it is not generally utilized along highway median strips or shopping plaza parking lots). However, *Z. integrifolia*, our native cycad, has shown itself to be hardy and resilient enough to be successfully utilized along such heavily trafficked or polluted sites.

Both of these cycads are readily used by the Atala butterfly in these 'semi-natural' urban environments, and often sustain considerable herbivory damage (**FIGURE 2**), sometimes to the extent that the cycad is unable to recover. Pesticides have been used repeatedly in urban arenas to control the Atala caterpillar damage [24-29], thereby triggering both an economic impact in chemicals as well as potentially detrimental environmental costs [29-31].

FIGURE 2: Atala larvae completely consuming a non-native *Zamia erosa* at a botanical garden in south Miami-Dade County (Photo credit: Koi S).

HOST PLANT CHOICE TESTS WITH THE ATALA BUTTERFLY

Larval and adult host plant choice tests were implemented to compare larval survival, development rates and subsequent adult choice between native *Z. integrifolia* and non-native *Z. vazquezii*. Oviposition choice by offspring has been shown to be at least partially affected by the adults' oviposition choices [31-34]. Understanding host preference, egg distribution, heritability of oviposition sites [33-40], colonization of non-native host plants [35], larval feeding strategies and choices and adult oviposition site selection [35-41] for the Atala butterfly and larvae will help lepidopterists, botanists and landscape managers develop feasible management protocol to protect the butterfly colonies while controlling herbivory by focusing on cultural practices and reducing pesticide use.

MATERIALS AND METHODS

BIOLOGICAL LIVESTOCK AND REARING

Atala larvae and pupae were collected from wild colonies in several southeast Florida stable colony sites that had been previously identified during on-going long-term surveys [24-26,42]. These choice test experiments were performed from 7 November 2012 to 30 January 2013.

All tests were completed under the same environmental conditions in a laboratory at the University of Florida in Gainesville, FL, USA (29° 39' 05" N, 82° 19' 29"W) in approximately 8:16 h photoperiod and stable ambient temperatures between 24°C (nocturnal) and 26°C (diurnal). Cages were sprayed twice daily with water misters to maintain normal relative humidity between 25-42%. Adult butterflies of both sexes were housed together in a large walk-in adult flight cage (LiveMonarch Foundation, 1.78 m2, Greenhouse Castle Cage, Boca Raton, FL, USA) to approximate a natural environment and to encourage mating. Cages were provided *ad libitum* with a nectar substitute (Gatorade®, The Gatorade Company, Chicago, IL, USA) in fruit punch, orange or watermelon-citrus flavors, hanging approximately 46 cm from the ceiling in a modified feeder (**FIGURE 3**) [43]. Water was also provided in feeders; host plants for larval consumption or adult oviposition were provided as described herein, depending on test. All feeders were cleaned and replenished daily and larval cages disinfected with 1% bleach solution weekly for disease control.

Larvae from all trials were reared with their brood mates on the host plant of adult or larval choice in square plastic boxes (19.05 × 19.05 × 10.16 cm); individuals that chose differently than brood mates were reared singly or with their congeners, depending on their choices. The box tops were covered with 1 mm tulle netting to prevent larval escape; the food of the larvae's choice was replaced daily as needed *ad libitum*. The rachides of all food fronds were housed in 15 cm floral aqua tubes (Aquapic, Syndicate Sales, and Kokomo, Indiana) to maintain plant freshness for the larvae (**FIGURE 4**). The larvae in all tests were maintained as described above until pupal emergence, when the eclosed adults were transferred to flight cages measuring approximately 30 × 30 × 61 cm (LiveMonarch Jumbo Cage, Boca Raton, Florida).

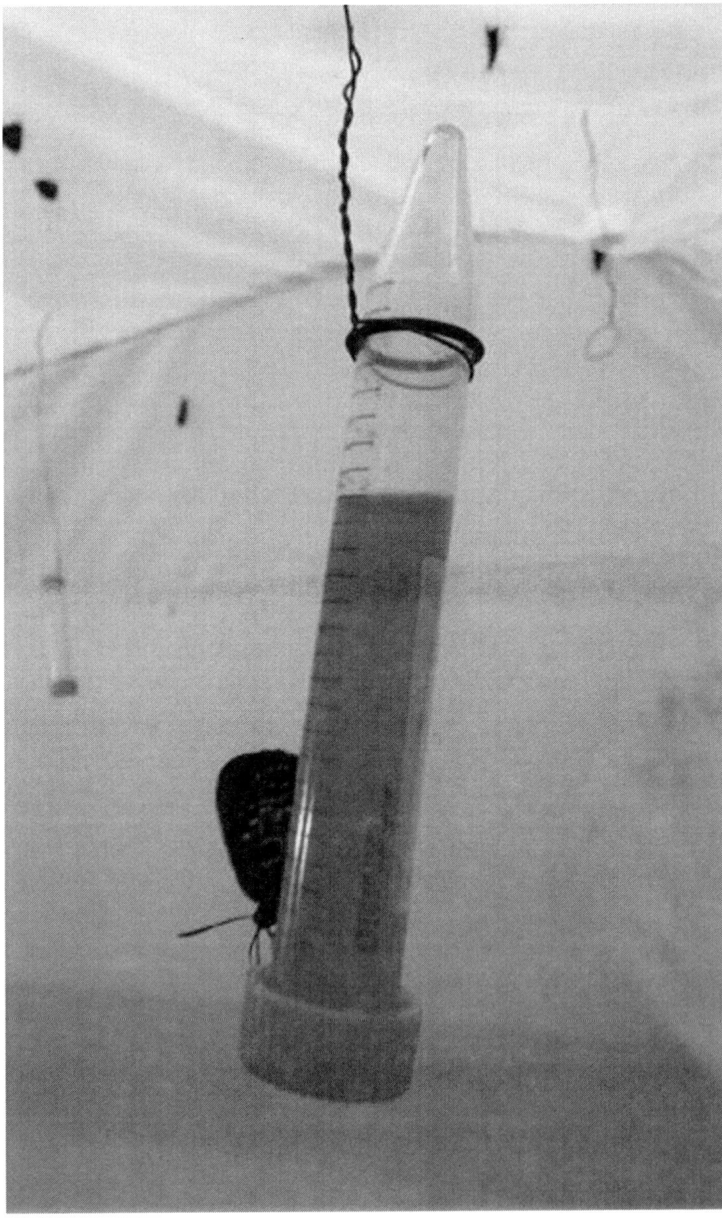

FIGURE 3: An economical, easily made liquid feeder adapted for butterflies (Photo credit: Koi S).

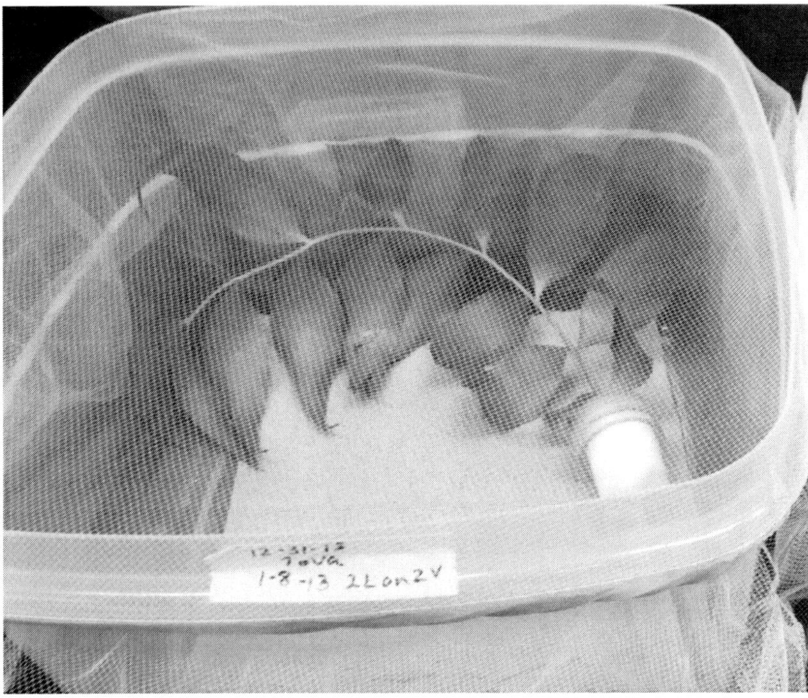

FIGURE 4: The larval box cage, with plant frond housed in an aqua tube, and tulle netting lid secured with a rubber band to prevent larval escape (Photo credit: Koi S).

ADULT CHOICE TEST I

For adult oviposition choice tests, one frond each from both *Z. integrifolia* and *Z. vazquezii*, closely matched in size, were installed in the flight cage. The host plants for ovipositioning were each encased in an aqua tube (Aquapic, Syndicate Sales, Kokomo, Indiana) to keep the leaves fresh, and were hung approximately 46 cm from the ceiling to make them more apparent to the adults (**FIGURE 5**). One leaf of each species was installed about two feet apart on the same side of the cage so that they were not touching. Leaves were collected daily from the previous day and number of eggs counted on each frond. Fresh leaves were installed and the positions of the fronds in the cage alternated daily to avoid position effect preference by the females. Trials were executed for forty-two uninterrupted days. Because the adults were free-flying in the flight cage, and both sex multiple mate [24-26], parentage was not followed for this test.

FIGURE 5: The adult breeding cage showing the host plant (and feeder tubes) suspended from the ceiling, where "Adult Host Plant Choice I" took place (Photo credit: Koi S).

The hatched larvae from the adult-choice test were maintained in larval boxes with chosen host plant as described above until pupal emergence. Those eclosed adults became First Generation Adult Test (**FIGURE 6** for experiment design).

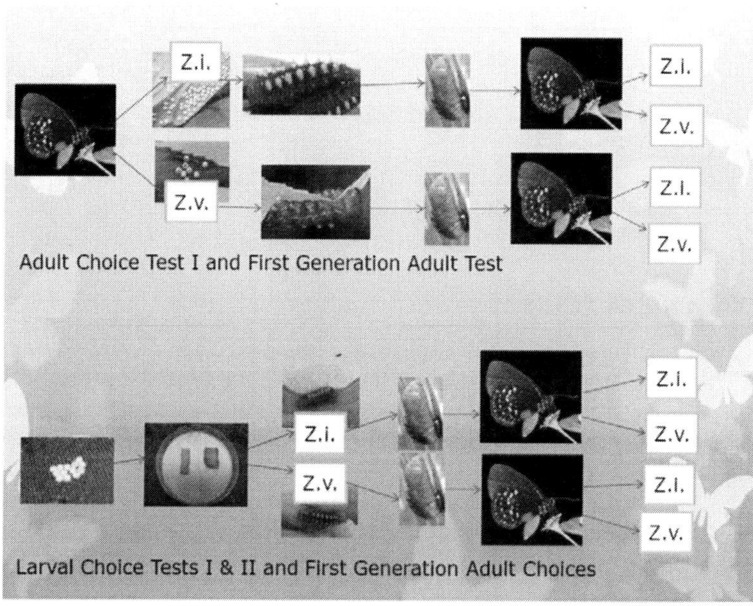

FIGURE 6: Graphic showing the adult choice test I-larval choice test I & II and first generation adult choice test design (Photo credit: Koi S).

FIRST GENERATION ADULT CHOICE TEST I

Larvae that successfully pupated and emerged as adults from "Adult Choice Test I" were housed together as they emerged in the smaller flight cages measuring approximately 30 × 30 × 61 cm (LiveMonarch Jumbo Cage, Boca Raton, FL, USA). Broods were housed together as much as possible, but because of the variable hatch rate and high mortality, the sample size was small. The cages were supplied with nectar and water feeders as previously described, as well as two fronds of host plant, one each *Z. integrifolia* and *Z. vazquezii*. Fronds were hung approximately 46 cm from the ceiling and installed about 15 cm apart. The position of the fronds was not alternated because the adults of both sexes were set free in the large flight cage as soon as the females chose an oviposition site (**FIGURE 6**). Only one cage had more than one female.

FIRST GENERATION ADULT CHOICE TEST II

Larvae that successfully pupated and emerged as adults from "Larval Choice Test II" were housed together as they emerged in the smaller fabric flight cages as previously described. Clutches were housed together as much as possible but because of the variable hatch rate and high mortality, the sample sizes are small for this test as well. The cages were supplied as described above and adults released to the larger flight cage as soon as the females made ovipositing choices (**FIGURE 6**).

LARVAL CHOICE TEST I

Eggs were collected from any non-plant substrate, such as the wall, floor, nectar feeders or roof, from inside the large flight cage. Multiple parents were most likely represented by the eggs in this test because of the nature of the free-flying adults in the flight cage. Sanitized 9 cm plastic petri dishes were prepared as choice test arenas by trimming Whatman filter papers to fit inside the petri dishes. A center line was drawn with a graphite pencil (**FIGURE 7**). The top of each half of the arena paper was labeled either "N" (designating native *Z. integrifolia*) or "NN" (designating non-native *Z. vazquezii*). The labels and plants were alternated in each arena to reduce possible side-preference effects by the larvae (i.e., 1="N-NN", 2="NN-N", 3="N-NN" etc.). One leaflet from each cycad species was pinched off at the petiole where it connected to the main stem of the compound leaf, and placed inside the arena on the appropriate side (**FIGURE 7**). Just before eggs were placed into the arena, the filter paper was dampened with deionized water and the dates of collection recorded so that expected hatch dates were known. Hatches were monitored immediately after eclosure and larval eclosure, timing, behavior and choice were photographed.

A single ovum from the day's collected eggs was placed in the center of each prepared petri dish and lightly sealed with string to prevent larvae loss, in case eclosure occurred earlier than expected. The arena and egg were allowed to rest undisturbed until the larvae hatched. On the expected day of larval hatch, the lid and string were removed from the petri dish, and fresh single leaflets were pinched off at the stems of each of the two host plants. The individual leaflets were installed on opposite sides of the center line approximately 1 cm from the center (**FIGURE 7**). If the larvae did not hatch when expected, the leaflets were changed daily until the larva did emerge.

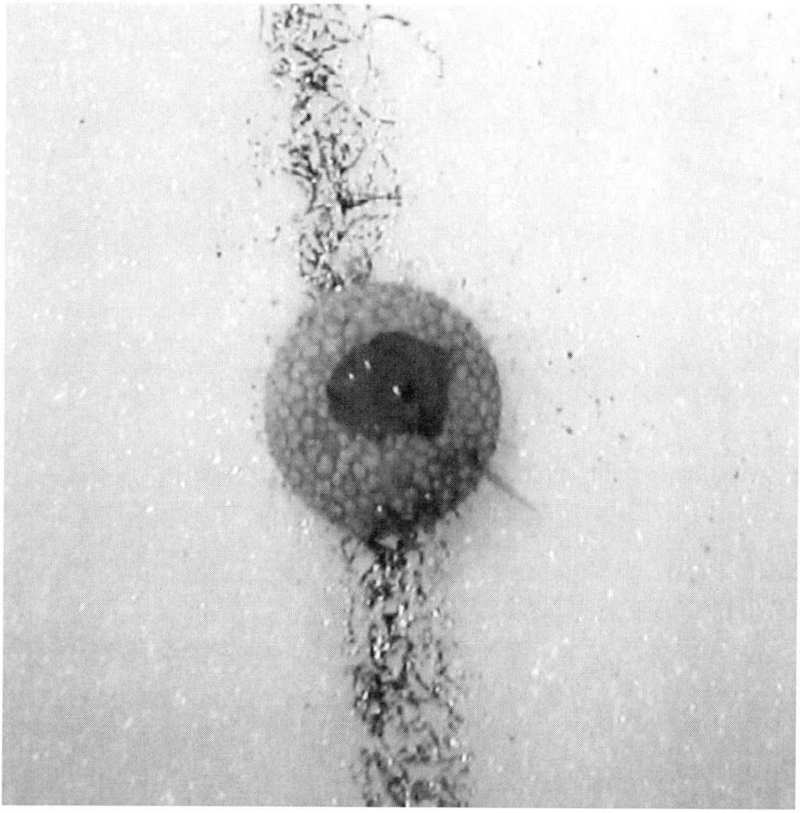

FIGURE 7: In Larval Host Choice Test I, a solitary egg was placed on the center line in the arena and the larva monitored to record choice and timing. The larva in this photograph is just beginning to hatch (Photo credit: Koi S).

The larvae were monitored to record behavior and choice immediately during and after eclosure. Choice was recorded when the larvae climbed onto the plant it chose and commenced feeding, i.e., "settling in" (**FIGURE 8A**) If the larvae did not remain on the leaf to continue feeding after an initial "natal test-bite," it was not recorded as a choice (**FIGURE 8B**). The time from eclosure to choosing its host plant was recorded. Larvae were reared singly on the host plant of their choice in standard square plastic boxes as described above.

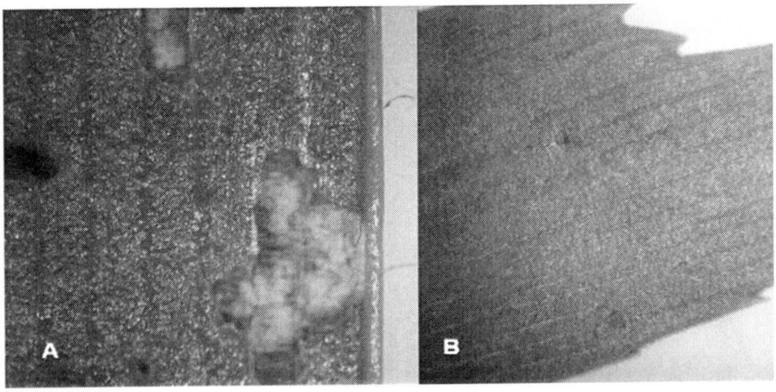

FIGURE 8: A) The native *Zamia integrifolia* leaflet has numerous natal test-bites indicating that the larva has "settled in" to consuming the plant. This kind of mark was counted as a choice (note the frass). B) The non-native *Z. vazquezii* shows shallow nibbles but no decisive bites and were not counted as choices (Photo credit: Koi S).

LARVAL CHOICE TEST II

The same basic protocol for this test was followed, with two changes: instead of one egg per arena, an entire brood of eggs from the same location was deployed in the arena. These small clusters were likely from the same female. Eggs were lined up on the center line in haphazard order. This test explored how and if siblings influenced each other in their choice of host plant (**FIGURE 9A-9D**). There were only 23 larvae successfully hatched in this experiment.

The second change in this test involved trimming both ends of the leaves with scissors (rather than pinched off at the petiole), to both simulate herbivory and release volatile plant chemical clues.

Larvae were reared with their brood mates on the host plant of their choice as described; individuals that chose differently than brood mates were reared singly or with congeners who chose likewise and maintained as described above until pupal emergence. The eclosed adults became First Generation Adult Test II (**FIGURE 6** is a graphical representation of the experimental design).

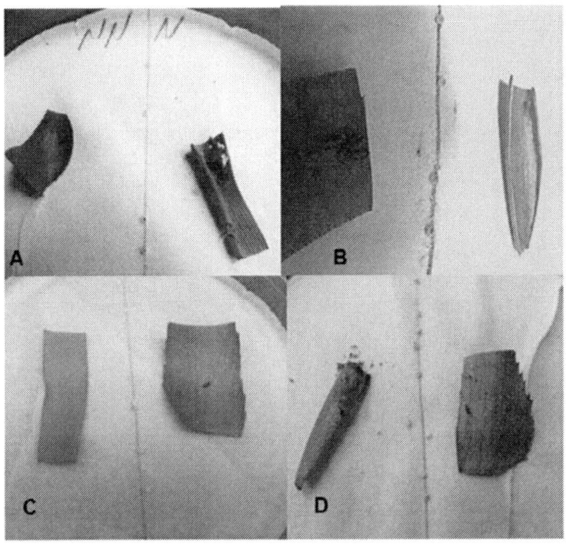

FIGURE 9: The arena for Larval Test II trials utilized the entire brood. A) Five larvae of five chose native *Zamia integrifolia.* B) The first eclosed larva chose non-native *Z. vazquezii,* but the remaining six eggs did not hatch. C) Five larvae in the brood chose non-native and one larva chose native. D) All five larvae chose native, but there was evidence of frass and nibbles on the non-native plant (Photo credit:Koi S).

RESULTS

ADULT CHOICE TEST

Eggs were collected from the leaves of the two cycads in the flight cage daily and counted. A total of 9055 eggs were laid during the fortytwo day trial. There was a significant difference in female host plant choice, with 8094 eggs laid on *Z. integrifolia* and 961 eggs laid on *Z. vazquezii* (p-value <0.000). Native *Z. integrifolia* (Zi) received 89% of the eggs and 11% were laid on *Z. vazquezii* (Zv). Although it is possible for all life stages to be completed successfully while feeding on the non-native plant Zv, the mortality rate of the larvae was high and many larvae feeding on both plants died before reaching pupation (**FIGURE 10**). Development times was not significantly different for any life stage between the two groups (in either the Adult or Larval Choice) or between the two plants ($p>0.05$), although there were obvious differences in sizes (**FIGURE 11**). The proportion larvae surviving on *Z. vazquezii* was lower, but not significantly so. Egg-to-adult survival was 0.021 for larvae consuming Zv and 0.153 for larvae feeding on native Zi. Pupa-to-adult eclosion survival was 0.666 for larvae that had consumed Zv, and 0.732 for larvae that had consumed Zi. Survival was markedly lower for larvae reared on Zv, but not significantly so (**FIGURE 12**).

FIGURE 10: Many of the larvae feeding on non-native *Zamia vazquezii* died before reaching pupation. This larva developed a fissure in the midgut caused by an erupted gut, likely associated with the sharper, tougher leaf structure of *Z. vazquezii* (Photo credit: Koi S).

In another example of the differences, the average number of days from oviposition to adult emergence in Adult Choice Test I was 46.80 days on non-native Zv and only 38.69 days on native Zi (**TABLE 1**). Only 2% of eggs laid on non-native Zv hatched and only seven out of fifteen adult butterflies reared on Zv successfully emerged from the pupa (66%). Only 15% of eggs laid on native Zi hatched, but 254 adults successfully emerged from 347 pupae (73%).

FIGURE 11: Comparisons of 6-day-old larvae in the Larval Choice Test I show obvious differences in development size; the larger larvae consumed native *Zamia integrifolia* while the smaller larvae ate the non-native *Z. vazquezii.* Note the surface scraping on the non-native, which has a much thinner epidermis than native plants (Photo credit: Koi S)

Immature development time and duration of life stages was not significantly different for larvae reared on either plant, but there was a significant difference between the number of eggs laid on *Zv* and *Zi* (χ^2=df14, 0.05, 169.00) (**TABLE 2**).

Although adults reared on *Zv* (Adult Choice) lived longer than other adults in the trials, longevity of the adults did not vary significantly between individuals reared on either plant; there were no significant differences between immature-to-adult development rates and lifespans (**TABLE 1**).

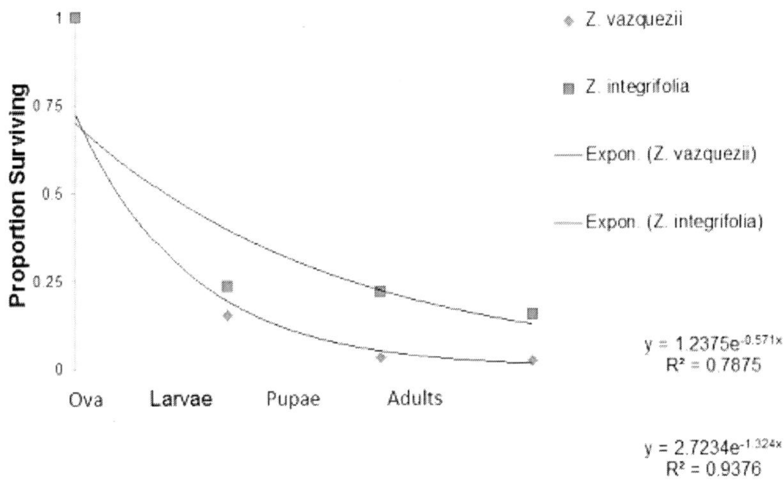

FIGURE 12: Proportion of Atala surviving to the next life stage was lower on non-native *Zamia vazquezii* host plant than on the native *Z. integrifolia* (Adult Choice Test I) (Photo credit: Koi S).

TABLE 1: Development times and life span were both different depending on whether the adult or larva made the choice of host plant, as well as which plant was chosen, but the differences were not statistically significant.

	Ova development (days)	Larval development (days)	Pupal development (days)	Total development (days)	Adult lifespan (days)	n
Larval choice: *Zamia integrifolia*	7.47	16.47	15.21	39.16	27.68	19
St. deviation	± 0.70	± 1.74	± 1.18	± 1.30	± 13.47	
Larval choice: *Zamia vazquezii*	7.00	14.00	19.00	42.75	25.75	4
St. deviation	± 1.15	± 2.00	± 2.06	± 3.69	± 16.32	
Adult choice: *Zamia integrifolia*	8.44	14.25	16.00	38.69	33.44	16
St. deviation	± 0.51	± 1.18	± 1.21	± 1.96	± 12.28	
Adult choice: *Zamia vazquezii*	8.87	22.20	15.73	46.80	22.13	16
St. deviation	± 0.35	± 2.76	± 3.65	± 4.83	± 15.32	

TABLE 2: Although nearly twice as many ova were laid on native *Zamia integrifolia* than on non-native *Zamia vazquezii*, the difference was not statistically significant.

	Ova observed	Ova expected	Chi-square	p-value >0.05
Zamia vazquezii	582	1122.50	260.26	3.841
Zamia integrifolia	1663	1122.50	260.26	
			520.52	

FIRST GENERATION ADULTS FROM "ADULT CHOICE" TEST

First generation adults that eclosed from "Adult Host Plant Choice Test" pupae chose to oviposit on native Zi in 100% of the trials (n=32, 11 females, 4 trials).

LARVAL HOST PLANT CHOICE TEST I

Most larvae chose native *Zi* (77%), but many nibbled the non- native plant first. Of 126 ova, 47 hatched to complete the choice trial. One neonate eclosed from an egg that had been laid on a non-host plant, Florida Keys Blackbead (*Pithecellobium bahamense*) Britton ex Britton & Rose, a southeast Florida native that occurs in coastal areas where there are extant Atala colonies. The plant had been installed in the flight cage as a roosting tree, and was not originally intended to be used for this experiment. However, it was an opportunity to observe any behavior or other changes that may take place if or when eggs are laid on non-host plants. When this larva eclosed, it moved toward the petri-dish sides and circled consistently, wandered to the native leaflet but left it again without feeding. It then wandered to the non-native leaflet in the same manner but did not commence feeding. After four hours without choosing a host, this larva was counted as the only "no choice" in all of the trials (the larva was removed from further testing in this trial but was placed with congeners in another cage). A third short trial within this test used 112 eggs which developed into 42 larvae, 93% of which chose native.

FIRST GENERATION ADULTS FROM "LARVAL HOST PLANT CHOICE" II

First generation adults that had been reared on both native and non-native plants in "Larval Host Plant Choice Test II" chose to oviposit on native *Z. integrifolia* 100% of the time (n=32 ova; 11 females).

LARVAL HOST PLANT CHOICE TEST II

The sample size was small (n=23) as individual mortality was high reared on *Zv*. Eggs collected from broods appeared to make independent choices, as not all of the larvae chose the same host plant. A higher percentage chose native over non-native (83%) than in the first trial and 17% chose the non-native host plant. Total development time from egg-to-adult (Larval Choice II) showed 42.75 days for larvae reared on *Zv*, and 39.16 days for larvae reared on *Zi*.

FIRST GENERATION ADULTS FROM "LARVAL HOST PLANT CHOICE" II

First generation adults that had been reared on both native and non-native plants in "Larval Host Plant Choice Test II" chose to oviposit on native *Z. integrifolia* 100% of the time (n=169 ova; 23 females).

DISCUSSION

All cycads contain variations of the same basic chemicals (cycasins, macrozamins, etc.) and are most likely recognized by the Atala butterfly as basal host plants. Some of the chemicals in cycads, such as cycasin and macrozamin, have been analyzed in some cycad species and in a few herbivorous insects, but there is a lot still unknown about the complex relationship between the insects and the plants. Many of the cycad species have not yet been chemically analyzed and the analyses of others have been performed using different methods or testing different parts making comparisons difficult to interpret.

Beta-methylazoxymethanol (BMAA) is stored in the idioblasts of the leaves of some cycads [20,44]; herbivory by insects has been shown to have beneficial effects regarding enhanced growth patterns as an agonist of glutamate receptors which induces increased hypocotyl growth, as evidenced in *Arabidopsis*. Teas was the first to record the metabolic pathways in *Seirarctia echo* by which the neurotoxic Methylazoxymethanol (MAM) in the coontie host plant was transformed into less toxic cycasin via beta-glucosidase activity, thereby avoiding autotoxicity. It is thought that the same metabolic detoxification methods are used by other the cycad insect associates. There is a question as to whether the BMAA neurotoxin may have evolved before it developed as an herbivore deterrent [44]. Brenner et al. [44] questions if the idioblasts may have evolved to protect the other cells because of the high concentrations of the chemical.

Azoxyglucosides are not released until leaf tissue injury occurs, which may provide a clue as to why the neonate larvae appeared to seek out the chemical volatiles released when the cycad leaflets were cut to simulate herbivory. This release of volatile organic chemicals from herbivory may in fact serve as a field signal for female Atala that are searching for a suitable host plant. This may also help explain why Atala larvae must often be introduced several times into new colony sites before the adults persist at the new location (Koi unpublished). Methylazoxyglucosidemethanol (MAM) is released via herbivory or as enzymes in the digestive tract of herbivores [21], but BMAA contents have not been analyzed in any of the herbivores yet [28]. Castillo- Guevara and Rico-Gray conjectured that macrozamins are the most primitive of azoxyglucosides, and may have evolved as a primeval method of storing nitrogen. They mention that because the plants are very slow growing, this added protection would have proven a beneficial to prevent competition for space in crowded tropical regions where many cycads grow. The sarcotesta of the *Z. integrifolia* seeds contain a growth inhibitor (Taylor) and must break down over time, usually germinating the next season unless the flesh is consumed by wildlife (some animals are capable of utilizing the toxic flesh). Taylor also indicated that the butterfly larvae associated with the cycads may break down the germination-inhibiting sarcotesta on the female seed cones, helping the seeds sprout faster than they do otherwise, and are therefore helpful herbivores.

Castillo-Guevara and Rico-Gray show a negative relationship between the types of herbivory and azoxyglucoside content of the plants, regardless of whether the herbivore was a butterfly or moth, a beetle, weevil, or midge, a mealy bug or scale insect. It does not matter what kind of herbivory occurs, it does not predict the neurotoxin concentration in the cycad.

However, both larvae and adults preferred native *Z. integrifolia* over non-native *Z. vazquezii*. Growth and survival were arrested to some extent on the non-native cycad, and development was influenced as well, although there was no significant difference in the growth rates. There was a significant difference in the number of eggs laid on the different plants, indicating that females choose the native plant more often than non-native. It was interesting that the first generation adult females from the larvae that had chosen the non-native cycad as its larval food chose the native cycad for ovipositing.

In general, the butterflies reared on *Z. vazquezii* lived longer than those reared on the native host, although not significantly so. It could be that insects that are tough enough to survive on the non-native plant simply have a more robust constitution. The surface epidermis of *Z. vazquezii* is much thinner than that of native *Z. integrifolia*, requiring larvae to feed longer and consequently it took the larvae longer to acquire enough nutrients to successfully complete their life cycle to adult emergence (**FIGURE 13**).

There were two unsuccessful attempts to rear individuals from eggs that had been laid on native and non-native plants by removing the eggs from their substrate and switching them to the other plant. However, because the hatch rate was very low, these trials were not used in any analyses.

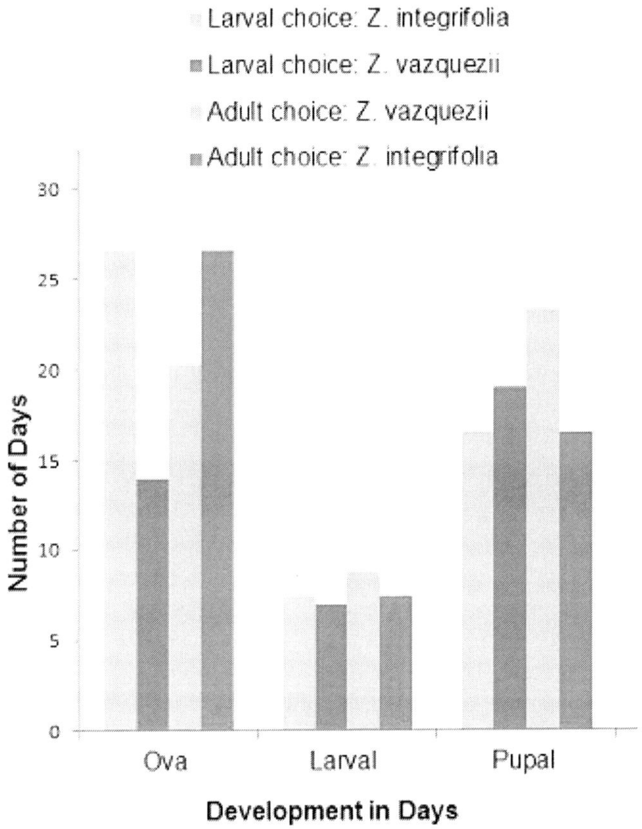

FIGURE 13: Comparisons of development time per life stage and adult lifespan between larval and adult choice trials. Difference was dependent on whether the adults or larvae chose the host plant, as well as which host plant was chosen (Photo credit: Koi S).

Switching from the pinched-off leaflet in Larval Test I to a clipped leaflet in Larval Test II was made based on observed behavior exhibited by the larvae in Larval Choice Test I that included apparent larval attempts to garner these volatiles from the atmosphere in the arena by rearing up on their prolegs and swaying back and forth with their "nose in the air," and then returning repeatedly to the egg, circling it and touching it (ostensibly looking for clues that it did not have because the eggs has been laid on non-plant substrates).

BENEFICIAL INSECT ASSOCIATIONS WITH CYCADS

There are some probable beneficial effects of insect-associated activity with cycads. Each cycad species has a mutualistic relationship with a specialist, sometimes obligate, pollinator weevil. Like butterflies, moths or other insects, including pests such as scale, they each play a part in the total ecology of the plants [1,2,15,19-21]. The co-relationships between the specialized insect herbivores and pollinators may be more recent than previously thought, as new molecular phylogenies show the relationships have co-evolved since the late Miocene [45].

Nagalingum et al. [45] notes that this may be one of the reasons for the low genetic diversity in the genera, which has made identification of the cycad species difficult [3,20,44,46-48]. It is thought now that the diversification of the six cycad genera may have first been triggered when the current continents were moved into their present locations as the tectonic plated shifted during the Miocene [45], although there is still some controversy [47]. A recent paper indicates that the seeds may have been ingested by large birds, mammals and other megafauna, and dispersed in clusters along with large amounts of 'fertilizer' in places some distance from the mother plant [49], explaining the growth habit of the plants, usually adapted to growing in high density. The large volume of frass left behind by small herbivores, especially butterflies and moths, acts as a fertilizer as well. Hall and Gimme [49] determined that opossums were primarily responsible for consuming the fleshy sarcotesta and dispersing the seeds of the Australian cycad *Macrozamia miquelii* (Zamiaceae). They found that few seeds were dispersed more than a meter from the maternal plant and that most juvenile seedlings did not establish within a few meters of the parent, even though most seeds fell beside the maternal plant. In Florida, squirrels are known to ingest the sarcotesta, but opossums and other wildlife may be responsible for dispersal of the seeds as well. Because of this growth habit, *Zamia* plants may be locally abundant, but rare; this is also true of *Z. integrifolia* in wild lands in Florida.

CONSERVATION CONCERNS: ENCOUNTERS BETWEEN WILDLIFE AND HUMANS

Both plants are pest-resistant under optimal conditions, but show signs of stress, such as scale infestations or sooty mold, when too crowded or poorly maintained. Recovery of the Atala butterfly from near-extinction is associated to the increased use of *Zamia* plants in landscaping, as well as increased conservation and restoration projects by biologists, botanists, park managers, scientists and concerned citizens. In 2008, Everglades National Park, for instance, grew additional coontie for the park from seed harvested four years prior in the park by volunteers, to prevent possible dilution of genetic material (Perry, pers. comm.). Most restoration endeavors include replanting native vegetation as well as removing non-native invasive plants.

Although it is generally accepted as good that the Atala has increased its range and distribution because of increased use of cycads for landscaping [24-27,42], the adult butterflies have expanded their oviposition choices during the past twenty years as well. That expansion includes many of the non-native introduced ornamental cycads found in south Florida's botanical and domestic gardens [22-27,42]. The larvae are able to successfully complete their life cycle on many, if not most, of these non-native plants [22-26].

This expansion into exotic Cycadales, many of which are extremely rare and valuable, has in turn increased potential struggles between home-owners, botanists, property managers and city planners with park managers, biologists, entomologists and conservationists as the Atala butterfly larvae attack these urban cycads [1,14,22,24-26]. Fortunately there have been no records of the Echo Moth using any of these non-native cycads as host plants at this time, although the moths do find their way to urban gardens occasionally.

Conservation methods are complicated because the often severe herbivory caused by the Atala larvae may damage the plant, but control of the larvae may impact the fragmented populations of the butterfly. The butterfly, like the coontie plant in the wild, is locally abundant but rare. The Atala is also currently listed as imperiled by the State of Florida, and the two natural ecosystems where coontie grows, the pine rocklands and tropical hardwood hammocks, are both now recognized as endangered (MDC undated; Myers). The question of how to address one endangered species' natural but possibly destructive interactions with another endangered species is a dilemma faced by conservationists worldwide [1,14,22].

In Eastern Oregon and Washington, for example, forestry managers face similar conflicts between balancing pest insect species, such as mosquitoes, while protecting scores of threatened insects that live in the tracts of forested area [50]. The paper points out that many insect species are ephemeral, and/or diminutive; that they are therefore difficult to monitor, challenging to find, may live in scattered fragmented habitats, but are nonetheless vital to overall health of the forest. Areas bordering urban localities are especially vulnerable; LaBonte et al. [50] point out the need for mosquito control spraying in recreational areas, controlled burns to maintain forest ecology, road building for access to locations within the forest, changes in soil chemistry from compaction and erosion, herbicide use to control unwanted invasive plant species; the possible extirpation of the *Polites mardon* butterfly and a flightless beetle, *Agonum belleri*, may have been caused because of these challenges.

Urban interface areas such as this become classic fields of controversy when conflicts between human wishes and endangered species are involved, in any country, in any state. The environmental impacts of pesticide use, herbicides and other chemical means of controlling pests while managing wildlife is extremely complex to manage [29-31].

CONCLUSION

HAND-MANAGEMENT AND IPM PRACTICES FOR ATALA BUTTERFLIES

Hand management practices must evolve as we learn more about the requirements of the Atala butterfly, including landscape architecture, specific nectar sources and host plants [42]. Recommendations involve a number of protocols, some of which have not yet been defined for this species. Pesticides to control unwanted insect pests such as mealy bugs should be used sparingly in butterfly gardens as non-target insects, such as Lepidoptera, are often impacted adversely by their use [30,51-53].

Planting native and non-native together may be another possible management protocol for protecting valuable cycads, as intercropping has been shown to be an effective pest management for conservation [40]. Because the Atala prefers native to non-native cycads, utilizing more native plants would be beneficial. One of the tricks to any successful butterfly garden is to move highly attractive host plants to areas where they will not be scrutinized by the public because complaints from people about herbivory is one of the biggest reasons the host plants get torn out of public venues [24,26,42]. It is a shame that rather than using the herbivory to teach the public about the insects, the plants are removed (often replaced by undesirable non-native vegetation). Atala butterflies use these pockets of coontie that are planted in public areas as stepping stones for dispersal [24,26,27,42,54], and it is certain that other butterflies need some form of connectively between fragmented suitable patches embedded within a concretized urban matrix.

Cultural management methods could be used to prevent herbivory, such as covering the most valuable plants while adult butterflies are active. A tomato cage covered with tulle mesh and buried into the ground an inch or so will prevent female ovipositioning on the plants. The adult Atala females will flatten their wings 'airplane style' to crawl into tight places, so the mesh must be securely closed on all sides and along the bottom.

Hand-removal of eggs on either *Z. integrifolia* or *Z. vazquezii* is easy with a thumb nail as they are not strongly adhered. The eggs can be frozen to humanely dispatch the embryos or simply left in the substrate where the ants and other insects will eat them. Although it can be time-consuming, hand-removal is certainly less expensive than buying a new cycad, especially since some valuable plants can cost thousands of dollars! If larvae hatch, they will most likely be unable to find food before dying as the larvae must find food within a few hours after eclosure or they will die.

Because it requires more time to develop on non-native *Z. vazquezii*, there is additional time for owners to remove the larvae from the plant. Contacting the local chapter of the North American Butterfly Association in southeast Florida, or the local Institute of Food and Agricultural Services (IFAS) Extension office Master Gardeners program will help find new 'foster homes' for unwanted or excess Atala larvae or pupae. Contacting the parks and natural areas management staff may welcome hosting this beautiful butterfly and be willing to take excess Atala. It is highly suggested not to carelessly introduce this butterfly outside of its native range (southeast Florida, south of Lake Okeechobee and east of the Everglades) [24,25]; the consequences of introducing even native species outside their normal range could have economically devastating results (a very destructive new herbivore attacking valuable nursery plants, for example). However, the butterfly is surviving in self-established colonies as far north as Martin County now, undoubtedly in response to changing climatic conditions as average temperatures increase.

AND LASTLY, A THOUGHT FROM THE WORLD-RENOWNED CYCAD BOTANIST, ROLF OBERPREILER: The natural insect fauna of cycads should not be brazenly dismissed as 'pests.' These insects are a natural and mostly vital component of the environment of the cycads, and their destruction can have severe impacts on the survival of the plants. The most obvious examples in this regard are the pollinators, but other insects may also play important roles in e.g., the disintegration of the cones and release of the seeds, the decomposition of old cones, leaves and stems, the recycling of nutrients, etc., their survival is so inextricably attached to that of the plants that extinction of their host plants will inevitably lead to the extinction of these insect species.

ACKNOWLEDGEMENTS

Staff, volunteers and students are sincerely thanked for help rearing stock in the laboratory. This work is part of the author's Master's thesis, funded by a research assistantship provided by the McGuire Center for Lepidoptera and Biodiversity, University of Florida, Gainesville, FL, USA.

REFERENCES

1. Oberprieler R (1995) The weevils (Coleoptera: Curculionoidea) associated with cycads, Proceedings, 3rd International Conference on Cycad Biology. Stellenbosch, Cycad Society of Pretoria, South Africa, pp: 295-365.
2. Chemnick J, Oberprieler R, Donaldson J, Terry I, Osborne R, et al. (2002) Insect pollinators of cycads. A report from a cycad pollination workshop held in Thailand, 2002 with a protocol for collecting and studying cycad pollinators. The Cycad Newsletter 27: 3-7.
3. Donaldson JS (ed.) (2003) Cycads. Status Survey and Conservation Action Plan. IUCN/SSC Cycad Specialist Group. International Union for Conservation of Nature and Natural Resources, Gland, Switzerland and Cambridge, UK, p: 86.
4. Cycads (2016) Official Newsletter of the IUCN SSC Cycad Specialists Group 1: 1-36.
5. http://www.iucnredlist.org/search
6. Thomas JA, Telfer MG, Roy DB, Preston CD, Greenwood JDD, et al. (2004) Comparative losses of British Butterflies, Birds and Plants and the Global Extinction Crisis. Science 303: 1879-1881.
7. Dirzo R, Young HS, Galetti M, Ceballos G, Isaac NJB, et al. (2014) Defaunation in the Anthropocene. Science 345: 401-406.
8. Ceballos G, García A, Ehrlich P (2010) The Sixth Extinction crisis loss of animal populations and species. J Cosmol 8: 1821-1831.
9. Ellis EC, Goldewijk KK, Siebert S, Lightman D, Ramankutty N (2010) Anthropogenic transformation of the biomes, 1700-2000. Glob Ecol Biogeogra 19: 5 89-606.
10. Gonzáles F (2004) Herbivoría en una gimnosperma endémica de Colombia, *Zamia encephalartoides* (Zamiaceae) por parte de *Eumaeus* (Lepidoptera:Lycaenidae). Revista de la Academia Colombiana de Ciencia 28: 233-243.
11. Coile NC, Garland MA (2003) Notes on Florida's endangered and threatened plants. Florida Department of Agriculture and Consumer Services, Division of Plant Industry. Bureau of Entomology, Nematology and Plant Pathology-Botany Section. Contribution No. 38 (4th edn.), p: 122.
12. Small JK (1921) Seminole Bread-The Conti. A history of the genus *Zamia* in Florida. Journal of the New York Botanical Garden 22: 121-137.
13. Oberprieler R (2004) Evil Weevils-the key to cycad survival and diversification. Proceedings, 6th International Conference on Cycad Biology, Chonburi, Thailand. Nong Nooch Tropical Garden, pp: 170-195.
14. Culbert DF (1994) An IPM approach for control of atala (*Eumaeus atala*) on Florida coonties (*Zamia floridana*). Proceedings of Florida State Horticulturalists107: 427-430.
15. Schneider D, Wink M, Sporer F, Lounibos P (2002) Cycads: their evolution, toxins, herbivores and insect pollinators. Naturewissenschaften 89: 281-294.
16. Klots AB (1951) A Field Guide to the Butterflies of North America, East of the Great Plains. Houghton Mifflin: Boston.
17. Rawson GW (1961) The recent re-discovery of *Eumaeus atala* (Lycaenidae) in southern Florida. J Lepidopterists' Soc 15: 237-244.
18. Deyrup M, Franz R (1994) Rare and Endangered Biota of Florida. Vol. IV. Invertebrates. In: Ashton RE (ed.) Jr. Series ed., University Press of Florida, Gainesville, p: 798.

19. Norstog KJ, Fawcett PKS (1989) Insect-cycad symbiosis and its relation to the pollination of *Zamia furfuracea* (Zamiaceae) by *Rhopalotria mollis* (Curculionidae). Am J Bot 76: 1380-1394.
20. Vovides AP (1991) Cone idioblasts of eleven cycad genera: morphology, distribution and significance. Botanical Gazette 152: 91-99.
21. Chaves R, Genaro J (2005) A new species of *Pharaxonotha* (Coleoptera: Erotylidae), probable pollinator of the endangered Cuban cycad, *Microcycas calocoma* (Zamiaceae). Insecta Mundi 19: 143-150.
22. Hubbuch C (1991) Wings of a dilemma: the Atala butterfly. Fairchild Tropical Botanic Garden Bulletin 46: 20-23.
23. Hammer RL (1996) New food plants for *Eumaeus atala* in Florida. News of the Lepidopterists' Society 38: 10.
24. Koi S (2013) Ecology and Conservation of *Eumaeus atala* Poey 1832 (Lepidoptera: Lycaenidae). MS thesis, University of Florida, FL, USA, p: 294.
25. Koi S, Daniels JC (2015) New and revised life history of the Florida hairstreak *Eumaeus atala* (Lepidoptera: Lycaenidae) with notes on its current conservation status. Florida Entomol 98: 1134-1147.
26. Koi S, Hall DW (2016) *Eumaeus atala,* Feature Creature, IFAS UF publication EENY-638.
27. Ramírez-Restrepo L, Koi S, MacGregor-Fors I (2015) Tales of Conservation in cities: The case of *Eumaeus* butterflies and their threatened cycad hostplants. Urban Ecosystems.
28. Prado A (2011) The cycad herbivores. Bulletin de la Société d'entomologie du QuébecAntennae 18: 3-6.
29. Pimentel D (2009) Pesticides and pest control. In: Integrated Pest Management: Innovation-Development Process. In: Peshin R and Dhawan A (eds.) Springer, Netherlands, pp: 83-87.
30. Pimentel D (2011) Environmental effects of pesticides on public health, birds and other organisms.
31. Pimentel D, Culliney TW, Bashore T (2013) Public health risks associated with pesticides and natural toxins in foods. In: Radcliffe EB, Hutchison WD, Cancelado RE (eds.) Radcliffe's IPM World Textbook, University of Minnesota, St. Paul, MN, Chemoecology 19: 117-154.
32. Vovides A (2010) *Zamia vazquezii*. The IUCN Red List of Threatened Species 2010: e.T42120A10655118.
33. Singer MC, Ng D, Thomas CD (1988) Heritability of oviposition preference and its relationship to offspring performance within a single insect population. Evolution 42: 977-985.
34. Mousseau TA, Dingle H (1991) Maternal effects in insect life histories. Ann Rev Entomol 36: 511-534.
35. Meagher RL, Nagoshi RN, Stuhl CJ (2011) Oviposition choice of two fall armyworm (Lepidoptera: Noctuidae) host strains. J Insec Behav 24: 337-347.
36. Mader BJ, Daoust SP, Cardinal-Aucoin M, Bauce E, Despland E (2012) Larval experience induces adult aversion to rearing host plants: a novel behavior contrary to Hopkins' host selection principle. Ecol Entomol 37: 204-211.
37. Singer MC (1983) Determinants of multiple host use by a phytophagous insect population. Evolution 37: 389-403.
38. Thompson JN, Pellmyr O (1991) Evolution of oviposition behavior and host preference in Lepidoptera. Ann Rev Entomol 36: 65-89.
39. Lindman L, Johansson B, Gotthard K (2013) Host plant relationships of an endangered butterfly, *Lopinga achine* (Lepidoptera: Nymphalidae) in northern Europe. J Insec Conserv 17: 375-383.
40. Floater GJ, Zalucki MP (2000) Habitat structure and egg distributions in the processionary caterpillar *Ochrogaster lunifer*: lessons for conservation and pest management. J App Ecol37: 87-99.

41. Dolek M, Freese-Hager A, Geyer A, Balletto E, Bonelli S (2013) Multiple oviposition and larval feeding strategies in Euphydras maturna (Linné , 1758) (Nymphalidae) at two disjoint European sites. J Insec Conserv 17: 357-366.
42. Koi S (2004) Establishing Naturally Sustainable Wild Populations of the *Eumaeus atala florida* Röber (Lycaenidae: Lepidoptera) and its Larval Host Plant, Coontie (Zamia species) within Historically Documented Florida Sites. National Fish and Wildlife Foundation, Budweiser Conservation Scholarship Program.
43. Hughes PR, Radke CD, Renwick JAA (1993) A simple, low-input method for continuous laboratory rearing of the monarch butterfly (Lepidoptera: Danaidae) for research. Amer Entomol 39: 109-111.
44. Brenner E, Stevenson D, Twigg R (2003) Cycads: evolutionary innovations and the role of plant-derived neurotoxins. TRENDS in Plant Science 8: 446-452.
45. Nagalingum N, Marshall C, Quental T, Rai H, Matthews S (2011) Recent synchronous radiation of a living fossil. Science 334: 796-799.
46. Vovides AP (1983) Systematic studies on the Mexican Zamiaceae. I. Chromosome numbers and karyotypes. Am J Bot 70: 1002-1006.
47. Nicolalde-Morejón F, Vergara-Silva F, González-Astorga J, Stevenson D (2011) Character-based, population-level DNA barcoding in Mexican species of *Zamia* L. (Zamiaceae: Cycadales). Mitochondrial DNA 21: 51-59.
48. Meerow AW, Francisco-Ortega J, Calonje M, Griffin MP, Ayala-Silva T, et al. (2012) Zamia (Cycadales: Zamiaceae) on Puerto Rico: Asymmetric genetic differentiation and the hypothesis of multiple introductions. Am J Bot 99: 1828-1839.
49. Hall JA, Gimme WH (2013) Seed dispersal of the Australian cycad *Macrozamia miquelii* (Zamiaceae): Are cycads megafauna-dispersed "grove forming" plants? Am J Bot 100: 1127-1136.
50. LaBonte JR, Scott DW, McIver JD, Hayes JL (2001) Threatened, endangered, and sensitive insects in eastern Oregon and Washington forests and adjacent lands. Northwest Sci 75: 185-198.
51. Salvato M (2001) Influence of mosquito control chemicals on butterflies (Nymphalidae, Lycaenidae, Hesperiidae) of the Lower Keys. J Lepidopterist's Soc 55: 8-14.
52. Hoang TC, Pryor RL, Rand GM, Frakes RA (2011) Use of butterflies as nontarget insect test species and the acute toxicity and hazard of mosquito control insecticides. Environ Toxicol Chem 30: 997-1005.
53. Bargar TA (2012) Risk assessment for adult butterflies exposed to the mosquito control pesticide naled. Environ Toxicol Chem 31: 885-891.
54. Lollar K (2004) Butterfly comes back from near extinction. Atalas being introduced in Florida.

CHAPTER 13

A Review of the Biology of Eucharitidae (Hymenoptera: Chalcidoidea) from Argentina

Javier Torréns

CRILAR-CONICET, Entre Ríos y Mendoza, 5301 Anillaco, La Rioja, Argentina

ABSTRACT

All the members of Eucharitidae are parasitoid of ants. Argentina has 14 genera and 41 species, but little is known about their biology. Herein are provided new data for host associations (host ant and/or host plant) of Galearia latreillei, Kapala spp., Latina rugosa, Orasema aenea, and Orasema sp. A revision of the most relevant biological aspects of Dicoelothorax platycerus, Latina rugosa, Neolirata alta, N. daguerrei, Lophyrocera variabilis, Orasema argentina, O. salebrosa, O. simplex, O. susanae, O. worcesteri, and O. xanthopus is included. New records of K. sulcifacies, Lo. plagiata, and Ob. semifumipennis in Argentina are presented. Galearia proseni is synonymized with G. latreillei.

1. INTRODUCTION

Eucharitidae parasitize the immature stages of Formicidae and are among the most diverse hymenopteran parasitoids of eusocial insects [1–8]. Females are oviparous and pro-ovigenic and lay their eggs inside or on plant tissues, either individually or in masses. They oviposit away from the host, with the active first instar larva (planidium) responsible for getting into the ant nest through various associations with foraging adult ants [9]. Once in contact with the larval ant host, the planidium either remains as an external parasite or burrows into the host. Upon pupation of the host, the larva migrates to the ventral region of the thorax, just posterior to the legs of the newly formed pupa, then resumes development through two additional instars [10, 11]. The adults emerge and leave the nest on their own or may be carried by the ants and deposited in the accumulation of colony waste [10, 12, 13].

Eucharitidae are present in all zoogeographic regions but most abundant in the tropics [8]. Fifty-four genera and about 420 species worldwide have been described. In Argentina, 14 genera and 41 species have been reported [8, 14–16].

Eucharitidae were well studied in a series of early taxonomic papers by Gemignani [17–19]; however, very little information was provided on their biology. This paper reviews our current understanding and contributes new data for some of the Argentinean species.

2. MATERIALS AND METHODS

Females were collected by sweep netting and provided twigs with leaves, fruits, and flowers of different species of plants in 10 × 3.5 cm plastic tubes to monitor oviposition habits. Host plants with eggs extracted from the field or oviposited by the captive females were placed into a cylindrical glass container of 10 × 10 cm with dampened cotton until emergence of the first instar (planidium).

Ant nests with adults, brood, and debris were collected into plastic containers. Adults and immature stages were then sorted from the debris, examined for parasitism, and in some cases returned to the containers to allow further development of immature ones. The immature stages were examined once daily until all parasitoids or ants emerged from the cocoons. In the cases where both parasitoid sexes emerged, they were put together in a cylindrical glass container of 10 × 10 cm containing different types of plants to allow for oviposition after mating.

A Leica MZ12 stereomicroscope was used for observations. Images were obtained using GT-Vision Ento-Vision software operating on a Leica M16 zoom lens linked to a JVC KY-F75U 3-CCD digital video camera or Leica Application Suite (version 3.5.0) software operating on a Leica MZ12 linked to a Leica DFC295 digital video camera. Images were enhanced with Corel Photopaint and Corel Draw (version 15). Some images were processed using Deep Focus (Stuart Ball).

The biogeographical distribution and classification of ecoregions in Argentina was taken from Morrone [20] and Bertonatti and Corcuera [21]. Geographic coordinates for eucharitid localities were estimated using Google Earth (version 6.2.2.6613).

3. GENERA AND SPECIES OF EUCHARITIDAE FROM ARGENTINA

Two of the four subfamilies of Eucharitidae are represented in Argentina, Oraseminae and Eucharitinae (Table 1). Oraseminae is represented only by Orasema Cameron. The Eucharitinae are comprised of 12 genera of Eucharitini with a dubious record of Psilocharis Heraty (Psilocharitini) from Déan Funes (Córdoba) [5].

3.1. Dicoelothorax Ashmead

This genus includes two species distributed in the Neotropical region: D. parviceps Cameron (Argentina, Brazil, Colombia, and Guyana) and D. platycerus Ashmead (Argentina, Bolivia, and Brazil) (Figures 1(a) and 1(b)) [8, 14, 22]. Biological information is only available for D. platycerus [22].

3.1.1. Dicoelothorax platycerus Ashmead

Habitat and location are as follows. Specimens were collected in San Vicente (Tucumán). The vegetation of this region is characterized by dry forests, dominated by deciduous, spiny, and small-leaves plants typical of the Chaco ecoregion [40] (Figures 1(c) and 1(d)). The host plant, Pseudabutilon virgatum (Cav.) (Malvaceae), is a ligneous shrub that occurs throughout the area and persists year round.

Life history and host ants are next. A single gravid female oviposited about 40 eggs per 1 mm^2 on the underside of leaves (Figure 1(e)), and eggs hatched within 10 days. First instars (planidia) (Figure 1(f)) are mobile and have a propensity to jump; larvae presumably attach phoretically to foraging ants under the host plant and get carried back to the ant nest where they attack the ant larvae [3]. Of two pupae of D. platycerus obtained from the host ants nest one male emerged 12 days after the nest was excavated, whereas the other pupa (female) did not emerge (Figure 2(d)).

Table 1: List of species of Eucharitidae in Argentina. Known biology is indicated for host ants (HAs), host plants (HPs), or immature stages (ISs).

Subfamilies/tribes/genera	Species	Biology	References
Eucharitinae			
Psilocharitini			
Psilocharis Heraty	*Psilocharis* sp.*	?	—
Eucharitini			
Colocharis Heraty	*Colocharis huggi* Torrèns	?	—
Dicoelothorax Ashmead	*Dicoelothorax parviceps* Cameron	?	—
	Dicoelothorax platycerus Ashmead	HP, HA, IS	[22]
Dilocantha Shipp	*Dilocantha bennetti* Heraty	?	—
	Dilocantha flavicornis (Walker)	?	—
Galearia Brullé	*Galearia latreillei* (Guérin-Méneville)	HP, HA, IS	***
	Kapala argentina Gemignani	?	—
	Kapala chacoensis Gemignani	?	—
Kapala Cameron	*Kapala furcata* (Fabricius)	HP	[2, 3]
	Kapala splendens Ashmead	?	—
	Kapala sulcifacies (Cameron)**	HP, IS	[23, 24]
Latina Koçak & Kemal	*Latina bonariensis* (Gemignani)	?	—
	Latina rugosa (Torrèns, Heraty & Fidalgo)	HP, HA, IS	[25], HA***
	Latina vianai (Gemignani)	?	—
Lophyrocera Cameron	*Lophyrocera daguerrei* (Gemignani)	?	—
	Lophyrocera plagiata (Walker)**	?	—
	Lophyrocera variabilis Torrèns, Heraty & Fidalgo	HP, HA, IS	[26]
	Neolirata alta (Walker)	HP, IS	[15]
Neolirata Torrèns & Heraty	*Neolirata daguerrei* (Gemignani)	HP, IS	[15]
	Neolirata furcula Torrèns & Heraty	?	—
Obeza Heraty	*Obeza maculata* (Westwood)	?	—
	Obeza nigriceps (Ashmead)	?	—
	Obeza semifumipennis (Girault)**	?	—
Parakapala Gemignani	*Parakapala decarloi* Gemignani	?	—
	Pseudochalcura alba Heraty & Heraty	?	—
	Pseudochalcura americana (Howard)	?	—
Pseudochalcura Ashmead	*Pseudochalcura frustrata* Heraty	?	—
	Pseudochalcura pauca Heraty	?	—
	Pseudochalcura prolata Heraty	?	—
Thoracantha Latreille	*Thoracantha spegazzinii* (Gemignani)	HP	[17]
	Thoracantha striata Perty	HP, IS	[8]
Oraseminae			
	Orasema aenea Gahan	HP, HA, IS	***
	Orasema argentina Gemignani	HA	[6, 17]
	Orasema deltae Gemignani	?	—
	Orasema freychei (Gemignani)	?	—
	Orasema gemignanii De Santis	?	—
Orasema Cameron	*Orasema salebrosa* Heraty	HA	[11, 27]
	Orasema simplex Heraty	HA, HP	[11, 28, 29]
	Orasema susanae Gemignani	HA	[6]
	Orasema vianai Gemignani	?	—
	Orasema worcesteri (Girault)	HA	[17]
	Orasema xanthopus (Cameron)	HP, HA, IS	[6, 11, 27, 28, 30–39]

Abbreviations: *doubtful record [5]; **new record of presence in Argentina; ***new biological record.

Figure 1: Dicoelothorax platycerus: (a) habitus (female); (b) habitus (male). Biology and immature stage of D. platycerus: (c) habitat; (d) Pseudabutilon virgatum; (e) underside of leaf of P. virgatum with eggs; (f) planidium (dorsal and ventral); (g) nest entrance of Ectatomma brunneum (opening indicated). Figures extracted from [22].

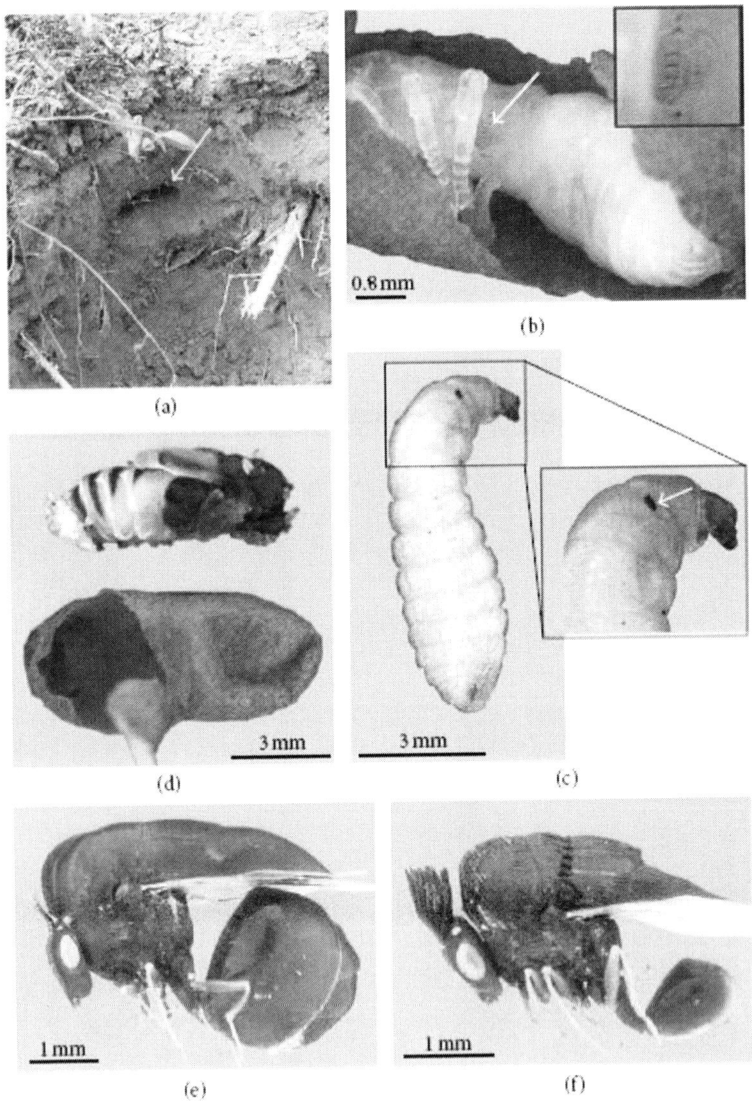

Figure 2: Biology and immature stage of D. platycerus: (a) brood chamber (indicated); (b) prepupa parasitized (2nd instar larva indicated and magnified); (c) ant larva parasitized (attached planidium magnified and indicated); (d) pupa extracted with ant cocoon (female, lateral). Galearia latreillei: (e) habitus (female); (f) habitus (male). Figures 2(a), 2(b), 2(c), and 2(d) are extracted from [22].

Ectatomma brunneum F. Smith (Ectatomminae) workers were observed and sampled from under the plants with Dicoelothorax. Of three ant nests found, immature ones were in two of them (H1 and H2). The disposition of chambers and general structure of nests are similar to those observed by Lapola et al. [41] (Figures 1(g) and 2(a)). Nest H1 contained 17 cocoons and 2 ant larvae, and nest H2 had 97 ant larvae and no cocoons. The percentage of parasitism ranged from 6.2% in H2 to 21% in H1. Of the 17 cocoons (H1) recovered, there were two cocoons each with one pupae of D. platycerus (1 female and 1 male) and 2 ant prepupae parasitized by second instars of D. platycerus (Figure 2(b)). In nest H2, 6 of the larvae were parasitized by externally located planidia (Figure 2(c)).

3.2. Galearia Brullé

The genus is comprised of two species, G. latreillei (Guérin-Méneville) and G. proseni Gemignani. Heraty [8] argued that the Argentinean male described as G. proseni by Gemignani [19] was likely the male of G. latreillei (Figures 2(e) and 2(f)). Based on the morphological similarity of a reared male with G. proseni (=Pseudokapala proseni), and its subsequent mating with a female of G. latreillei, infer that the suggestion by Heraty is correct and propose here a new synonymy of G. proseni with G. latreillei. The one species has a widespread Neotropical distribution, being present in Argentina, Bolivia, Brazil, and Venezuela [8, 14].

The only known biological record was from Gemignani [17] in which he mentioned that an adult of G. latreillei (=Thoracantha latreillei) was collected from the waste pile of a nest of Pogonomyrmex cunicularius Mayr (=P. carnivora), but this ant association is likely invalid [8].

Galearia latreillei was collected in northcentral and northwestern Argentina, and information on life history, immature stages, and host association is included.

3.2.1. Galearia latreillei (Guérin-Méneville)

Habitat and locations are as follows. Specimens were collected in Cabeza de Buey (Salta), Campo Gallo, Suncho Corral, and Tintina (Santiago del Estero). The Cabeza de Buey locality consists of mixed yunga (humid mountain forest) and xeric lowland Chaco vegetation. In the two localities in Santiago del Estero, located in the center and north of the province, the vegetation is typical of the chaco ecoregion (Figure 3(a)). The host plant, Sida cordifolia L. (Malvaceae), is a perennial, herbaceous plant with stems that are yellow-green, hairy, long, and slender, and their leaves are oblong-ovate, covered with hairs (Figure 3(b)).

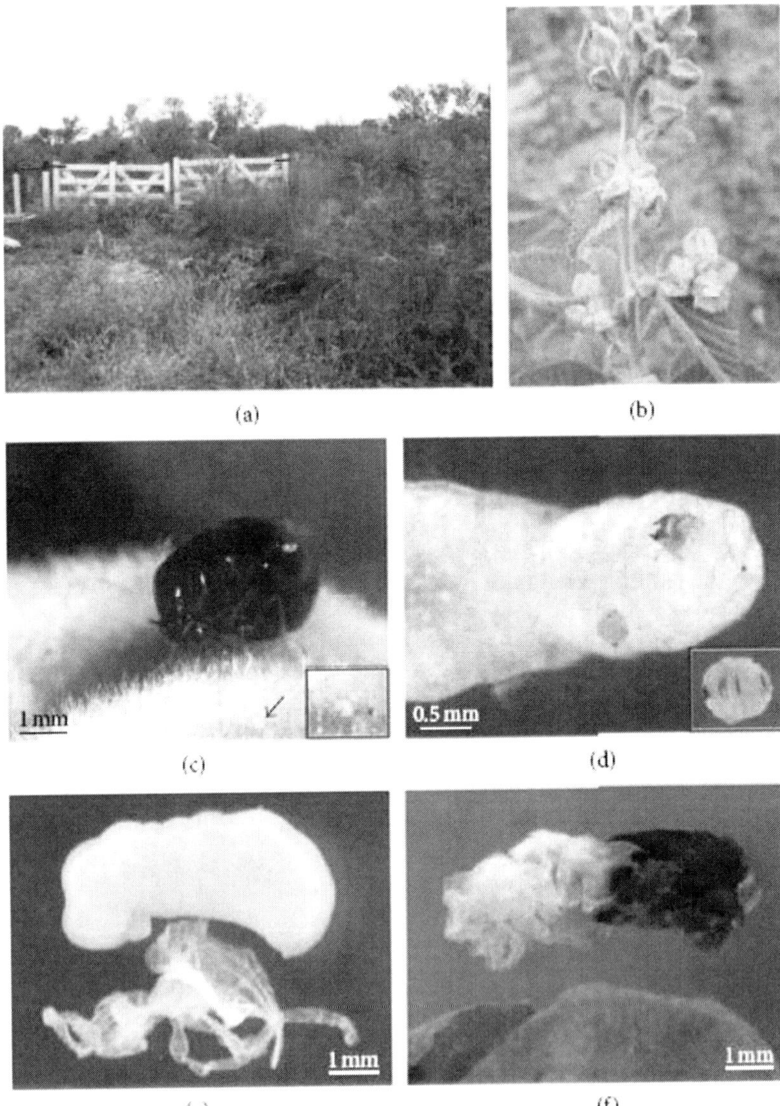

Figure 3: Biology and immature stage of Galearia latreillei: (a) habitat; (b) Sida cordifolia; (c) female of G. latreillei ovipositing on leaf of S. cordifolia (eggs indicated and magnified); (d) prepupa parasitized (1st instar larva indicated and magnified); (e) third instar (with remains of ant pupa); (f) pupa extracted from ant cocoon (male, lateral).

Life history and host ants are next. Both sexes of G. latreillei were obtained from a nest of Ectatomma brunneum. The adult wasps were together for two days before mating occurred. The female then oviposited about 400 eggs that were dispersed among the spicules forming the pubescence on the stem of S. cordifolia near to the leaves or in the underside of leaves near the base (Figure 3(c)). Eggs hatched within 11 days. The planidia were very mobile and had a propensity to jump.

Nests of Ectatomma brunneum were excavated from near to the host plant, with immature ones found at a depth of 6 to 8 cm. From 50 cocoons, we extracted 10 pupae of G. latreillei. One male and one female emerged about 4 days after the nest was excavated, whereas the other pupae did not emerge (Figure 3(f)). Three other cocoons yielded one second-instar and two third-instars (Figures 3(d) and 3(e)). Of the 50 cocoons recovered, 13 were attacked giving a percentage parasitism of 26%.

Discussion. Ectatomma brunneum has also been reported as the ant host for Dicoelothorax platycerus [22] and for an unidentified species of Kapala Cameron (Eucharitidae: Eucharitini) in French Guiana [42]. Similarly, another species of the same ant genus, E. tuberculatum (Olivier), is known to be attacked by three different eucharitid genera, Dilocantha Shipp, Isomerala Shipp, and Kapala [43].

3.3. Kapala Cameron

Kapala includes 16 species, but there are many undescribed species in the Neotropical region. It is widespread and diverse in both the Nearctic and Neotropical regions and also includes one widespread afrotropical species, Kapala ivorensis Risbec [8].

In Argentina, 5 species were recorded: K. argentina Gemignani, K. chacoensis Gemignani, K. furcata (Fabricius), K. splendens Ashmead, and K. sulcifacies (Cameron) [8, 14]. Partial biological information is available for K. furcata and K. sulcifacies (summarized later). New data is also added for two unidentified species.

3.3.1. Kapala furcata (Fabricius)

This species was observed ovipositing on floral buds of Mikania sp. (Asteraceae) [2] that were infested with aphids [3].

3.3.2. Kapala sulcifacies (Cameron)

This species has been reported as ovipositing in floral buds of Cordia curassavica (Jacq.) Roem. & Schult. (Boraginaceae) (=Cordia macrostachya), Gossypium hirsutum L. (Malvaceae), and in a flowering asclepiad [23], with eggs laid in clusters of 200–300 eggs [24].

3.3.3. Kapala spp.

A species sampled in Campo Gallo (Santiago del Estero) oviposited into flower buds of Sphaeralcea bonariensis (Cav.) Griseb. (Malvaceae), with the planidia emerging 9 days after oviposition. Another species was collected in Rosario de la Frontera (Salta) over an unidentified Sapindaceae, but no oviposition was observed.

3.4. Latina Koçak and Kemal

Latina (=Laurella Heraty) includes four species distributed in the Neotropical region: Latina bonariensis (Gemignani) (Argentina), L. guriana (Heraty) (Venezuela), L. rugosa (Torréns, Heraty and Fidalgo) (Argentina) (Figures 4(a) and 4(b)), and L. vianai (Gemignani) (Argentina) [8, 25].

Latina rugosa was collected in northwestern Argentina and the taxonomic and biological aspects provided by Torréns, Heraty, and Fidalgo [25].

3.4.1. Latina rugosa (Torréns, Heraty, and Fidalgo)

Habitat and location are as follows. Specimens were collected at Rosario de la Frontera (Salta); the collection site was a forest of Piptadenia macrocarpa Benth. (Cebil Colorado) (Fabaceae). The vegetation of this region corresponds to the Yungas and Chaco ecoregions [40]. The host plants, Serjania glabrata (Sapindaceae), are perennial shrubs with pubescent and serrated leaves, with the plants dispersed between trees in the collection area (Figures 4(c) and 4(d)).

Life history and host ants are next. Adults of L. rugosa were collected in the same location, mainly close to or on the host plant. A single gravid female oviposited about 25 eggs per 1 mm^2 on the underside of leaves (Figure 4(d)). Eggs hatched within 6 days. The planidia (Figure 4(e)) were mobile and able to jump.

Figure 4: Latina rugosa: (a) habitus (female); (b) habitus (male). Biology and immature stage of L. rugosa: (c) habitat; (d) underside of leaf of Serjania glabrata with eggs (eggs represented in white area and magnified); (e) planidium (dorsal and ventral view); (f) ant larva parasitized (attached planidia magnified and indicated). Figures 4(d) and 4(e) are extracted from [25].

Odontomachus chelifer (Latreille) (Ponerinae) workers were observed and collected under the host plants from which L. rugosa were collected. One O. chelifer nest was identified only by a small ground opening. The ant nests were excavated and the cocoons and ants larvae extracted at a depth of 16 cm; however, the nest appeared to be much deeper, and it was difficult to tell whether the entire brood was extracted. Of the five ant larvae extracted, one had three planidia externally attached (Figure 4(f)), while of the 19 cocoons only one planidium was found attached externally to a prepupa. From this sample, the percentage of parasitism was 8.3% of 24 immature ones.

Discussion. Data presented here confirm the ant host association of Latina rugosa as Odontomachus chelifer. This ant genus is also the host of other eucharitids genera as Ancylotropus Cameron, Chalcura Kirby, Schizaspidia Westwood, and Kapala Cameron [8, 42].

3.5. Lophyrocera Cameron

Lophyrocera Cameron includes seven species distributed across South and Central America and the western United States (Neotropical and Nearctic): L. apicalis Ashmead (USA), L. daguerrei (Gemignani) (Argentina), L. chilensis (Brèthes) (Chile), L. plagiata (Walker) (Argentina and Brazil), L. pretendens (Walker) (Brazil), L. stramineipes Cameron (Panama), and L. variabilis Torréns et al. (Argentina) (Figures 5(a)–5(c)) [8, 14, 26].

Lophyrocera variabilis was collected in northwestern Argentina, with information available on life history, immature stages, and host association [26].

3.5.1. Lophyrocera variabilis Torréns, Heraty, and Fidalgo

Habitat and location are as follows. The habitat consists of mixed Yungas and Chaco vegetation in Los Chorrillos (Tucumán) (Figure 5(d)). The host plant, Vassobia breviflora (Sendtn) Hunz. (Solanaceae), common name "Chalchal de la gallina", is a spiny shrub with globe-shaped fruits, which are red in color when mature (Figure 5(e)) [44].

Life history and host ants are next. Females were observed ovipositing in the immature (green) fruit of V. breviflora, with eggs deposited in small masses within the fruit (Figure 5(f)). Only undeveloped eggs were obtained from immature fruits while mature fruits taken from the ground had mature eggs and larvae. The planidia (Figure 5(g)) crawl and do not have the ability to jump.

Figure 5: Lophyrocera variabilis: (a) and (b) habitus (female); (c) habitus (male). Biology and immature stages of L. variabilis: (d) habitat; (e) Vasobia breviflora; (f) saggittal section of unripe fruits of V. breviflora with eggs (egg mass magnified); (g) planidium (before hatching; dorsal and ventral view). Figures 5(a)–5(c), 5(f), and 5(g) are extracted from [26].

In the field, a species of Camponotus Mayr (Formicinae: Camponotini) visited and foraged below the host plant. Camponotus are known to collect fruit pulp and small seeds [45], and a direct interaction of foragers with the ripe fruit and planidia is very likely, as proposed for Pseudochalcura [9].

Nests of Camponotus were located under host plants or within a few meters (Figure 6(a)). In total, 35 Lophyrocera pupae were found in 7 of the 13 nests excavated, and of these, three had two pupae of L. variabilis in the same cocoon (Figure 6(b)). No larvae were found. The parasitism rate ranged from 0 to 6.21%.

Figure 6: Biology and immature stage of Lophyrocera variabilis: (a) ant nest of Camponotus sp.; (b) two Lophyrocera pupae extract from same ant cocoon (females). Neolirata alta: (c) habitus (female). Biology and immature stage of N. alta: (d) habitat; (e) underside of leaf of P. virgatum with eggs (magnified area with eggs); (f) planidia. Figure 6(b) is extracted from [26], and Figures 6(c), 6(d), 6(e), and 6(f) are extracted from [15].

3.6. Neolirata Torréns and Heraty

This genus includes three species distributed in the Neotropical region: N. alta (Walker) (Argentina, Brazil, and Uruguay) (Figure 6(c)), N. daguerrei (Gemignani) (Argentina, and Brazil) (Figures 7(a) and 7(b)), and N. furcula Torréns and Heraty (Brazil) [15].

Figure 7: Neolirata daguerrei: (a) habitus (female); (b) habitus (male). Biology and immature stage of N. daguerrei: (c) habitat; (d) underside of leaf of Urvillea chacoensis with eggs (eggs represented in white area and magnified); (e) egg; (f) planidia. Figures extracted from [15].

Neolirata alta and N. daguerrei were collected in northwestern Argentina, and their taxonomic and biological information is given in Torréns and Heraty [15].

3.6.1. Neolirata alta (Walker)

Habitat and location are as follows. Specimens were collected in Los Baños and Rosario de la Frontera (Salta) and Tapia and San Vicente (Tucumán). In Los Baños, the vegetation corresponds to the transition Yungas and Chaco ecoregions, while the others are typical of the Chaco ecoregion (Figure 6(d)). The host plant, Pseudabutilon virgatum, was widely distributed in all four areas (Figure 6(e)).

Life history and host ants are next. The female oviposited about 32 eggs per mm^2 at random between the spicules on the underside of a leaf (Figure 6(e)). Eggs hatched within 14 days. The planidia (Figure 6(f)) were mobile and have the ability to jump.

The host ant remains unknown. A few meters from where the female was collected in San Vicente (Tucumán), there was a nest of Ectatomma brunneum. This nest was excavated, but no immature stages were found.

3.6.2. Neolirata daguerrei (Gemignani)

Habitat and location are as follows. Most specimens were collected in Tapia (Tucumán) (Figure 7(c)); the vegetation corresponds to the Chaco ecoregion [40]. The host plant, Urvillea chacoensis Hunz. (Sapindaceae), is a climbing vine distributed throughout the collection area; its leaves are marginally serrate and pubescent [46] (Figure 7(d)).

Life history and host ants are next. Females were observed ovipositing on the underside of leaves of U. chacoensis. A single gravid female oviposited about 28 eggs per mm^2 (Figure 7(d)). Eggs hatched within 9 days (Figure 7(e)). Planidia (Figure 7(f)) are very mobile and jump.

Host ant unknown.

3.7. Orasema Cameron

Orasema is composed of 57 species, but many are still undescribed. Their distribution is Neotropical, Nearctic, and Paleotropical [8]. In Argentina, Orasema is widely distributed, with 11 species documented: O. aenea Gahan, O. argentina Gemignani, O. deltae Gemignani, O. freychei (Gemignani), O. gemignanii De Santis, O. salebrosa Heraty, O.

simplex Heraty, O. susanae Gemignani, O. vianai Gemignani, O. worcesteri (Girault), and O. xanthopus (Cameron).

Several authors have studied the biological aspects of Orasema [5, 7, 11, 27–29, 47–56]. Among the genera of ants recorded as attacked by Orasema are Formica Linnaeus, Monomorium Mayr, Pheidole Westwood , Solenopsis Westwood, Temnothorax Mayr, Tetramorium Mayr, Wasmanni a Forel, and a dubious case of Eciton Latreille [5–8, 11, 27, 47, 54]. Immature stages were described by several authors [5–8, 11, 24, 27, 47, 48, 53, 54, 56, 57].

Herein are summarized the most relevant data on the biology of the species found in Argentina, with new data for O. aenea and some records of an unidentified species.

3.7.1. Orasema aenea Gahan

Habitat and location are as follows. Specimens of O. aenea (Figures 8(a) and 8(b)) were collected in Caimancito (Jujuy). The vegetation and geographic location corresponds to the foothills of the Yungas ecoregion. The host plant, Tecoma stans (L.) Juss. ex Kunth (Bignoniaceae) (common name, Guarán amarillo), is a shrub or small tree that grows 3–6 m tall, with leaves decussate with elliptic-lanceolate and serrated edges, and it blooms from August to October [40] (Figure 8(c)).

Life history and host ants are next. Females were observed ovipositing on the undersides of leaves of T. stans by creating an incision and laying a single egg in short linear rows (Figures 8(d) and 8(e)). Eggs hatched within 9 days. Planidia (Figure 8(f)) crawl and leave the incision but do not have the ability to jump.

Although host ants were not located in the area, the host has been reported as Solenopsis quinquecuspis Forel (Myrmicinae) [27].

Discussion. Plants used for oviposition also include Ilex paraguayensis A.St.-Hil. (Aquifoliaceae) (Yerba Mate) and Olea europapaea L. (Oleaceae) (Olive) for which Orasema is considered as a potential pest [50, 57]. Orasema aenea was found on both T. stans and Vaccinium corymbosum L. (Ericaceae) (blueberry), with the latter association recorded by Varone and Briano [29].

Figure 8: Orasema aenea: (a) habitus (female); (b) habitus (male). Biology and immature stage of O. aenea: (c) habitat; (d) female of O. aenea ovipositing on leaf of Tecoma stans; (e) underside of leaf of T. stans with incisions (magnified area with egg extracted from incision); (f) planidia (lateral and ventral).

3.7.2. Orasema argentina Gemignani

It is associated with Pheidole nitidula Emery (Myrmicinae) [6, 17].

3.7.3. Orasema salebrosa Heraty

It is associated with Solenopsis invicta Buren, S. macdonaghi Santschi and S. richteri Forel (Myrmicinae) [11, 27].

3.7.4. Orasema simplex Heraty

It is associated with Solenopsis richteri, S. invicta, and S. quinquecuspis (Myrmicinae) [11, 27]. Varone and Briano reported Zea mays L., Glycine max L., Vinca rosae L., Citrus limon (L.) Burn, Capsicum annuum L., Smilax campestris Griseb, Paspalum unispicatum (Scribn. & Merr.)
Nash, P. denticulatum Trin., P. notatum Fluegge, P. dilatatum Poir, Grindeli a pulchella Dann., Stevia aff. entreriensis Hieron, Eupatorium aff. laevigatum L., Sesbania virgata (Cav.) Pers., Asclepias curassavica L., Verbena montevidensis Spreng., Sida rhombifolia L., and Stemodia aff. lanceolata Benth. with oviposition marks of Orasema simplex in nonchoice laboratory tests and in field surveys [29].

3.7.5. Orasema susanae Gemignani

It is associated with Pheidole near tetra Creighton [6].

3.7.6. Orasema worcesteri (Girault)

It is associated with Pheidole radoszkowskii Mayr (Myrmicinae) (=P. nitidula) [17].

3.7.7. Orasema xanthopus (Cameron)

Various aspects related to its biology were recorded by several authors [11, 28, 30–39]. Orasema xanthopus is associated with several species of Solenopsis, such as S. invicta [11, 27, 32–35, 39, 58], S. quinquecuspis [27], S. richteri [33], and the S. saevissima (Smith) complex [6, 11, 35].

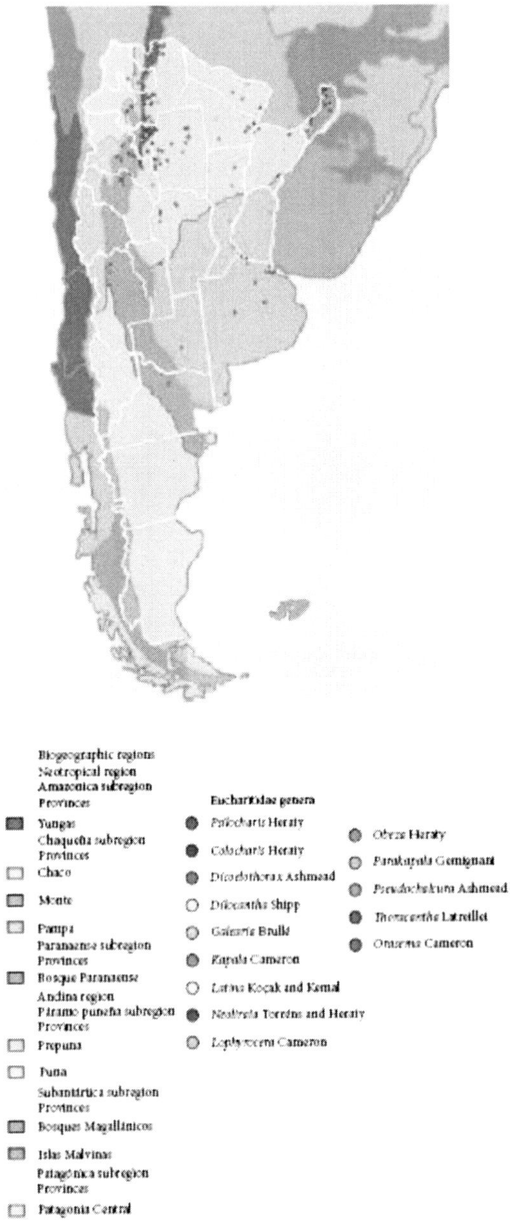

Figure 9: Distribution of genera of Eucharitidae in Argentina.

3.7.8. Orasema sp.

Several females were collected in Villa Vil (Catamarca) ovipositing into the stem tissue below the flower buds and along the petiole and midrib of leaves of Lantana xenica Moldenke (Verbenaceae).

3.8. Thoracantha Latreille

This genus is comprised of three species, Thoracantha anchura Walker (Brazil), T. spegazzinii (Gemignani) (Argentina), and T. striata Perty (Argentina and Brazil) [8].

3.8.1. Thoracantha spegazzinii (Gemignani)

A single female was collected on a flower of a Malvaceae. This data was included on the holotype label but not used in the original description of the species by Gemignani [17].

3.8.2. Thoracantha striata Perty

Heraty observed females ovipositing in patches on the underside of leaves of Lantana sp. (Verbenaceae); oviposition took place over 1-2 hours. Eggs and planidia were obtained [8].

4. CONCLUSION

Eucharitidae are found in almost all biogeographic regions in northern Argentina (Figure 9). Most genera are distributed in the Chaco ecoregion and the transition between Chaco and Yungas, but more surveys are necessary in the Monte and Pampa ecoregions, and in those it was areas never surveyed for eucharitids including Central Patagonia and Prepuna. Herein we presented a new record for Kapala sulcifacies (Cameron) from Salta (Rosario de la Frontera, 20/03/2003; one female and two males deposited in Instituto Fundación Miguel Lillo, Tucumán, Argentina), Lophyrocera plagiata (Walker) from Misiones (Mado, Puerto Magdalena, 23/10/1964; one female deposited in American Museum of Natural History, NY, USA) and Obeza semifumipennis (Girault) from

Formosa (Pirané, 31/12/1948; two males deposited in Instituto Fundación Miguel Lillo, Tucumán, Argentina).

Although we have detailed information for most genera, little or nothing is known about the biology of many species. Host relationships were summarized by Heraty [8], Lachaud and Pérez-Lachaud [59], and Lachaud et al. [42]. Herein we presented a new host association for Galearia latreillei from Ectatomma brunneum and confirm the association of Latina rugosa with Odontomachus chelifer suggested by Torréns et al. [25]. Of the remaining genera present in Argentina, ant host relationships can be inferred from species found elsewhere in South and Central America. Generally, it is expected that in Argentina, Orasema (Oraseminae) are exclusively found on Myrmicinae, the genera Lophyrocera, Obeza, and Pseudochalcura attack Camponotini (Formicinae), and the remaining genera in the Kapala clade all attack either Ectatomminae or Ponerinae [8, 9, 11, 12, 23, 26, 27, 29, 42, 43, 59].

Eucharitidae utilize a variety of distinct methods for oviposition. Oraseminae oviposit into incisions made in leaf tissues [2, 5, 49, 52]. Damage to the leaves can be caused by scaring of the plant tissue [7] or through secondary infections caused by the punctures [50]. Because of this, Orasema have been considered as potential pests of banana, citrus, olive, tea, and yerba mate [30, 31, 48, 50, 52, 53]. However, they are never regarded as a continuing pest problem. In contrast, Eucharitinae oviposit either on the undersides of the leaves, into fruits or into the bracts of flower buds, without causing cosmetic damage to the plants. However, as parasitoids of Ectatomma, they might have a negative impact on ants that are potential biological control agents [60]. Importantly, various details of the oviposition behavior, plant and ant host choice, behavior of the planidia both within and outside of the nest, and development within the nest are all key pieces of information to provide a better understanding of how the eucharitids gain access and specialize on their particular ant host group.

ACKNOWLEDGMENTS

The author would like to thank J. M. Heraty and P. Fidalgo for their help in all his studies and Elizabeth Murray for constructive comments and suggestions of this paper. This investigation was made possible through funding by the Projects PIP6361 from CONICET and PICT 01238 provided by Agencia Nacional de Promoción Científica y Tecnológica. He dedicates this paper to the memory of Emma "Chongui" Wuscovi who supported him at all times in his Ph.D. degree.

REFERENCES

1. C. P. Clausen, "The immature stages of the Eucharidae," Proceedings of the Entomological Society of Washington, vol. 42, no. 8, pp. 161–170, 1940.
2. C. P. Clausen, "The oviposition habits of the Eucharidae (Hymenoptera)," Journal of the Washington Academy of Sciences, vol. 30, no. 12, pp. 504–516, 1940.
3. C. P. Clausen, "The habits of the Eucharidae," Psyche, vol. 48, pp. 57–69, 1941.
4. J. M. Heraty, "A revision of the Nearctic Eucharitinae (Hymenoptera: Chalcidoidea: Eucharitidae)," Proceedings of the Entomological Society of Ontario, vol. 85, pp. 61–103, 1985.
5. J. M. Heraty, "Classification and evolution of the Oraseminae in the Old World, with revisions of two closely related genera of Eucharitinae (Hym: Eucharitidae)," Life Sciences Contributions (Royal Ontario Museum), vol. 157, pp. 1–174, 1994.
6. J. M. Heraty, "Biology and importance of two eucharitid parasites of Wasmannia and Solenopsis," in Exotic Ants: Biology, Impact and Control of Introduced Species,, D. Williams, Ed., pp. 104–120, Westview Press, Boulder, Colo, USA, 1994.
7. J. M. Heraty, "Phylogenetic relationships of Oraseminae (Hymenoptera: Eucharitidae)," Annals of the Entomological Society of America, vol. 93, no. 3, pp. 374–390, 2000.
8. J. M. Heraty, "A revision of the genera of Eucharitidae (Hymenoptera: Chalcidoidea) of the world," Memoirs of the American Entomological Institute, vol. 68, pp. 1–367, 2002.
9. J. M. Heraty and K. N. Barber, "Biology of Obeza floridana (Ashmead) and Pseudochalcura gibbosa (Provancher) (Hymenoptera: Eucharitidae)," Proceedings of the Entomological Society of Washington, vol. 92, no. 2, pp. 248–258, 1990.
10. C. P. Clausen, "The biology of Schizaspidia tenuicornis Ashm., a eucharid parasite of Camponotus," Annals of the Entomological Society of America, vol. 16, pp. 195–217, 1923.
11. J. M. Heraty, D. P. Wojcik, and D. P. Jouvenaz, "Species of Orasema parasitic on the Solenopsis saevissima-complex in South America (Hymenoptera: Eucharitidae, Formicidae)," Journal of Hymenoptera Research, vol. 2, no. 1, pp. 169–182, 1993.
12. G. Pérez-Lachaud, J. M. Heraty, A. Carmichael, and J.-P. Lachaud, "Biology and behavior of Kapala (Hymenoptera: Eucharitidae) attacking Ectatomma, Gnamptogenys, and Pachycondyla (Formicidae: Ectatomminae and Ponerinae) in Chiapas, Mexico," Annals of the Entomological Society of America, vol. 99, no. 3, pp. 567–576, 2006.
13. S. C. Buys, R. Cassaro, and D. S. Alomon, "Biological observations on Kapala Cameron 1884 (Hymenoptera Eucharitidae) in parasitic association with Dinoponera lucida Emery 1901 (Hymenoptera Formicidae) in Brazil," Tropical Zoology, vol. 23, no. 1, pp. 29–34, 2010.
14. J. Torréns, Estudios de Biología y Taxonomía de Eucharitidae (Hymenoptera: Parasitica) de Argentina, parasitoides de Formicidae (Hymenoptera: Aculeata) [Ph.D. thesis], Universidad Nacional de Tucumán, Tucumán Province, Argentina, 2010.
15. J. Torréns and J. M. Heraty, "A new genus of Eucharitidae (Hymenoptera: Chalcidoidea), with notes on life history and immature stages," Zootaxa, vol. 3630, no. 2, pp. 347–358, 2013.
16. J. Torréns, "Primer registro de Colocharis (Chalcidoidea: Eucharitidae) en la Argentina y descripción de una nueva especie," Revista de la Sociedad Entomológica Argentina, vol. 71, no. 3-4, pp. 215–218, 2012.
17. E. V. Gemignani, "La familia " Eucharidae" (Hymenoptera: Chalcidoidea) en la República Argentina," Anales del Museo Nacional de Historia Natural, vol. 37, pp. 477–493, 1933.

18. E. V. Gemignani, "Nueva nota sobre la familia Eucharidae (Hymenoptera: Chalcidoidea)," Anales del Museo Nacional de Historia Natural, vol. 39, pp. 159–166, 1937.
19. E. V. Gemignani, "Nuevas especies de la familia Eucharidae (Insecta, Hymenop. Chalcid.)," Comunicaciones del Museo Argentino de Ciencias Naturales, "Bernardino Rivadavia", vol. 1, pp. 1–15, 1947.
20. J. J. Morrone, "Biogeografía de América Latina y el Caribe, vol. 3 of M&T-Manuales & Tesis SEA, Sociedad Entomológica Aragonesa, Zaragoza, Spain, 2001.
21. C. Bertonatti and J. Corcuera, Situación Ambiental Argentina, Fundación Vida Silvestre Argentina, Buenos Aires, Argentina, 2000.
22. J. Torréns and J. M. Heraty, "Description of the species of Dicoelothorax Ashmead (Chalcidoidea, Eucharitidae) and biology of D. platycerus Ashmead," ZooKeys, vol. 165, pp. 33–46, 2012.
23. J. M. Heraty and J. B. Woolley, "Separate species or polymorphism: a recurring problem in Kapala (Hymenoptera: Eucharitidae)," Annals of the Entomological Society of America, vol. 86, no. 5, pp. 517–531, 1993.
24. J. M. Heraty and D. C. Darling, "Comparative morphology of the planidial larvae of Eucharitidae and Perilampidae (Hymenoptera: Chalcidoidea)," Systematic Entomology, vol. 9, no. 3, pp. 309–328, 1984.
25. J. Torréns, J. M. Heraty, and P. Fidalgo, "Biology and description of a new species of Laurella Heraty (Hymenoptera: Eucharitidae) from Argentina," Proceedings of the Entomological Society of Washington, vol. 109, no. 1, pp. 45–51, 2007.
26. J. Torrens, J. M. Heraty, and P. Fidalgo, "Biology and description of a new species of Lophyrocera Cameron (Hymenoptera: Eucharitidae) from Argentina," Zootaxa, no. 1871, pp. 56–62, 2008.
27. L. Varone, J. M. Heraty, and L. A. Calcaterra, "Distribution, abundance and persistence of species of Orasema (Hym: Eucharitidae) parasitic on fire ants in South America," Biological Control, vol. 55, no. 1, pp. 72–78, 2010.
28. R. K. Vander Meer, D. P. Jouvenaz, and D. P. Wojcik, "Chemical mimicry in a parasitoid (Hymenoptera: Eucharitidae) of fire ants (Hymenoptera: Formicidae)," Journal of Chemical Ecology, vol. 15, no. 8, pp. 2247–2261, 1989.
29. L. Varone and J. Briano, "Bionomics of Orasema simplex (Hymenoptera: Eucharitidae), a parasitoid of Solenopsis fire ants (Hymenoptera: Formicidae) in Argentina," Biological Control, vol. 48, no. 2, pp. 204–209, 2009.
30. A. Tocchetto, "Bicho costureiro," Revista Agronomica, Porto Alegre, vol. 6, pp. 587–588, 1942.
31. F. S. Roberts, "Insects affecting banana production in central America," in Proceedings of the 10th International Congress of Entomology, vol. 3, pp. 411–415, 1958.
32. A. Silveira-Guido, "Investigations on the biology and biological control of the fire ant, Solenopsis saevissima richteri Forel in Uruguay," Annual Report, Departamento de Sanidad Vegetal, Facultad de Agronomía, Universidad de la República, Montevideo, Uruguay, 1962.
33. A. Silveira-Guido, P. San-Martin, C. Crisci-Pisano, and J. Carbonnell-Bruhn, "Investigations on the biology and biological control of the fire ant, Solenopsis saevissima richteri Forel in Uruguay," Third Report, Departamento de Sanidad Vegetal, Facultad de Agronomía, Universidad de la República, Montevideo, Uruguay, 1964.
34. R. N. Williams, "Insect natural enemies of fire ants in South America with several new records," in Proceedings of the Tall Timbers Conference on Ecology, Animal Control and Habitat Management, vol. 7, pp. 123–134, 1980.
35. R. N. Williams and W. H. Whitcomb, "Parasites of fire ants in South America," in Proceedings of the Tall Timbers Conference on Ecology, Animal Control and Habitat Management, vol. 5, pp. 49–59, 1973.
36. D. P. Wojcik, "Status and search for natural control agents in South America," in Proceedings of the Annual Fire Ant Conference, vol. 5, pp. 29–37, 1986.

37. D. P. Wojcik, "Survey for biocontrol agents in Brazil: a final report, with comments on preliminary research in Argentina," in Proceedings of the Imported Fire Ant Conference, pp. 50–62, Athens, Ga, USA, 1988.
38. D. P. Wojcik, "Behavioural interactions between ants and their parasites," The Florida Entomologist, vol. 72, no. 1, pp. 43–51, 1989.
39. D. P. Wojcik, D. P. Jouvenaz, W. A. Banks, and A. C. Pereira, "Biological control agents of fire ants in Brazil," in Chemistry and Biology of Social Insects, J. Eder and H. Rembold, Eds., pp. 627–628, J. Peperny, München, Germany, 1987.
40. A. P. L. Digilio and P. R. Legname, "Los árboles indígenas de la provincia de Tucumán," Opera Lilloana, vol. 15, pp. 1–283, 1966.
41. D. M. Lapola, W. F. Aantonalli Jr., and E. Giannotti, "Arquitetura de ninhos da formiga neotropical Ectatomma brunneum F. Smith, 1858 (Formicidae: Ponerinae) em ambientes alterados," Revista Brasileira de Zoociências, vol. 5, no. 2, pp. 177–188, 2003.
42. J.-P. Lachaud, P. Cerdan, and G. Pérez-Lachaud, "Poneromorph ants associated with parasitoid wasps of the genus Kapala Cameron (Hymenoptera: Eucharitidae) in French Guiana," Psyche, vol. 2012, Article ID 393486, 6 pages, 2012.
43. G. Pérez-Lachaud, J. A. López-Méndez, and J. P. Lachaud, "Eucharitid parasitism of the Neotropical ant Ectatomma tuberculatum: parasitoid co-occurrence, seasonal variation, and multiparasitism," Biotropica, vol. 38, no. 4, pp. 574–576, 2006.
44. J. Pensiero, J. D. Muñoz, and V. Martinez, "Alternativas de sustentabilidad del bosque nativo del Espinal. Área Etnobotánica," Proyecto de Investigación Aplicada a los Recursos Forestales Nativos (PIARFON), January 2013, http://www.ambiente.gov.ar/?idarticulo=949.
45. D. W. Davidson, "Ecological studies of neotropical ant gardens," Ecology, vol. 69, no. 4, pp. 1138–1152, 1988.
46. M. S. Ferrucci, "Sapindaceae Juss. Guías ilustradas de clases," Aportes Botánicos de Salta. Serie Flora, vol. 7, no. 4, pp. 1–44, 2004.
47. W. M. Wheeler, "The polymorphism of ants, with an account of some singular abnormalities due to parasitism," Bulletin of the American Museum of Natural History, vol. 23, pp. 1–93, 1907.
48. G. C. Wheeler and E. W. Wheeler, "New hymenopterous parasites of ants (Chalcidoidea: Eucharidae)," Annals of the Entomological Society of America, vol. 30, no. 1, pp. 163–175, 1937.
49. H. L. Parker, "Oviposition habits and early stages of Orasema sp.," Proceedings of the Entomological Society of Washington, vol. 44, no. 7, pp. 142–145, 1942.
50. J. C. Nicolini, "La avispa costurera y la tuberculosis del olivo," Revista Agronomica, Porto Alegre, vol. 20, pp. 35–36, 1950.
51. A. F. van Pelt, "Orasema in nests of Pheidole dentata Mayr (Hymenoptera: Formicidae)," Entomological News, vol. 61, no. 6, pp. 161–163, 1950.
52. G. M. Das, "Preliminary studies on the biology of Orasema assectator Kerrich (Hym., Eucharitidae), parasitic on Pheidole and causing damage to leaves of tea in Assam," Bulletin of Entomological Research, vol. 54, no. 3, pp. 373–379, 1963.
53. G. J. Kerrich, "Descriptions of two species of Eucharitidae damaging tea, with comparative notes on other species (Hym., Chalcidoidea)," Bulletin of Entomological Research, vol. 54, no. 3, pp. 365–372, 1963.
54. J. B. Johnson, T. D. Miller, J. M. Heraty, and F. W. Merickel, "Observations on the biology of two species of Orasema (Hym.: Eucharitidae)," Proceedings of the Entomological Society of Washington, vol. 88, no. 3, pp. 542–549, 1986.
55. B. Carey, K. Visscher, and J. Heraty, "Nectary use for gaining access to an ant host by the parasitoid Orasema simulatrix (Hymenoptera, Eucharitidae)," Journal of Hymenoptera Research, vol. 27, pp. 47–65, 2012.
56. J. M. Heraty, Classification and evolution of the Oraseminae (Hymenoptera: Eucharitdae) [Ph.D. thesis], Texas A&M University, College Station, Tex, USA, 1990.

57. A. B. Gahan, "A contribution to the knowledge of the Eucharidae (Hymenoptera: Chalcidoidea)," Proceedings of the United States National Museum, vol. 88, pp. 425–458, 1940.
58. L. de Santis, "Una nueva especie de "Orasema" del Uruguay (Hymenoptera: Eucharitidae)," Revista Sociedad Entomológica Argentina, vol. 7, pp. 1–3, 1968.
59. J.-P. Lachaud and G. Pérez-Lachaud, "Diversity of species and behavior of hymenopteran parasitoids of ants: a review," Psyche, vol. 2012, Article ID 134746, 24 pages, 2012.
60. J.-P. Lachaud and G. Pérez-Lachaud, "Impact of natural parasitism by two eucharitid wasps on a potential biocontrol agent ant in southeastern Mexico," Biological Control, vol. 48, no. 1, pp. 92–99, 2009.

CHAPTER 14

Biological Control Strategies for Mosquito Vectors of Arboviruses

Yan-Jang S. Huang [1,2], Stephen Higgs [1,2] and Dana L. Vanlandingham [1,2,*]

[1]*Department of Diagnostic Medicine/Pathobiology, College of Veterinary Medicine, Kansas State University, Manhattan, KS, 66506, USA*

[2]*Biosecurity Research Institute, Kansas State University, Manhattan, KS 66506, USA*

ABSTRACT

Historically, biological control utilizes predatory species and pathogenic microorganisms to reduce the population of mosquitoes as disease vectors. This is particularly important for the control of mosquito-borne arboviruses, which normally do not have specific antiviral therapies available. Although development of resistance is likely, the advantages of biological control are that the resources used are typically biodegradable and ecologically friendly. Over the past decade, the advancement of molecular biology has enabled optimization by the manipulation of genetic materials associated with biological control agents. Two significant advancements are the discovery of cytoplasmic incompatibility induced by *Wolbachia* bacteria, which has enhanced replacement programs, and the introduction of dominant lethal genes into local mosquito populations through the release of genetically modified mosquitoes. As various arboviruses continue to be significant public health threats, biological control strategies have evolved to be more diverse and become critical tools to reduce the disease burden of arboviruses.

Keywords: arboviruses; mosquitoes; predators; entomopathogenic microorganisms; *Wolbachia*; genetically modified mosquitoes

1. INTRODUCTION

Although significant advancements have been made in developing therapeutics and vaccines for mosquito-borne pathogens over the last few decades, efficient vector control strategies are still the primary method used for control and prevention of mosquito-borne diseases. Mosquitoes are important vector species for various arboviruses belonging to three different families: Flaviviridae, Togaviridae, and Bunyaviridae. Unlike the *Anopheles* species mosquitoes, which are responsible for transmission of human malaria, the majority of mosquito species responsible for transmission of arboviruses belong to the genera of *Culex* and *Aedes*. Among all the medically important mosquito species, the most significant advancement in formulating control strategies has taken place in *Aedes aegypti*, a highly anthropophilic species known for its competence for several emerging and re-emerging arboviruses with high human public health significance including: dengue virus (DENV), chikungunya virus (CHIKV), yellow fever virus (YFV), and Zika virus (ZIKV) [1]. In addition to *Ae. aegypti*, other mosquito species—for example, the highly invasive *Ae. albopictus* mosquito, which has become an important vector species for multiple arboviruses, especially DENV and CHIKV—have become targets for novel control strategies [2,3]. Other zoophilic mosquito species, especially *Culex* species mosquitoes, are also important targets for vector control and for the prevention of arboviruses, including Japanese encephalitis virus (JEV) and West Nile virus (WNV) [4]. As various mosquito species and arboviruses continue to expand to new areas, there will be a continuing need for new biological control strategies. Although the goal of completely eliminating disease vectors is impractical, it has always been true that the reduction in the numbers of competent vectors is an important part of disease control. This is particularly true for control of many mosquito-borne arboviruses, which lack specific antiviral therapeutics or efficacious vaccines. Previously, vector control approaches could be divided into three general categories: chemical control, biological control, and environmental management, regardless of the pathogens to be targeted. All control strategies share a common goal of reducing the sizes of vector populations. The fundamental approach for chemical control is through the release of insecticidal chemical compounds, which inevitably faces the challenge of insecticide resistance and off-target effects on other arthropod species. Biological control depends on the use of predatory or parasitic organisms targeting disease vectors. Additionally, the removal of habitats in the environment can aid in the population reduction of disease vectors. In order to transiently eliminate mosquito larvae or adults, these various vector control strategies require knowledge of the biology of the targeted mosquito species, knowledge of the pharmacological mechanisms of the insecticides used, or the behaviors of the introduced predatory or competing species. For example, the use of indoor residual spray is more likely to be effective in controlling indoor species such as *Ae. aegypti* but has limited impact on the populations of various *Culex* species mosquitoes, which rest outdoors [5].

During the past two decades, integrated vector management (IVM) for control of arboviruses has gradually replaced the use of individual vector control strategies and promotes the development of control strategies based on vector biology. In addition to transient suppression of vector populations through interventions, IVM also focuses on rigorous monitoring of outcomes, and identification and correction of problems encountered [6]. Maintaining the positive outcomes of vector control during interepidemic periods is a critical factor in integrated vector control [7]. Another positive aspect of evidence-based vector control programs is the termination of existing strategies that are found to be ineffective and the development of new methods for vector control based on a better understanding of mosquito biology. The development of new techniques often leads to more efficient tools for the control of mosquito vectors for arboviruses.

This review will focus on summarizing available control strategies for mosquito vectors of public health importance. New tools for biological control using new technologies, recent advances, and optimization of existing methods will be discussed.

2. PREDATORS FOR MOSQUITO CONTROL—HISTORICAL AND ONGOING STORIES OF SUCCESS

Historically, successful biological control of mosquitoes has frequently depended on the introduction of mosquito predators. Theoretically, predaceous animal species can result in the decline of mosquito populations in nature. However, there have been limited evidence-based biological control programs which show that introducing certain species can significantly reduce mosquito populations and ultimately lead to the actual reduction of disease burden in humans and animals. As some of these species can be exotic, depending on the region they are introduced to, a critical analysis comparing the benefit of disease control and the risk associated with the use of exotic species is needed prior to use in the environment. Three predators used in biological control of mosquitoes in recent history are listed in Table 1 and discussed below.

Table 1. Predators of mosquito larvae as tools for biological control of mosquitoes.

Organisms	Species	Targeting Species	Limitations
Fish	Mainly *Gambusia affinis*	Nonspecific due to diet and predatory behaviors	Other off-target arthropod species in the same water body can be affected. Potential damage to the ecological system can occur.
Larvae of *Toxorhynchites* species mosquitoes	*Tx. splendens, Tx. brevipalpis, Tx. moctezuma, Tx. Amboinensis, and Tx. rutilus*	Mainly *Ae. aegypti*	Sylvatic species cannot be readily adapted to human environment.
Copepods	Mainly *Mesocyclops* and *Macrocyclops* species	Mainly *Ae. aegypti*	Most effective against first instar larvae.

2.1. Gambusia affinis, the Mosquito Fish

The release of predatory species has played a significant role in biological control for mosquitoes. For example, the use of *Gambusia affinis*, the mosquito fish, first occurred over 100 years ago and showed high compatibility with the simultaneous use of other chemical or biological control tools [8]. Because they are tolerant to insecticides, particularly larvicides, they are ideal to use with chemical control methods [9]. Additionally, when *G. affinis* are used with other biological control tools, control of mosquitoes can be further sustained. *G. affinis* can thrive in permanent water bodies, are relatively long-lived, have a slower rate of reproduction, and have been found to be effective in the transient suppression or extinction of targeted mosquito species.

A trial in the rice fields of California, mainly targeting *Culex tarsalis*, demonstrated the release of *G. affinis* followed by the treatment of *Bacillus thuringiensis israelensis* (*Bti*) prevented the anticipated resurgence of the vector population, which was observed in control groups treated only with *Bti* [10]. As larvivorous fish species often have a broad range of organisms in their diet, off-target effects on other arthropod species were also reported [11]. As the discovery of additional larvivorous fish species continues, it is evident that it is impossible to generate a comprehensive list of larvivorous fish species because of their nonspecific nature of predation. It has become clear that predation by larvivorous fish species can be exploited as biological control strategies when appropriate indigenous species are used. Another example of using fish species to control arboviral diseases, especially dengue, was reported in Taiwan, where multiple species—including *G. affinis*, *Poecilia reticulata*, *Tilapia mossambica*, and *Sarotherodon niloticus*—were simultaneously released in household water containers as a method of control. The strategy resulted in a reduced container index for the larvae of *Aedes* species mosquitoes, especially *Ae. aegypti* and *Ae. albopictus* [12]. As biological control strategies using *G. affinis* became popular, characterization of the behaviors of mosquitoes in relation to the presence of *G. affinis* has revealed complexity in the ecology of mosquito vectors. In a study utilizing outdoor breeding sites, it was demonstrated that female *Cx. pipiens* mosquitoes can sense the presence of *G. affinis* and selectively lay eggs in water bodies where the mosquito fish are not present or present in low quantities [13]. Similar observations were made in *Cx. tarsalis* and *Cx. quinquefasciatus*, but not *Ae. aegypti*, indicating that different species of mosquitoes have evolved to sense the predation risks in their natural habits [14]. Because oviposition of *Ae. aegypti* and perhaps the majority of *Ae. albopictus* occurs in artificial containers, where *G. affinis* normally is not present, the efficacy for the control of both vector species by the mosquito fish remains questionable. Another factor that must be considered is that the introduction of an exotic species can adversely affect the ecology and biodiversity of the aquatic environment; for example, the release of *G. affinis* was dispersed further than intended following flooding conditions in India [15]. Therefore, the use of this species for mosquito control requires further assessment between the cost and benefit [16].

2.2. Predatory Toxorhynchites Species Mosquitoes

Historically, larvae belonging to multiple species of mosquitoes in the genus *Toxorhynchites* have been reported to be predators for larvae of medically important mosquitoes. *Toxorhynchites* are autogenous, and lack the need for blood feeding, which makes this species of mosquito ideal for release in the environment for biological control without increasing the risk of disease transmission. The predatory behavior by *Toxorhynchites splendens*, *Tx. brevipalpis*, *Tx. moctezuma*, and *Tx. rutilus* on larvae of *Ae. aegypti* is also an important factor in its use as a biological control species [17,18,19,20]. However, since predation occurs at the larval stage, the coexistence of *Toxorhynchites* species mosquitoes and the prey population need to be in the same water body. This can substantially limit the application of *Toxorhynchites* as a biological control species in the urban environment, especially for some sylvatic species. However, there has been evidence suggesting that *Tx. amboinensis* can successfully locate oviposition sites in cryptic inaccessible water bodies and successfully reduce the quantity of larvae of *Ae. aegypti* [21].

The use of *Toxorhynchites* for biological control in the urban and sylvatic environments has also been examined. For example, *Tx. splendens* has been found to have a preference for oviposition in artificial containers with wider openings over narrow breeding sites [22]. Other species are not as well suited for release in the urban environment; the sylvatic *Tx. rutilus* may not be adapted to the urban environment, where the majority of transmission of arboviruses by *Ae. aegypti* takes place [23]. However, when used to control other tree-dwelling species such as *Ae. triseriatus* and *Ae. hendersoni*, *Tx. rutilus*, and *Tx. brevipalpis* showed high spatial aggregation in breeding sites with their potential prey population [24]. This agrees with the finding that the preferred oviposition locations mainly require the presence of trees [25]. In contrast to the sylvatic species, weekly release of *Tx. amboinensis* successfully reduced the population of *Ae. aegypti* and *Cx. quinquefasciatus* in New Orleans, Louisiana, USA [26]. Similar predatory behavior of *Tx. amboinensis* and *Tx. moctezuma* on *Ae. aegypti* was reported in a field release on the Fiji and Caribbean islands, respectively [27,28,29]. Other studies further indicated the predation can be further expected in other medically important *Aedes* species mosquitoes such as *Ae. albopictus* and *Ae. polynesiensis* [30,31].

With the release of transgenic mosquitoes expressing toxic proteins controlled by tetracycline, one critical issue for the continuous use of *Toxorhynchites* species mosquitoes is whether or not any adverse effects among *Toxorhynchites* species mosquitoes can be caused by ingesting the larvae of transgenic mosquitoes. Predation of *Tx. splendens* and *Tx. amboinensis* on transgenic lines of *Ae. aegypti* did not show any significant difference in longevity, which supports the continuous use of the *Toxorhynchites* species for biological control in combination with the use of transgenic mosquitoes [32]. Similar conclusions were drawn from the evaluation of the predatory behavior of *Tx. speciosus* on uninfected and *Wolbachia*-infected *Ae. aegypti* [33].

2.3. Copepods to Control Aedes Larvae

Copepods, mainly *Mesocyclops* and *Macrocyclops* species, are crustaceans that have been used for biological control for *Aedes* species mosquitoes by mainly feeding on first instar larvae. With no known harmful effects to human health, release of cyclopoid copepods became a popular method for mosquito control after the initial success in reducing the populations of *Aedes* species mosquitoes in French Polynesia [34]. In Queensland, Australia, significant reduction of larvae in larval breeding sites was reported by introduction of *Mesocyclops* species. In the same study, the dispersal of copepods to neighboring water bodies was observed after flooding and led to more extended results in controlling the population of larvae [35]. Simple protocols for culturing copepods species have been established for maintenance and mass propagation prior to release [36]. In laboratory studies, *Mesocyclops* species have also been demonstrated to prey on larvae of *Anopheles* and *Culex* species mosquitoes [37,38,39,40]. However, the majority of applications are still focused on the control of *Ae. aegypti* in DENV- and CHIKV-endemic countries. Predation behavior by the *Mesocyclops* species was characterized by scanning electron microscope indicating the anal segment, the siphon, and the abdomen are preferred over the head segment [41]. Another interesting determinant for the predation behavior is the interactions between copepods and the size of containers as mosquito breeding sites. The increase in predatory efficiency associated with the decrease in the size of containers was demonstrated by a higher larval mortality rate [42].

In addition to the laboratory studies and field surveys on predatory behaviors of copepods, application of copepods in public health control programs has also provided valuable information on how the introduction of copepods can be a control strategy for arboviral diseases. One of the most well-established evidence-based programs using copepods for mosquito control was established in Vietnam with the primary goal of controlling DENV transmission in the communities selected for the control programs combining introduction of *Mesocyclops* and health education. From initial surveys, *Mesocyclops* have been prevalent in multiple water bodies used as breeding sites of *Aedes* larvae and acting as a control agent for *Aedes* species mosquitoes [43]. Although transmission of DENV still led to asymptomatic infections in children, significant reduction in the populations of *Ae. aegypti* and *Ae. albopictus* has been reported in the communities where the treatment of *Mesocyclops* was implemented in the second year of the program [44]. Significant decline in disease incidence was achieved at the end of the program in 2002 and 2003 [45]. The results were sustained for several years with high prevalence of *Mesocyclops* in water bodies and low abundance of *Ae. aegypti* larvae [46].

3. APPLICATION OF MICROORGANISMS TO MOSQUITO CONTROL

Naturally occurring infections of microorganisms can occur at different stages of the mosquito life cycle and lead to different physiological outcomes. Early development of entomopathogenic microorganisms as control strategies focused on suppressing the overall population of vectors through larvicidal mechanisms of action. Several microorganisms including fungi, bacteria, and nematodes have been reported to show the initial success in causing lethality in mosquitoes, especially at the larval stage. The history of using microorganisms for biological control can be traced back to the 1960s, when entomopathogenic fungus was first demonstrated to cause mortality in larvae of the malaria vector *Anopheles gambiae* [47]. The larvicidal toxins produced by *Bti* and *Lysinibacillus sphaericus* (*Ls*) have been used in biological control since their discovery and have been extensively commercialized. Penetration of mermithid nematodes, especially *Romanomermis* species, in mosquito larvae has also been evaluated for the development of biological control strategies [48]. However, it has proven difficult to commercialize because the viability of the microorganisms is difficult to maintain in storage and transportation. Recently, evaluation of *Wolbachia* endosymbiotic bacterium in controlling arboviral diseases, especially dengue, utilizes cytoplasmic incompatibility to reduce the vector population as a sterilization technique. Moreover, the unexpected benefit of introducing *Wolbachia* to arthropods by reducing the vector competence and life expectancy of disease vectors further provides additional advantage of *Wolbachia*-based biological control.

Candidate microorganisms are constantly screened for their entomopathogenic capabilities in mosquitoes, so it is impossible to provide a comprehensive list of these microorganisms. However, to meet the operational needs for vector control in the field and the scale of production in the industry, there have only been a few microorganisms deployed as tools for biological control of mosquitoes. In the past two decades, pathogenic mechanisms of these organisms in mosquitoes have been further characterized. With the advancement of recombinant DNA technologies, genetic manipulations of microorganisms are used for the optimization of existing tools and the formulation of control strategies. Whilst the further application of insecticidal activity of several entomopathogenic microorganisms can be helpful in the control of pests with agricultural and public health importance [49], the off-target effects to other arthropod species should not be ignored in the process of formulating microorganism-based biological control strategies.

3.1. Endotoxins of Bacillus thuringiensis israelensis and Lysinibacillus sphaericus

Following the discovery of toxins as an effective mosquito control over three decades ago, the toxins produced by *Bti* and *Ls* have the advantage of being highly effective as larvicides for various vector species of arboviruses. *Bti* toxicity is broadly effective against medically important mosquitoes in the genera of *Aedes, Culex*, and *Anopheles*, whereas, toxicity of *Ls* is mainly limited to *Culex* and *Anopheles* species mosquitoes [50]. The molecular mechanisms of *Bti* toxins in mosquitoes and the general mechanisms of killing by disrupting the membranous structures in other insect species have continued to be studied. Ingestion of the sporulated cells by larvae leads to the ingestion of four toxin proteins. Three of the four toxins (Cry4A, Cry4B, and Cry11A) produced by *Bti* are homologous in their structures and utilize similar mechanisms of killing with other members in the Cry toxin family [51]. Cell death is triggered by osmotic shock through the pore structure formed by the oligomerization of toxin monomers [52]. Interestingly, transcriptome analysis of *Ae. aegypti* larvae provided the complimentary knowledge that the physiological responses to Cry toxins can only be initiated in the presence of functional wild-type toxins. Disruption in the formation of pore structures by monomers of Cry toxins removes the toxicity and fails to trigger the physiological responses [53]. The other toxin, Cyt1A, provides a synergistic mechanism for all three Cry toxins by serving as a receptor to induce the binding between the Cry toxins and brush border membrane [54]. The synergistic mechanism is particularly important to overcome the resistance developed towards the Cry toxins among mosquito larvae [55]. Similar to the mechanism of lysis induced by *Bti*, strains of *Ls* mainly utilize a highly conserved binary (Bin) toxin. Bin is proteolytically processed into two individual proteins, BinA and BinB, in the larvae and is followed by the formation of pore structures prior to cell lysis [56,57]. The toxicity to specific species of mosquitoes is mainly determined at the stage of binding. Binding of the Bin toxin is highly specific to cells in the gut of larvae belonging to the genera of *Culex* and *Anopheles* with close affinity to the cells in *Ae. aegypti* [58].

Although the approach of *Bti* and *Ls* is considered biodegradable and relatively environmentally friendly without the use of insecticidal compounds, the first significant challenge for its application in the field is still the emerging resistance first observed in *Culex* species mosquitoes to toxins produced by *Ls* [59,60]. The advancement of recombinant DNA technology allows the expression of a combination of toxins in the same strain of bacteria. This was particularly helpful in resolving the resistance of *Cx. quinquefasciatus* to *Ls* by introducing toxins from *Bti* [61]. The development of recombinant strains of *Ls* also enhances its application as the introduced *Bti* toxins increase the toxicity of *Ls* to *Aedes* species mosquitoes [62,63]. In addition to the resistance to the bacterial toxins, another active research area in evaluating the use of *Bti* and *Ls* in vector control is to determine the phenotypic changes of the populations that survived the treatment. An important question one must ask is whether or not selection by *Bti* at the larval stage will lead to the emergence of adult populations with higher vector competence for arboviruses. From the experimental evidence gathered from *Ae. aegypti*, the presence of *Bti* at various concentrations mainly led to the accelerated rate of development at the larval stage without significantly altering the phenotypes or increasing the susceptibility to DENV [64]. However, as the development rate of larvae increases under the selective pressure of *Bti*, it may potentially alter the overall vectorial capacity of a given vector population by the fitness gain in development. Changes in physiology and vector competence for other arboviruses in different medically important mosquito species, subjected to similar control strategies, remains to be determined. It is anticipated the commercialization of new *Bti* and *Ls* products and their field use will be continued as new technologies allow the development of more broadly toxic bacterial strains. A critical but unaddressed aspect in IVM is to evaluate the reduction of disease burden caused by various arboviruses after the intervention is introduced [65].

3.2. Entomopathogenic Fungus

The concept of using entomopathogenic fungus for mosquito control was first published on *An. gambiae* through the use of fungus belonging to the genus of *Coelomomyces* [47]. Although it achieved the goal of inducing high mortality in local *An. gambiae* populations, its further development was largely hindered by the lack of a robust in vitro culture system [50]. The use of fungal agents as a biological control strategy for mosquitoes can also be successful by changing the physiology of the insect to decrease the likelihood of blood feeding and, ultimately, disease transmission. For example, infection by *Beauveria bassiana* reduces the survival rate, blood-feeding success, and fecundity in *Ae. aegypti* [66]. Another species used in the field was *Lagenidium giganteum* in California targeting *Cx. tarsalis*, the vector for Western equine encephalitis virus and WNV [67]. The same species was later reported to suppress the populations of *Cx. quinquefasciatus* and *Ae. aegypti* in artificial containers [68,69]. As zoospores normally invade the larvae in water, the constant recycling of fungus in the aquatic environment effectively results in population suppression for extended periods without the requirement of follow-up treatment [70]. Further studies have found that the susceptibility to fungal infection is possibly age-dependent [71]. The establishment of fungal infection was generally found to be highly efficient in larvae of *Ae. aegypti* and *Cx. pipiens* but not *An. gambiae* because of the differential efficiency in melanization and encapsulation of fungus as immune mechanisms [72]. Subsequent colonization of the larvae is initiated after entering the hemocoel [73]. Proliferation of the fungus is generally temperature-dependent and results in high mortality within a few days [74]. In spite of the success of one previously commercialized product, production and stability of larvicidal zoospores remains challenging for large-scale manufacturers [75].

The other important genus being actively investigated for mosquito control is the *Metarhizium* species. These fungus species are normally found in the soil and can be highly virulent to various terrestrial insects. Although mosquitoes are not the natural hosts of the *Metarhizium* species, the pathogenic mechanisms of two species, *Metarhizium anisopliae* and *M. brunneum*, have been recently determined. Both species can establish infections which ultimately lead to lethality by either conidia or blastospores, the latter generally considered more virulent [76]. Infection by *M. anisopliae* is found to be pathogenic to a wide range of mosquitoes in the genera of *Aedes* and *Culex* [76,77]. Histological examination initially suggested ingestion of conidia ultimately leads to the blockage of anatomic structures [78]. However, the pathogenic mechanisms were later determined to be associated with apoptosis rather than the fungal colonization. A clear distinction in its infection process between mosquito larvae and terrestrial insects is the conidia of *M. anisopliae* does not attach to the cuticle of the mosquito larvae, as it normally does to the surface of terrestrial insects, through hydrophobic interactions. The difference is likely due to the fact that *M. anisopliae*, a fungus species normally found in soil, has not evolved to interact with mosquito larvae through the same hydrophobic interactions in water. The most likely route of entry is the ingestion of conidia as the gut of larvae showed high focal concentrations of conidia [75]. Mortality of larvae is the result of stress-induced apoptosis, as demonstrated by the experimental evidence that treatment with protease inhibitors substantially increased the survival of larvae [79]. In contrast to the pathway of ingestion utilized by conidia, blastospores of *M. brunneum* readily adhere to the cuticle of larvae by multiple mechanisms, including the hydrophilic surface, upregulated expression of adhesion molecules, and secretion of insoluble mucilage [80]. In fact, blastospores are able to detect the presence of larvae and increase the expression levels of adhesion molecules required for the invasion of mosquito larvae. Because of the higher affinity to mosquito larvae and rapid killing after penetration, blastospores of the *Metarhizium* species have a high potential for field application.

3.3. Endosymbiotic Wolbachia Bacterium as A Tool for Biological Control

The first discovery of endosymbiotic *Wolbachia pipientis* in *Cx. pipiens* was published in 1924 [81]. For several decades, there was very limited advancement on the characterization of this Gram-negative intracellular bacterium because of the difficulty in its cultivation, which was later improved by the availability of continuous insect cell lines [82,83]. Therefore, the nomenclature of *Wolbachia*, which was previously based on examination of insect tissues, was difficult until molecular genetics tools became available. Phylogenetic analyses on the sequences of 16S ribosome RNA genes have shown a lack of correlation between the *Wolbachia* symbionts and the phylogenetics of their hosts. Further interpretation of the more likely case scenario is acquisition of one single species by multiple insects through horizontal transfer [84]. Further work on its surface protein, *Wsp*, is used as the standard of current nomenclature, by dividing the clade of *Wolbachia* naturally found in insects into two major groups [85,86]. Infection of *Wolbachia* in mosquitoes can lead to several desired phenotypic changes that reduce or block the transmission of arboviruses and parasites.

One of the rationales of using *Wolbachia* for biological control of disease vectors is formulated on cytoplasmic incompatibility as a result of its infection in the reproductive organs. The infected males and females utilize different mechanisms to maintain their advantage in reproduction. This mechanism is not just limited to mosquitoes, but shared by the symbiotic relations between *Wolbachia* and all the other infected arthropod species. Successful production of viable progenies can only take place when an infected female mates with an infected male. Mating between an infected male and an uninfected female leads to abnormality in embryonic development with no or limited numbers of viable progenies. In such a transmission cycle, males are the dead-end host of *Wolbachia* and can sterilize the uninfected females. In contrast to the role of the dead-end host observed in male mosquitoes, an infected female can achieve the goal of reproduction by mating either with an infected male or an uninfected male. Importantly, all the progeny from an infected female will carry the bacteria through maternal transmission [87]. With greater frequencies of producing viable progenies by infected adults, it is anticipated the spread of *Wolbachia* in a given vector population can ultimately lead to a high prevalence of infection. This is particularly helpful in the development of a gene drive system to replace the wild-type competent vector populations for human or agriculture pathogens with refractory populations.

The earliest publication on cytoplasmic incompatibility to suppress the population of competent vectors was first proposed and tested in controlling mosquitoes in the *Cx. pipiens* complex as a potential control strategy for filariasis transmission [88]. Whilst cytoplasmic incompatibility was described in the publication and evaluated as a biological control strategy in a subsequent field trial [89], the mechanism of cytoplasmic incompatibility by *Wolbachia* infection was not proposed until the electron micrographs of infected eggs became available [90]. After treatment with tetracycline, the complementary evidence suggests that cytoplasmic incompatibility in mosquitoes was due to infection of *Wolbachia* [91]. Another critical factor that shows the feasibility of introducing *Wolbachia* to medically important arthropod species was confirmed by using microinjection technique to transinfect the other uninfected arthropod species at the embryonic stage [92]. This finding is particularly important because *Ae. aegypti*—a vector species for DENV, YFV, CHIKV, and ZIKV—is known to be rarely infected by *Wolbachia* [93]. The microinjection method was later used by many others to establish *Wolbachia*-infected *Ae. aegypti*, *Ae. albopictus*, and *Ae. polynesiensis* lines, providing the foundation knowledge for the population suppression trials currently being evaluated [94,95,96]. With the increase of global incidence of dengue and other arboviral diseases depending on the transmission of DENV [97], the physiological and behavioral changes associated with *Wolbachia* infection in *Ae. aegypti* have been extensively characterized.

After the first experiment demonstrating that transinfection of *Wolbachia* from *Ae. albopictus* to *Drosophila simulans* can be achieved, infection in laboratory colonies of *Ae. aegypti* was first artificially induced by microinjection of *Wolbachia* from *Ae. albopictus* [95]. In the same study, it was also shown that a fitness cost associated with *Wolbachia* infection may be a critical factor to consider in releasing *Wolbachia*-infected mosquitoes for population replacement in the field. Although the bacterium has been evolved to ensure its selective advantage, a minimum starting ratio between the infected and uninfected mosquitoes is still required in order to achieve fixation at 100% (i.e., a complete replacement of the wild-type competent population). Introducing a virulent strain of *Wolbachia*, *w*MelPop, into *Ae. aegypti* from *Drosophila simulans* showed additional promising aspects for utilizing *Wolbachia* for mosquito control by significantly reducing the life span of infected females and altering the feeding behavior of females in addition to the sterility induced in uninfected females. The observation is particularly important as transmission of arboviruses normally cannot take place after the extrinsic incubation period (EIP) has passed. For several medically important flaviviruses, the EIPs can sometimes be as long as 14 days [98,99]. The significantly shortened lifespan theoretically can substantially limit transmission by a female mosquito and the vectoring potential of a female mosquito simultaneously infected by arboviruses and *Wolbachia*. Transmission may not occur because some of the mosquitoes may not live for a period longer than EIPs of arboviruses [100]. The change of female feeding behavior caused by *Wolbachia* may also provide synergistic effect towards the goal of biological control. In the laboratory, infected females suffer from tissue damage in the proboscis and showed a decline in feeding success on vertebrate hosts and a smaller quantity of blood ingested among older mosquitoes [101].

As the decrease in successful feedings in older adults is well-correlated with reduced egg production, one must ask: What is the advantageous fitness gain, which compensates to such fitness cost, among *Wolbachia*-infected *Ae. aegypti*? One potential underlying mechanism is the protection against other pathogenic microorganisms invading the insect species [102]. This was later considered as another desirable outcome of using *Wolbachia* in vector control, because this pathogen protection mechanism ultimately can be used for blocking arbovirus infections in mosquitoes. Although this was initially observed in naturally infected *Drosophila melanogaster* and *Cx. quinquefasciatus*, the similar finding in transinfected *Ae. aegypti* was soon reported in other experiments. Infection of *Wolbachia* was first demonstrated to block the infection process of several flaviviruses in *Ae. aegypti*, interestingly except for WNV [103,104,105,106,107]. This further provides an additional mechanism to suppress transmission of DENV and CHIKV by introducing *Wolbachia* to *Ae. aegypti* in nature [108]. A strong degree of resistance to DENV was also observed in *Ae. polynesiensis* infected by *Wolbachia* [109]. Evidence of resistance induced by *Wolbachia* infection in *Ae. albopictus* has been limited. However, in vitro experiments did show the interference with CHIKV replication and production [110]. Characterization of pathogen-resistant phenotypes in arthropod vectors following infection of arthropods with *Wolbachia* has provided important knowledge with regards to how this mechanism operates. It has been demonstrated that the viral resistance caused by *Wolbachia* infection cannot be induced by introducing the bacteria at the adult stage, although its infection can be artificially established. An experiment conducted by intrathoracically inoculating *Wolbachia* in adult *Cx. tarsalis* did not successfully suppress the replication of WNV [111]. In addition to the discovery showing viral resistance can only be produced through its introduction at a certain life stage of mosquitoes, another important discovery has focused on how *Wolbachia* modulates the physiological responses of mosquitoes to limit arbovirus infections.

Detailed molecular mechanisms for resistance have been determined through several experiments. In an in vitro model using *Drosophila melanogaster* cells and Semliki Forest virus, it has been shown that the onset of resistance induced by *Wolbachia* occurs shortly after infection and targets various stages of the RNA virus replication cycle [112]. Surprisingly, as an obligatory intracellular microorganism, the characterization of antimicrobial responses in arthropods infected by *Wolbachia* did not show a universal upregulation of immune responses and cannot sufficiently explain how resistance can be induced and which effector genes are really involved in the development of resistance [111,113,114,115,116]. It is important to keep in mind characterization of immune responses to *Wolbachia* has largely been studied in *Drosophila*, a model organism in genetics study that is not normally infected by arboviruses.

In contrast to the extensive work on *Ae. aegypti*, progress in introducing *Wolbachia* to reduce the populations of *Ae. albopictus* is relatively slow. Because there is naturally occurring infection of *Wolbachia* in *Ae. albopictus*, introducing *Wolbachia* to *Ae. albopictus* has been found even more challenging. The superinfection of *Wolbachia* in *Ae. albopictus* has been reported to be caused by two strains of *Wolbachia* belonging to group A and B [99,117]. The superinfection status in nature further complicates the formulation of *Wolbachia*-based control strategies, as males infected by one strain of *Wolbachia* cannot sterilize dually infected females in nature. Females infected by one strain of *Wolbachia* may even be sterilized by dually infected males in nature. Characterization of cytoplasmic incompatibility caused by naturally occurring superinfection suggests release of triple-infected individuals may be the only approach to reduce the population of superinfected *Ae. albopictus* in nature [96]. The approach has been proven to be feasible in *Drosophila simulans* but has not been described in any other mosquito species but *Ae. albopictus* [118,119].

Following the establishment of several lines of *Aedes* species mosquitoes infected by *Wolbachia* in the laboratory, trials of releasing infected males were launched in the field to evaluate whether the sterilization caused by mating between infected males and uninfected females can be used as a control strategy. These trials have mainly targeted the population of *Ae. aegypti* in DENV-endemic regions. However, the first open release of *Wolbachia*-infected *Aedes* species mosquitoes occurred in French Polynesia targeting *Ae. polynesiensis*. With comparable level of fertilization in the trial and control site, cytoplasmic incompatibility caused by the incompatible males resulted in the significant reduction in the hatching rate of eggs collected in the field and the decline of local population. However, the evidence-based evaluation of this strategy is still challenged by the relatively large numbers of uninfected males in the field. Whether the release of infected males is truly the cause of population reduction still has to be confirmed [120]. Introduction of *Wolbachia*-infected *Ae. aegypti* in the field did not take place until 2011. The release of *Ae. aegypti* infected by the *w*Mel strain reached stable and high frequencies of infection, indicating the feasibility of completely replacing disease-transmitting vector populations with *Wolbachia*-infected refractory populations [121]. More importantly, in a follow-up study performed after release, the field population still maintained a high degree of refractoriness to DENV, as the infection of *Wolbachia* still persisted [122,123]. Ongoing trials and evaluations of releasing *Wolbachia*-infected *Ae. aegypti* are currently being conducted in Vietnam, Indonesia, Columbia, Brazil, and Singapore. One can certainly expect *Wolbachia* remaining a popular biological control method for mosquitoes because of its success.

4. GENETICALLY MODIFIED MOSQUITOES FOR VECTOR CONTROL

Biological control for mosquito-borne diseases by genetically engineered mosquitoes was proposed even before the techniques of stably transforming mosquitoes became available [124]. The concept of using genetically engineered mosquitoes for biological control is based on the goal of replacing the population of competent mosquitoes in nature with engineered mosquitoes. There have been two major approaches being investigated for genetically engineered mosquitoes as biological control strategies. The first approach focuses on directly inducing the resistance to arboviruses by "intracellular immunization" through the expression of antiviral genes. Although it does not directly reduce the numbers of mosquito vectors like other biological control strategies, disruption of transmission is achieved by lowering the numbers of competent mosquitos in nature. The genetic products elevate the required dosage for the establishment of infection in arthropods and subsequently reduce the likelihood of transmission. This approach has been often pathogen-specific as most of the successful examples are based on the expression of antiviral RNA molecules for arboviruses or single-chain immunoglobulins for parasites [125,126,127,128,129]. Alternatively, the decrease of transmission can also be achieved by introducing genetically modified populations that disrupt the life cycles of wild populations and achieve the reduction of population. This approach normally cannot target a specific arbovirus or pathogen but acts against specific species, which may carry one or multiple pathogens. A particularly good example is the sterile insect technique (SIT) that disrupts the insemination step in mating. The release of irradiated sterile males, which carry sperm with chromosomal damage, mate with females but fail to successfully inseminate them; this can ultimately drive the species to extinction. The elimination program of the New World screwworm *Cochliomyia hominivorax* in the United States was achieved by weekly release of 50 million sterile males [130]. The success of the SIT–based elimination program is based on the monogamous mating behavior of female arthropods because the mating of a wild-type female will only take place once with a genetically modified male and preclude the production of viable wild-type progenies. This strategy becomes the fundamental theory for the formulation of biological control based on SIT and other similar strategies. Despite the tremendous success in eradicating agricultural pests, the development of SIT was very limited. Successful reduction of populations was only achieved in a few instances [131]. Irradiation of pupae to produce sterile males proved to be technically challenging because of the lack of mating success later at the adult stage [132]. Sterilization using chemical compounds such as thiotepa was successful but also led to the potential of environmental contamination [133]. A modern adapted version of SIT has been successfully developed to target *Ae. aegypti* for the control of DENV, CHIKV, and ZIKV [134]. The technique does not rely on the early-acting lethality as SIT, but depends on the late-acting lethality to prevent the production of viable progenies. The approach is based on "release of insects with dominant lethality" (RIDL), which introduces the expression of a dominant lethal gene in the vector populations by releasing male transgenic mosquitoes carrying the dominant lethal gene. The development of RIDL will be further discussed in Section 4.2. Comparison of SIT and RIDL is listed in Table 2.

Table 2. Comparison of sterile insect technique (SIT) and release of insects with dominant lethality (RIDL).

Technique	Mechanism of Population Suppression	Introduction of Lethality by Genetically Modified Arthropods	Requirement of Sex Separation
SIT	Suppression of population by lethality at embryo stage	Sterilization of males at pupae stage prevents the successful insemination in female adults after mating	Yes, manual separation of males and females is required
RIDL	Suppression of population by lethality at larval stage in the absence of selectable antibiotics	Introduction of dominant lethal genes is achieved by releasing transgenic males	No, sex-specific promoters can allow the separation of males and females

The first critical issue in genetic manipulation is to deliver the genetic elements into the genomes of medically important mosquito species. The technique was initially described in three pioneering studies utilizing microinjection methods to deliver DNA molecules into eggs of mosquitoes [135,136,137]. Although the species described in the publications are different, the experimental approach independently taken by three groups was similar. Insertion of foreign genes into mosquito genomes was achieved by the use of *Drosophila* transposable P elements. Because of the limited capability of carrying large genetic elements and lack of site-specific integration, modified methods with bacteriophage Cre, phiC31, and yeast flippase (FLP) recombinases were further developed [138,139,140,141]. Other transposable elements isolated from other arthropod species remain as popular options for developing methods for transforming mosquitoes [142]. Among all the transposable elements selected for genetic engineering, *piggybac* transposon from lepidopteran insects is the most advanced and versatile tool that was later shown to transform several species of medically important mosquitoes [141,143,144,145]. Other methods such as transduction by retroviruses have been evaluated in vitro but have not been successfully applied in vivo [128,146]. Recently, the advancement of gene-editing technologies further allow the genetic manipulation of mosquitoes to specifically disrupt and edit individual genes in vivo. Using the transcription activator-like elements nuclease (TALEN), initial success was reported in disrupting the function of kynurenine 3–monooxygenase gene [147]. The rapid development of CRISPR/Cas9 utilizes a similar double-strand break–repair mechanism shared by other homing endonuclease genes' (HEGs') mechanism, with a significant improvement of specificity and throughput through the use of short guide nucleotide sequences [148,149,150].

Another key aspect of generating genetically engineered mosquitoes is the expression of introduced effector genes. The first promoter successfully used for transgenic expression in mosquitoes was the constitutive *Drosophila* heat shock protein 70 (Hsp70) promoter. However, the nature of constitutive expression raises the concern of increased fitness cost while being used in vivo [151]. How to precisely express the transgenes in the correct tissues at the desired timing subsequently became an important topic for research. Inducible elements from *Drosophila* have been tested in transformed mosquitoes, but the application is limited due to the difficulty in the delivery of inducers in nature [152,153]. A more desirable mechanism is to precisely drive gene expression after the blood meal because the infection of arboviruses in mosquitoes is always initiated by the ingestion of blood meals. There have been several promoters that have been demonstrated to induce the gene expression precisely after ingestion of blood meals

[154,155]. One of the most promising applications is the carboxypeptidase A promoter in transgenic DENV-resistant *Ae. aegypti* [125]. The gene expression driven by this promoter is relatively robust and responds specifically to blood meals [156].

Selection of transformed mosquitoes is a critical step for establishing genetically engineered mosquitoes. The first available platform of selectable markers used compounds that are toxic to eukaryotic organisms, such as neomycin and hygromycin. The advancement of reporter gene techniques enables the selection of stably transformed individuals by visual selection without the use of antibiotic selectable markers [157,158,159].

When a disease-resistant or sterile line of genetically engineered mosquito is established, another crucial factor for its success is the presence of a powerful gene drive system for the introduction of effector genes [160]. The goal of replacing wild-type competent mosquitoes with genetically engineered mosquitoes will be highly unlikely because unfeasibly large numbers will be required for replacement in the absence of gene drive systems. Additionally, the large number of genetically engineered mosquitoes must also be continuously released for extended periods, thus failing to meet the reasonably efficient timescale needed for disease control. An example of this challenge was the large number of irradiated sterile males needed during the screwworm eradication program in the United States [161]. Good candidates for gene drive systems not only must be powerful for the spread of the effector genes in the target population, but also have high specificity to minimize the spread to the nontarget species. Preferably, a reverse system should be available in the event that unanticipated damage to the environment or ecological system occurs. Several systems, mainly based on the selfish genes spreading at the cost of arthropod hosts, have been developed and evaluated including transposable elements and *Wolbachia*. With the high specificity conferred by the CRISPR/Cas9 system and the catalytic activity of Cas9 nuclease, the CRISPR/Cas9 system has also shown its feasibility in spreading the effector genes in model organisms [162]. In this section of the review, genetically modified mosquitoes as biological control tools will be discussed.

4.1. Genetically Modified Mosquitoes with Viral Resistance

One of the earliest concepts to disrupt the transmission cycles of arboviruses is to introduce genes that lead to viral resistance in competent vectors. Although several antiviral genes have been isolated and characterized, there has not been a single gene demonstrated to be fully responsible for the susceptibility or resistance to arboviruses. There has been a significant amount of evidence showing arthropod immune responses can modulate the outcomes of infection. Although the techniques have been developed, overexpression of antiviral genes in mosquitoes still fail to completely block replication of arboviruses precluding the possibility of developing disease-resistant mosquitoes by overexpressing antiviral genes [163,164]. A more feasible approach was to incorporate viral genomic sequences into the antiviral RNA interference (RNAi) pathway to specifically inhibit the replication of arboviruses [165,166,167]. Expression of antiviral inverted repeat (IR) sequences against various arboviruses was first achieved by the expression of Sindbis virus (SINV) vector. Because of the pathogenicity of SINV in humans, the approach still did not sufficiently fulfill the goal of biological control of arboviruses. A much-improved platform was later developed by the expression of IR sequence against DENV-2 driven by the carboxypeptidase A promoter in *Ae. aegypti* transformed by the mariner transposable element [125]. However, the resistance phenotype was only transiently maintained for several generations in the laboratory, precluding the possibility of conducting field trials to demonstrate its efficacy in disease control [168]. A similar approach was later used to generate additional lines of genetically modified *Ae. aegypti*, which had increased stability in mosquitoes [169]. The other platform was similarly constructed by driving the expression of IR sequences with the female salivary gland-specific 30K b promoter [170]. However, neither the transgene controlled by the carboxypeptidase A promoter nor the transgene controlled by the 30K b promoter successfully resulted in the phenotype of complete resistance to DENV. Since more than one serotype of DENV may be simultaneously transmitted in endemic regions and there is no cross-protection against other serotypes [125], a critical but unaddressed gap in this "proof-of-principle" study is how to generate and stably maintain genetically modified *Ae. aegypti*, which can be resistant to multiple serotypes of DENV in nature.

There have also been two other innovative strategies being evaluated to introduce resistance to arboviruses, especially DENV. Both strategies utilize catalytic RNA molecules, which are designed to recognize specific viral sequences. The first strategy is to express hammerhead ribozymes derived from plant viroids. The engineered hammerhead ribozymes were designed to target the genomic RNA of DENV-2 and able to cause greater than 100-fold reduction in transduced *Ae. albopictus* C6/36 cell line [128]. The other strategy is to express the trans-splicing group-I intron in mosquitoes. The catalytic RNA molecule only requires nine nucleotides in length to initiate the cleavage of viral genome, indicating the likelihood of identifying a highly conserved region among four serotypes of DENV as its target is higher than hammerhead ribozymes. The results from the transformed C6/36 cell lines have shown initial success in reducing the titer of DENV-2 in vitro. Because of the conserved sequence also found in three other serotypes of DENV, the antiviral activity is presumably similar, but was not confirmed in the same study [126]. An improved version of the same antiviral effector transgene was developed subsequently by coupling the *trans*-splicing group-I intron with the proapoptosis gene Bax and shown to be effective against all four serotypes of DENV in vitro [127]. Further expansion of the antiviral activity against both DENV and CHIKV was achieved by developing a dual acting *trans*-splicing group-I intron in a similar experimental approach [171].

Several other significant gaps still remain before further trials of these strategies in nature should be conducted. The rapid evolution of RNA viruses often results in diverse sequences, especially in the envelope protein. Whether or not the hammerhead ribozymes or *trans*-splicing group-I introns can induce similar level of resistance against other DENV or CHIKV strains that belong to different genotypes and contain diverse sequences in nature is still unclear. As the molecule often recognizes sequences less than 20 nucleotides in length as its target, the tolerance of mismatch will have to be considered for the likelihood of selecting escape mutants to emerge.

4.2. Release of Insects Carrying a Dominant Lethal Gene (RIDL)

Originally constructed in *Drosophila melanogaster* as a proof-of-concept technique, populations of insects can be controlled by dominant lethal genes repressed by tetracycline [172]. The system was developed with the goal to replace the laborious procedure of removing the sterile females, which essentially have no contribution or are even detrimental to SIT. It generally consists of a transcriptional control system, which can be repressed by the presence of tetracycline [173], a female-specific expression system for the tetracycline-sensitive transcription factor, and a cytotoxic gene lethal to females. First developed as an autocidal control system of the Mediterranean fruit fly, *Ceratitis capitata*, the approach of RIDL is based on a transcriptional control system regulating the expression of tetracycline-controlled transactivator (tTA) [174]. The addition of tetracycline in the diet under laboratory rearing conditions prevents overexpression of tTA and the downstream positive feedback to induce larval lethality. In nature, the absence of tetracycline triggers the overexpression of tTA through its positive feedback mechanism. tTA serves as its own transcription activator, binding to its promoter element tetO in the transgenic mosquito and driving its overexpression. High concentrations of tTA is lethal to larval development, presumably due to the interactions of its VP16 domain with other critical transcription factors required for larval development.

The genetic system was further developed for mosquito control and shares a similar mechanism of killing as the Mediterranean fruit fly, except the expression of tTA is controlled by a combination of a tetracycline operator and the minimal promoter of *Drosophila* Hsp70. Because of the lack of a female-specific promoter element, the transgenic mosquito line OX513A achieved the goal of biological control by producing lethality in both males and females. Overexpression of tTA in the absence of tetracycline results in high mortality, whereas, exposure to tetracycline removes the lethality [134]. Manual selection of male mosquitoes prior to the release in the field is required due to the fact that transgenic female *Ae. aegypti* can still be competent for various arboviruses. In the subsequent study determining its performance in the laboratory, the transgenic OX513A line demonstrated compromised fitness by showing higher larval mortality rate and reduced longevity and flight capacity in adults compared to the wild-type counterpart, but was still able to maintain comparable insemination efficiency in a subsequent mating study, indicating its capability of introducing the lethal gene into the wild-type population [175,176,177,178]. A subsequent release–recapture study in Malaysia showed a similar finding to the transgenic line, showing a decreased mean distance traveled compared to the other wild-type laboratory line of *Ae. aegypti* [179]. The 4-week-long release experiment on Grand Cayman further showed that transgenic males are able to mate with wild-type females, as the marker of transgenic mosquitoes was detected from the subsequent hatching of eggs collected from the ovitraps [180]. In Brazil, an even longer release program that lasted 6 weeks was continuously monitored for over a year and suggested that sustained release of the transgenic OX513A line in relative large numbers can effectively reduce the population of *Ae. aegypti* in nature; however, it took several months to achieve significant suppression after the initial release [181,182]. Similar observation was also reported in the release study in Panama, as significant reduction of *Ae. aegypti* was not reported until 109 days after the initial release [183]. In the study conducted in Brazil, several key factors determining the success of such a strategy were discovered, including the extent of infestation by *Ae. aegypti* in the region, the mass rearing capacity to produce sufficient quantities for release, and the area covered by the release program.

The discovery of the female-specific *Aedes Actin-4* (*AeAct-4*) promoter from the *Ae. aegypti* pupal mRNA provided the ideal candidate promoter for the development of a female-specific RIDL system [184]. The expression of *AeAct-4* is predominantly found in the indirect flight muscles in females, whereas the presence of multiple start and stop codons as a result of alternative splicing in males prevents the expression. The *AeAct-4* promoter was subsequently used to govern the expression of tTA in pupae, and limits its expression only in females. As a consequence of tTA expressed at high abundance in the indirect flight muscles, female-specific flightless phenotype was produced because of the abnormality in development. The inability to fly impairs several steps in the life cycle of female mosquitoes by preventing escape from water after eclosion, mating, acquisition of blood meals from vertebrate hosts, and, ultimately, production of viable eggs [185]. It is immediately apparent that flightless females should not be able to transmit arboviruses because of the lack of their ability to locate susceptible hosts. Elimination of *Ae. aegypti* in laboratory studies was achieved as egg production significantly declined after the release [186]. This method has been further developed, and the same concept has been applied to *Ae. albopictus*. Isolation and characterization of *Ae. albopictus Actin-4* (*AealbAct-4*) shows the feasibility of reproducing the same flightless phenotype as described in *Ae. aegypti* [187].

The advancement in the molecular genetics of mosquitoes has created tools that can be used to either specifically increase the resistance to arboviruses or reduce the population of competent vectors in nature. With the success of RIDL in reducing *Ae. aegypti* population in Grand Cayman and Brazil, elimination of mosquitoes by release of genetically engineered mosquitoes can be achieved. However, the commitment to the approach and the logistics for continuously producing genetically engineered mosquitoes is required to achieve and maintain suppression of the vector population. Although the evidence gathered from the study in Panama did not show a significant increase of *Ae. albopictus* population after the decline in the population of *Ae. aegypti* [183], it seems inevitable that migration of *Ae. aegypti* from neighboring infested areas can potentially serve as a source of reintroduction after the majority of the population is eliminated by RIDL in given areas. Similar observation and concern was also reported in the study of releasing *Wolbachia*-infected *Ae. aegypti* to reduce the local vector populations in Australia [121]. Although it is compelling to conclude that limiting the transmission of CHIKV, DENV, and ZIKV by technology will be beneficiary and perhaps cost-effective to the community by preventing significant economic loss and expenditures on healthcare caused by the diseases, a significant amount of funding will still need to be located, followed by the question of how long the RIDL program must last to sustain the results of vector control in different endemic areas. The analysis between the costs of vector controls by RIDL and the costs of vaccination, if vaccines are available, will still have to be conducted. It is more likely that RIDL will be an important approach in combination with other methods for disease control before vaccines become available, especially for CHIKV and ZIKV.

5. CONCLUSIONS

Biological control strategies for mosquito populations are undoubtedly more ecologically friendly than the use of insecticides. Because the advancement of molecular biological technologies has allowed the genetic manipulation of different organisms, we have seen the improvement of existing tools and methods by genetic engineering of entomopathogenic microorganisms and releasing male mosquitoes containing dominant lethal genes. One can surely anticipate the larger variety of innovative approaches will be developed in the future. As the inevitable increase in human and cargo movement across political and geographical borders has led to the reemergence or introduction of arboviruses in several locations, we can certainly anticipate biological control will maintain its importance in vector control. However, a more significant challenge will also remain as to how the results of biological control of mosquitoes can be sustained, as we have seen from the field experiments of *Wolbachia*-infected *Ae. aegypti* and RIDL are still challenged by reinfestation caused by the dispersal of neighboring populations.

AUTHOR CONTRIBUTIONS

Yan-Jang S. Huang, Stephen Higgs and Dana L. Vanlandingham wrote the paper.

CONFLICTS OF INTEREST

All authors declare that no competing financial interests exist.

REFERENCES

1. Tabachnick, W.J. Evolutionary genetics and arthropod-borne disease: The yellow fever mosquito. Am. Entomol. 1991, 37, 14–26.

2. Reiter, P.; Fontenille, D.; Paupy, C. Aedes albopictus as an epidemic vector of chikungunya virus: Another emerging problem? Lancet Infect. Dis. 2006, 6, 463–464.
3. Benedict, M.Q.; Levine, R.S.; Hawley, W.A.; Lounibos, L.P. Spread of the tiger: Global risk of invasion by the mosquito Aedes albopictus. Vector-Borne and Zoonotic Diseases 2007, 7, 76–85.
4. Solomon, T.; Vaughn, D.W. Pathogenesis and Clinical Features of Japanese Encephalitis and West Nile Virus Infections. In Japanese Encephalitis and West Nile Viruses; Springer: Berlin, Germany, 2002.
5. World Health Organization. Pesticides and Their Application; World Health Organization: Geneva, Switzerland, 2006.
6. World Health Organization. Global Strategic Framework for Integrated Vector Management; World Health Organization: Geneva, Switzerland, 2004.
7. Achee, N.L.; Gould, F.; Perkins, T.A.; Reiner, R.C., Jr.; Morrison, A.C.; Ritchie, S.A.; Gubler, D.J.; Teyssou, R.; Scott, T.W. A critical assessment of vector control for dengue prevention. PLoS Negl. Trop. Dis. 2015, 9, e0003655.
8. Bay, E.C. Mosquito control by fish: A present-day appraisal. WHO Chron. 1967, 21, 415–423.
9. Tietze, N.S.; Hester, P.G.; Hallmon, C.F.; Olson, M.A.; Shaffer, K.R. Acute toxicity of mosquitocidal compounds to young mosquitofish, Gambusia affinis. J. Am. Mosq. Control Assoc. 1991, 7, 290–293.
10. Kramer, V.L.; Garcia, R.; Colwell, A.E. An evaluation of Gambusia affinis and Bacillus thuringiensis var. israelensis as mosquito control agents in California wild rice fields. J. Am. Mosq. Control Assoc. 1988, 4, 470–478.
11. Linden, A.L.; Cech, J.J., Jr. Prey selection by mosquitofish (Gambusia affinis) in California rice fields: Effect of vegetation and prey species. J. Am. Mosq. Control Assoc. 1990, 6, 115–120.
12. Wang, C.H.; Hwang, J.S.; Lay, J.R. Preliminary study on the biological control of dengue vectors by fish in Liouchyou Prefecture, Pingtung County, Taiwan. Gaoxiong Yi Xue Ke Xue Za Zhi 1990, 6, 382–388.
13. Angelon, K.A.; Petranka, J.W. Chemicals of predatory mosquitofish (Gambusia affinis) influence selection of oviposition site by Culex mosquitoes. J. Chem. Ecol. 2002, 28, 797–806.
14. Van Dam, A.R.; Walton, W.E. The effect of predatory fish exudates on the ovipostional behaviour of three mosquito species: Culex quinquefasciatus, Aedes aegypti and Culex tarsalis. Med. Vet. Entomol. 2008, 22, 399–404.
15. Haq, S.; Prasad, R.N.; Prasad, H.; Shukla, R.P.; Sharma, V.P. Gambusia affinis: Dispersal due to floods and its failure to colonize new water bodies in Shahjahanpur District (U.P.). Indian J. Malariol. 1992, 29, 113–118.
16. Azevedo-Santos, V.M.; Vitule, J.R.S.; Garcia-Berthou, E.; Pelicice, F.M.; Simberloff, D. Misguided strategy for mosquito control. Science 2016, 351.
17. Yasuno, M.; Tonn, R.J. Bionomics of Toxorhynchites splendens in the larval habitats of Aedes aegypti in Bangkok, Thailand. Bull. World Health Organ. 1970, 43, 762–766.
18. Trpis, M. Interaction between the predator Toxorhynchites brevipalpis and its prey Aedes aegypti. Bull. World Health Organ. 1973, 49, 359–365.
19. Padgett, P.D.; Focks, D.A. Laboratory observations on the predation of Toxorhynchites rutilus rutilus on Aedes aegypti (Diptera: Culicidae). J. Med. Entomol. 1980, 17, 466–472.
20. Sherratt, T.N.; Tikasingh, E.S. A laboratory investigation of mosquito larval predation by Toxorhynchites moctezuma on Aedes aegypti. Med. Vet. Entomol. 1989, 3, 239–246.
21. Focks, D.A.; Sackett, S.R.; Dame, D.A.; Bailey, D.L. Ability of Toxorhynchites amboinensis (Doleschall) (Diptera: Culicidae) to locate and oviposit in artificial containers in an urban environment. Environ. Entomol. 2014, 12, 1073–1077.

22. Mohamad, N.; Zuharah, W.F. Influence of container design on predation rate of potential biocontrol agent, Toxorhynchites splendens (Diptera: Culicidae) against dengue vector. Trop. biomed. 2014, 31, 166–173.
23. Focks, D.A.; Sackett, S.R.; Dame, D.A.; Bailey, D.L. Toxorhynchites rutilus rutilus (Diptera: Culicidae): Field studies on dispersal and oviposition in the context of the biocontrol of urban container-breeding mosquitoes. J. Med. Entomol.1983, 20, 383–390.
24. Schuler, T.C.; Beier, J.C. Oviposition dynamics of two released species of Toxorhynchites (Diptera: Culicidae) and potential prey species. J. Med. Entomol. 1983, 20, 371–376.
25. Jordan, S. Influence of tree trunks on the spatial distribution of Toxorhynchites r. rutilus ovipositions in a coastal oak/palm hammock in Florida. J. Am. Mosq. Control Assoc. 1991, 7, 452–455.
26. Focks, D.A.; Sackett, S.R.; Dame, D.A.; Bailey, D.L. Effect of weekly releases of Toxorhynchites amboinensis (Doleschall) on Aedes aegypti (l.) (Diptera: Culicidae) in New Orleans, Louisiana. J. Econ. Entomol. 1985, 78, 622–626.
27. Toohey, M.K.; Goettel, M.S.; Takagi, M.; Ram, R.C.; Prakash, G.; Pillai, J.S. Field studies on the introduction of the mosquito predator Toxorhynchites amboinensis (Diptera: Culicidae) into Fiji. J. Med. Entomol. 1985, 22, 102–110.
28. Rawlins, S.C.; Clark, G.G.; Martinez, R. Effects of single introduction of Toxorhynchites moctezuma upon Aedes aegypti on a Caribbean island. J. Am. Mosq. Control Assoc. 1991, 7, 7–10.
29. Tikasingh, E.S.; Eustace, A. Suppression of Aedes aegypti by predatory Toxorhynchites moctezuma in an island habitat. Med. Vet. Entomol. 1992, 6, 272–280.
30. Mercer, D.R.; Wettach, G.R.; Smith, J.L. Effects of larval density and predation by Toxorhynchites amboinensis on Aedes polynesiensis (Diptera: Culicidae) developing in coconuts. J. Am. Mosq. Control Assoc. 2005, 21, 425–431.
31. Nyamah, M.A.; Sulaiman, S.; Omar, B. Field observation on the efficacy of Toxorhynchites splendens (Wiedemann) as a biocontrol agent against Aedes albopictus (Skuse) larvae in a cemetery. Trop. Biomed. 2011, 28, 312–319.
32. Nordin, O.; Donald, W.; Ming, W.H.; Ney, T.G.; Mohamed, K.A.; Halim, N.A.; Winskill, P.; Hadi, A.A.; Muhammad, Z.S.; Lacroix, R.; et al. Oral ingestion of transgenic RIDL Ae. aegypti larvae has no negative effect on two predator Toxorhynchites species. PLoS ONE 2013, 8, e58805.
33. Hurst, T.P.; Pittman, G.; O'Neill, S.L.; Ryan, P.A.; Nguyen, H.L.; Kay, B.H. Impacts of Wolbachia infection on predator prey relationships: Evaluating survival and horizontal transfer between wMelPop infected Aedes aegypti and its predators. J. Med. Entomol. 2012, 49, 624–630.
34. Lardeux, F.; Riviere, F.; Sechan, Y.; Kay, B.H. Release of Mesocyclops aspericornis (Copepoda) for control of larval Aedes polynesiensis (Diptera: Culicidae) in land crab burrows on an atoll of French Polynesia. J. Med. Entomol. 1992, 29, 571–576.
35. Kay, B.H.; Lyons, S.A.; Holt, J.S.; Holynska, M.; Russell, B.M. Point source inoculation of Mesocyclops (Copepoda: Cyclopidae) gives widespread control of Ochlerotatus and Aedes (Diptera: Culicidae) immatures in service manholes and pits in north Queensland, Australia. J. Med. Entomol. 2002, 39, 469–474.
36. Suarez, M.F.; Marten, G.G.; Clark, G.G. A simple method for cultivating freshwater copepods used in biological control of Aedes aegypti. J. Am. Mosq. Control Assoc. 1992, 8, 409–412.
37. Pernia, J.; De Zoppi, R.E.; Palacios-Caceres, M. Prey-predator relationship between the cyclopoids Mesocyclops longisetus and Mesocyclops meridianus with Anopheles aquasalis larvae. J. Am. Mosq. Control Assoc. 2007, 23, 166–171.
38. Soumare, M.K.; Cilek, J.E.; Schreibers, E.T. Prey and size preference of Mesocyclops longisetus (Copepoda) for Aedes albopictus and Culex quinquefasciatus larvae. J. Am. Mosq. Control Assoc. 2004, 20, 305–310.

39. Calliari, D.; Sanz, K.; Martinez, M.; Cervetto, G.; Gomez, M.; Basso, C. Comparison of the predation rate of freshwater cyclopoid copepod species on larvae of the mosquito Culex pipiens. Med. Vet. Entomol. 2003, 17, 339–342.
40. Micieli, M.V.; Marti, G.; Garcia, J.J. Laboratory evaluation of Mesocyclops annulatus (Wierzejski, 1892) (Copepoda: Cyclopidea) as a predator of container-breeding mosquitoes in Argentina. Mem. I. Oswaldo. Cruz. 2002, 97, 835–838.
41. Chaper, S.; Hernandez-Chavarria, F. Scanning electron microscopy of damage caused by Mesocyclops thermocyclopoides (Copepoda: Cyclopoidea) on larvae of the dengue fever vector Aedes aegypti (Diptera: Culicidae). Rev. Biol. Trop. 2006, 54, 843–846.
42. Dieng, H.; Boots, M.; Tuno, N.; Tsuda, Y.; Takagi, M. A laboratory and field evaluation of Macrocyclops distinctus, Megacyclops viridis and Mesocyclops pehpeiensis as control agents of the dengue vector Aedes albopictus in a peridomestic area in Nagasaki, Japan. Med. Vet. Entomol. 2002, 16, 285–291.
43. Nam, V.S.; Yen, N.T.; Holynska, M.; Reid, J.W.; Kay, B.H. National progress in dengue vector control in Vietnam: Survey for Mesocyclops (Copepoda), Micronecta (Corixidae), and fish as biological control agents. Am. J. Trop. Med. Hyg. 2000, 62, 5–10.
44. Kay, B.H.; Nam, V.S.; Tien, T.V.; Yen, N.T.; Phong, T.V.; Diep, V.T.; Ninh, T.U.; Bektas, A.; Aaskov, J.G. Control of Aedes vectors of dengue in three provinces of Vietnam by use of Mesocyclops (Copepoda) and community-based methods validated by entomologic, clinical, and serological surveillance. Am. J. Trop. Med. Hyg. 2002, 66, 40–48.
45. Vu, S.N.; Nguyen, T.Y.; Tran, V.P.; Truong, U.N.; Le, Q.M.; Le, V.L.; Le, T.N.; Bektas, A.; Briscombe, A.; Aaskov, J.G.; et al. Elimination of dengue by community programs using Mesocyclops (Copepoda) against Aedes aegypti in central Vietnam. Am. J. Trop. Med. Hyg. 2005, 72, 67–73.
46. Kay, B.H.; Tuyet Hanh, T.T.; Le, N.H.; Quy, T.M.; Nam, V.S.; Hang, P.V.; Yen, N.T.; Hill, P.S.; Vos, T.; Ryan, P.A. Sustainability and cost of a community-based strategy against Aedes aegypti in northern and central Vietnam. Am. J. Trop. Med. Hyg. 2010, 82, 822–830.
47. Muspratt, J. Destruction of the larvae of Anopheles gambiae giles by a Coelomomyces fungus. Bull. World Health Organ.1963, 29, 81–86.
48. Sanad, M.M.; Shamseldean, M.S.M.; Elgindi, A.-E.Y.; Gaugler, R. Host penetration and emergence patterns of the mosquito-parasitic mermithids Romanomermis iyengari and Strelkovimermis spiculatus (Nematoda: Mermithidae). J. Nematol. 2013, 45, 30–38.
49. Creighton, C.S.; McFadden, T.L.; Cuthbert, R.B.; Onsager, J.A. Control of four species of caterpillars on cabbage with Bacillus thuringiensis var. Alesti, 1969–70. J. Econ. Entomol. 1972, 65, 1399–1402.
50. Bellows, T.S.; Fisher, T.W. Handbook of Biological Control: Principles and Applications of Biological Control; Academic Press: San Diego, CA, USA, 1999.
51. Boonserm, P.; Davis, P.; Ellar, D.J.; Li, J. Crystal structure of the mosquito-larvicidal toxin Cry4Ba and its biological implications. J. Mol. Biol. 2005, 348, 363–382.
52. Sriwimol, W.; Aroonkesorn, A.; Sakdee, S.; Kanchanawarin, C.; Uchihashi, T.; Ando, T.; Angsuthanasombat, C. Potential prepore trimer formation by the Bacillus thuringiensis mosquito-specific toxin: Molecular insights into a critical prerequisite of membrane-bound monomers. J. Biol. Chem. 2015, 290, 20793–20803.
53. Canton, P.E.; Cancino-Rodezno, A.; Gill, S.S.; Soberon, M.; Bravo, A. Transcriptional cellular responses in midgut tissue of Aedes aegypti larvae following intoxication with Cry11Aa toxin from Bacillus thuringiensis. BMC Genomics2015, 16, 1–17.
54. Pérez, C.; Fernandez, L.E.; Sun, J.; Folch, J.L.; Gill, S.S.; Sóberon, M.; Bravo, A. Bacillus thuringiensis subsp. israelensis Cyt1Aa synergizes Cry11Aa toxin by functioning as a membrane-bound receptor. Proc. Natl. Acad. Sci. USA 2005, 102, 18303–18308.

55. Wirth, M.C.; Georghiou, G.P.; Federici, B.A. CytA enables CryIV endotoxins of Bacillus thuringiensis to overcome high levels of CryIV resistance in the mosquito, Culex quinquefasciatus. Proc. Natl. Acad. Sci. USA 1997, 94, 10536–10540.
56. Cokmus, C.; Davidson, E.W.; Cooper, K. Electrophysiological effects of Bacillus sphaericus binary toxin on cultured mosquito cells. J. Invertebr. Pathol. 1997, 69, 197–204.
57. Schwartz, J.L.; Potvin, L.; Coux, F.; Charles, J.F.; Berry, C.; Humphreys, M.J.; Jones, A.F.; Bernhart, I.; Dalla Serra, M.; Menestrina, G. Permeabilization of model lipid membranes by Bacillus sphaericus mosquitocidal binary toxin and its individual components. J. Membr. Biol. 2001, 184, 171–183.
58. Davidson, E.W. Variation in binding of Bacillus sphaericus toxin and wheat germ agglutinin to larval midgut cells of six species of mosquitoes. J. Invertebr. Pathol. 1989, 53, 251–259.
59. Rodcharoen, J.; Mulla, M.S. Cross-resistance to Bacillus sphaericus strains in Culex quinquefasciatus. J. Am. Mosq. Control Assoc. 1996, 12, 247–250.
60. Nielsen–Leroux, C.; Pasquier, F.; Charles, J.F.; Sinegre, G.; Gaven, B.; Pasteur, N. Resistance to Bacillus sphaericus involves different mechanisms in Culex pipiens (Diptera: Culicidae) larvae. J. Med. Entomol. 1997, 34, 321–327.
61. Poncet, S.; Bernard, C.; Dervyn, E.; Cayley, J.; Klier, A.; Rapoport, G. Improvement of Bacillus sphaericus toxicity against dipteran larvae by integration, via homologous recombination, of the Cry11A toxin gene from Bacillus thuringiensis subsp. israelensis. Appl. Environ. Microbiol. 1997, 63, 4413–4420.
62. Federici, B.A.; Park, H.W.; Bideshi, D.K.; Wirth, M.C.; Johnson, J.J. Recombinant bacteria for mosquito control. J. Exp. Biol. 2003, 206, 3877–3885.
63. Federici, B.A.; Park, H.W.; Bideshi, D.K.; Wirth, M.C.; Johnson, J.J.; Sakano, Y.; Tang, M. Developing recombinant bacteria for control of mosquito larvae. J. Am. Mosq. Control Assoc. 2007, 23, 164–175.
64. Alto, B.W.; Lord, C.C. Transstadial effects of Bti on traits of Aedes aegypti and infection with dengue virus. PLoS Negl. Trop. Dis. 2016, 10, e0004370.
65. Boyce, R.; Lenhart, A.; Kroeger, A.; Velayudhan, R.; Roberts, B.; Horstick, O. Bacillus thuringiensis israelensis (Bti) for the control of dengue vectors: Systematic literature review. Trop. Med. Int. Health 2013, 18, 564–577.
66. Darbro, J.M.; Johnson, P.H.; Thomas, M.B.; Ritchie, S.A.; Kay, B.H.; Ryan, P.A. Effects of Beauveria bassiana on survival, blood-feeding success, and fecundity of Aedes aegypti in laboratory and semi-field conditions. Am. J. Trop. Med. Hyg.2012, 86, 656–664.
67. Kerwin, J.L.; Washino, R.K. Ground and aerial application of the asexual stage of Lagenidium giganteum for control of mosquitoes associated with rice culture in the central valley of California. J. Am. Mosq. Control Assoc. 1987, 3, 59–64.
68. Rueda, L.M.; Patel, K.J.; Axtell, R.C. Efficacy of encapsulated Lagenidium giganteum (Oomycetes: Lagenidiales) against Culex quinquefasciatus and Aedes aegypti larvae in artificial containers. J. Am. Mosq. Control Assoc. 1990, 6, 694–699.
69. Guzman, D.R.; Axtell, R.C. Population dynamics of Culex quinquefasciatus and the fungal pathogen Lagenidium giganteum (Oomycetes: Lagenidiales) in stagnant water pools. J. Am. Mosq. Control Assoc. 1987, 3, 442–449.
70. Hallmon, C.F.; Schreiber, E.T.; Vo, T.; Bloomquist, A. Field trials of three concentrations of Laginex as biological larvicide compared to Vectobac-12as as a biocontrol agent for Culex quinquefasciatus. J. Am. Mosq. Control Assoc. 2000, 16, 5–8.
71. Orduz, S.; Axtell, R.C. Compatibility of Bacillus thuringiensis var. Israelensis and Bacillus sphaericus with the fungal pathogen Lagenidium giganteum (Oomycetes: Lagenidiales). J. Am. Mosq. Control Assoc. 1991, 7, 188–193.
72. Golkar, L.; LeBrun, R.A.; Ohayon, H.; Gounon, P.; Papierok, B.; Brey, P.T. Variation of larval susceptibility to Lagenidium giganteum in three mosquito species. J. Invertebr. Pathol. 1993, 62, 1–8.

73. Brey, P.T.; Lebrun, R.A.; Papierok, B.; Ohayon, H.; Vennavalli, S.; Hafez, J. Defense reactions by larvae of Aedes aegypti during infection by the aquatic fungus Lagenidium giganteum (Oomycete). Cell Tissue Res. 1988, 253, 245–250.
74. Patel, K.J.; Rueda, L.M.; Axtell, R.C.; Stinner, R.E. Temperature-dependent development of the fungal pathogen Lagenidium giganteum (Oomycetes: Lagenidiales) in larvae of Culex quinquefasciatus (Diptera: Culicidae). J. Med. Entomol. 1991, 28, 95–100.
75. Scholte, E.J.; Knols, B.G.; Samson, R.A.; Takken, W. Entomopathogenic fungi for mosquito control: A review. J. Insect Sci. 2004, 4, 19.
76. Riba, G.; Keita, A.; Soares, G.G., Jr.; Ferron, P. Comparative studies of Metarhizium anisopliae and Tolypocladium cylindrosporum as pathogens of mosquito larvae. J. Am. Mosq. Control Assoc. 1986, 2, 469–473.
77. Agudelo-Silva, F.; Wassink, H. Infectivity of a Venezuelan strain of Metarhizium anisopliae to Aedes aegypti larvae. J. Invertebr. Pathol. 1984, 43, 435–436.
78. Lacey, C.M.; Lacey, L.A.; Roberts, D.R. Route of invasion and histopathology of Metarhizium anisopliae in Culex quinquefasciatus. J. Invertebr. Pathol. 1988, 52, 108–118.
79. Butt, T.M.; Greenfield, B.P.; Greig, C.; Maffeis, T.G.; Taylor, J.W.; Piasecka, J.; Dudley, E.; Abdulla, A.; Dubovskiy, I.M.; Garrido-Jurado, I.; et al. Metarhizium anisopliae pathogenesis of mosquito larvae: A verdict of accidental death. PLoS ONE 2013, 8, e81686.
80. Alkhaibari, A.M.; Carolino, A.T.; Yavasoglu, S.I.; Maffeis, T.; Mattoso, T.C.; Bull, J.C.; Samuels, R.I.; Butt, T.M. Metarhizium brunneum blastospore pathogenesis in Aedes aegypti larvae: Attack on several fronts accelerates mortality. PLoS Pathog. 2016, 12, e1005715.
81. Hertig, M.; Wolbach, S.B. Studies on Rickettsia-like micro-organisms in insects. J. Med. Res. 1924, 44, 329–374.
82. O'Neill, S.L.; Pettigrew, M.M.; Sinkins, S.P.; Braig, H.R.; Andreadis, T.G.; Tesh, R.B. In vitro cultivation of Wolbachia pipientis in an Aedes albopictus cell line. Insect Mol. Biol. 1997, 6, 33–39.
83. Dobson, S.L.; Marsland, E.J.; Veneti, Z.; Bourtzis, K.; O'Neill, S.L. Characterization of Wolbachia host cell range via the in vitro establishment of infections. Appl. Environ. Microbiol. 2002, 68, 656–660.
84. O'Neill, S.L.; Giordano, R.; Colbert, A.M.; Karr, T.L.; Robertson, H.M. 16S rRNA phylogenetic analysis of the bacterial endosymbionts associated with cytoplasmic incompatibility in insects. Proc. Natl. Acad. Sci. USA 1992, 89, 2699–2702.
85. Zhou, W.; Rousset, F.; O'Neil, S. Phylogeny and PCR-based classification of Wolbachia strains using wsp gene sequences. Proc. Biol. Sci. 1998, 265, 509–515.
86. Van Meer, M.M.; Witteveldt, J.; Stouthamer, R. Phylogeny of the arthropod endosymbiont Wolbachia based on the wsp gene. Insect Mol. Biol. 1999, 8, 399–408.
87. Yen, J.H. Transovarial transmission of Rickettsia-like microorganisms in mosquitoes. Ann. N. Y. Acad. Sci. 1975, 266, 152–161.
88. Laven, H. Eradication of Culex pipiens fatigans through cytoplasmic incompatibility. Nature 1967, 216, 383–384.
89. Curtis, C.F.; Adak, T. Population replacement in Culex fatigens by means of cytoplasmic incompatibility. Bull. World Health Organ. 1974, 51, 249–255.
90. Yen, J.H.; Barr, A.R. New hypothesis of the cause of cytoplasmic incompatibility in Culex pipiens L. Nature 1971, 232, 657–658.
91. Yen, J.H.; Barr, A.R. The etiological agent of cytoplasmic incompatibility in Culex pipiens. J. Invertebr. Pathol. 1973, 22, 242–250.
92. Braig, H.R.; Guzman, H.; Tesh, R.B.; O'Neill, S.L. Replacement of the natural Wolbachia symbiont of Drosophila simulans with a mosquito counterpart. Nature 1994, 367, 453–455.
93. Coon, K.L.; Brown, M.R.; Strand, M.R. Mosquitoes host communities of bacteria that are essential for development but vary greatly between local habitats. Mol. Ecol. 2016, 25, 5806–5826.

94. Brelsfoard, C.L.; Sechan, Y.; Dobson, S.L. Interspecific hybridization yields strategy for South Pacific filariasis vector elimination. PLoS Negl. Trop. Dis. 2008, 2, e129.
95. Xi, Z.; Khoo, C.C.; Dobson, S.L. Wolbachia establishment and invasion in an Aedes aegypti laboratory population. Science 2005, 310, 326–328.
96. Xi, Z.; Dean, J.L.; Khoo, C.; Dobson, S.L. Generation of a novel Wolbachia infection in Aedes albopictus (Asian tiger mosquito) via embryonic microinjection. Insect Biochem. Mol. Biol. 2005, 35, 903–910.
97. Gubler, D.J. Epidemic dengue/dengue hemorrhagic fever as a public health, social and economic problem in the 21st century. Trends Microbiol. 2002, 10, 100–103.
98. McElroy, K.L.; Girard, Y.A.; McGee, C.E.; Tsetsarkin, K.A.; Vanlandingham, D.L.; Higgs, S. Characterization of the antigen distribution and tissue tropisms of three phenotypically distinct yellow fever virus variants in orally infected Aedes aegypti mosquitoes. Vector Borne Zoonotic. Dis. 2008, 8, 675–687.
99. Salazar, M.I.; Richardson, J.H.; Sanchez-Vargas, I.; Olson, K.E.; Beaty, B.J. Dengue virus type 2: Replication and tropisms in orally infected Aedes aegypti mosquitoes. BMC microbiol. 2007, 7, 9–21.
100. McMeniman, C.J.; Lane, R.V.; Cass, B.N.; Fong, A.W.; Sidhu, M.; Wang, Y.F.; O'Neill, S.L. Stable introduction of a life-shortening Wolbachia infection into the mosquito Aedes aegypti. Science 2009, 323, 141–144.
101. Turley, A.P.; Moreira, L.A.; O'Neill, S.L.; McGraw, E.A. Wolbachia infection reduces blood-feeding success in the dengue fever mosquito, Aedes aegypti. PLoS Negl. Trop. Dis. 2009, 3, e516.
102. Hedges, L.M.; Brownlie, J.C.; O'Neill, S.L.; Johnson, K.N. Wolbachia and virus protection in insects. Science 2008, 322, 702.
103. Walker, T.; Johnson, P.H.; Moreira, L.A.; Iturbe-Ormaetxe, I.; Frentiu, F.D.; McMeniman, C.J.; Leong, Y.S.; Dong, Y.; Axford, J.; Kriesner, P.; et al. The wMel Wolbachia strain blocks dengue and invades caged Aedes aegypti populations. Nature 2011, 476, 450–453.
104. Moreira, L.A.; Iturbe-Ormaetxe, I.; Jeffery, J.A.; Lu, G.; Pyke, A.T.; Hedges, L.M.; Rocha, B.C.; Hall-Mendelin, S.; Day, A.; Riegler, M.; et al. A Wolbachia symbiont in Aedes aegypti limits infection with dengue, chikungunya, and plasmodium. Cell 2009, 139, 1268–1278.
105. Van Den Hurk, A.F.; Hall-Mendelin, S.; Pyke, A.T.; Frentiu, F.D.; McElroy, K.; Day, A.; Higgs, S.; O'Neill, S.L. Impact of Wolbachia on infection with chikungunya and yellow fever viruses in the mosquito vector Aedes aegypti. PLoS Negl. Trop. Dis. 2012, 6, e1892.
106. Aliota, M.T.; Walker, E.C.; Uribe Yepes, A.; Velez, I.D.; Christensen, B.M.; Osorio, J.E. The wMel strain of Wolbachia reduces transmission of chikungunya virus in Aedes aegypti. PLoS Negl. Trop. Dis. 2016, 10, e0004677.
107. Hussain, M.; Lu, G.; Torres, S.; Edmonds, J.H.; Kay, B.H.; Khromykh, A.A.; Asgari, S. Effect of Wolbachia on replication of West Nile virus in a mosquito cell line and adult mosquitoes. J. Virol. 2013, 87, 851–858.
108. Glaser, R.L.; Meola, M.A. The native Wolbachia endosymbionts of Drosophila melanogaster and Culex quinquefasciatus increase host resistance to West Nile virus infection. PLoS ONE 2010, 5, e11977.
109. Bian, G.; Zhou, G.; Lu, P.; Xi, Z. Replacing a native Wolbachia with a novel strain results in an increase in endosymbiont load and resistance to dengue virus in a mosquito vector. PLoS Negl. Trop. Dis. 2013, 7, e2250.
110. Raquin, V.; Valiente Moro, C.; Saucereau, Y.; Tran, F.H.; Potier, P.; Mavingui, P. Native Wolbachia from Aedes albopictus blocks chikungunya virus infection in cellulo. PLoS ONE 2015, 10, e0125066.
111. Dodson, B.L.; Hughes, G.L.; Paul, O.; Matacchiero, A.C.; Kramer, L.D.; Rasgon, J.L. Wolbachia enhances West Nile virus (WNV) infection in the mosquito Culex tarsalis. PLoS Negl. Trop. Dis. 2014, 8, e2965.
112. Rainey, S.M.; Martinez, J.; McFarlane, M.; Juneja, P.; Sarkies, P.; Lulla, A.; Schnettler, E.; Varjak, M.; Merits, A.; Miska, E.A.; et al. Wolbachia blocks viral genome replication

early in infection without a transcriptional response by the endosymbiont or host small RNA pathways. PLoS Pathog. 2016, 12, e1005536.

113. Wong, Z.S.; Brownlie, J.C.; Johnson, K.N. Oxidative stress correlates with Wolbachia-mediated antiviral protection in Wolbachia-Drosophila associations. Appl. Environ. Microbiol. 2015, 81, 3001–3005.

114. Chrostek, E.; Marialva, M.S.; Yamada, R.; O'Neill, S.L.; Teixeira, L. High anti-viral protection without immune upregulation after interspecies Wolbachia transfer. PLoS ONE 2014, 9, e99025.

115. Martinez, J.; Longdon, B.; Bauer, S.; Chan, Y.S.; Miller, W.J.; Bourtzis, K.; Teixeira, L.; Jiggins, F.M. Symbionts commonly provide broad spectrum resistance to viruses in insects: A comparative analysis of Wolbachia strains. PLoS Pathog. 2014, 10, e1004369.

116. Bourtzis, K.; Pettigrew, M.M.; O'Neill, S.L. Wolbachia neither induces nor suppresses transcripts encoding antimicrobial peptides. Insect Mol. Biol. 2000, 9, 635–639.

117. Sinkins, S.P.; Braig, H.R.; O'Neill, S.L. Wolbachia superinfections and the expression of cytoplasmic incompatibility. Proc. Biol. Sci. 1995, 261, 325–330.

118. Rousset, F.; Braig, H.R.; O'Neill, S.L. A stable triple Wolbachia infection in Drosophila with nearly additive incompatibility effects. Heredity 1999, 82, 620–627.

119. Fu, Y.; Gavotte, L.; Mercer, D.R.; Dobson, S.L. Artificial triple Wolbachia infection in Aedes albopictus yields a new pattern of unidirectional cytoplasmic incompatibility. Appl. Environ. Microbiol. 2010, 76, 5887–5891.

120. O'Connor, L.; Plichart, C.; Sang, A.C.; Brelsfoard, C.L.; Bossin, H.C.; Dobson, S.L. Open release of male mosquitoes infected with a Wolbachia biopesticide: Field performance and infection containment. PLoS Negl. Trop. Dis. 2012, 6, e1797.

121. Hoffmann, A.A.; Montgomery, B.L.; Popovici, J.; Iturbe-Ormaetxe, I.; Johnson, P.H.; Muzzi, F.; Greenfield, M.; Durkan, M.; Leong, Y.S.; Dong, Y.; et al. Successful establishment of Wolbachia in Aedes populations to suppress dengue transmission. Nature 2011, 476, 454–457.

122. Frentiu, F.D.; Zakir, T.; Walker, T.; Popovici, J.; Pyke, A.T.; Van Den Hurk, A.; McGraw, E.A.; O'Neill, S.L. Limited dengue virus replication in field-collected Aedes aegypti mosquitoes infected with Wolbachia. PLoS Negl. Trop. Dis.2014, 8, e2688.

123. Ye, Y.H.; Carrasco, A.M.; Dong, Y.; Sgro, C.M.; McGraw, E.A. The effect of temperature on Wolbachia-mediated dengue virus blocking in Aedes aegypti. Am. J. Trop. Med. Hyg. 2016, 94, 812–819.

124. Higgs, S. Alternative approaches to control dengue and chikungunya. Bayer Public Health 2013, 24, 35–42.

125. Franz, A.W.; Sanchez-Vargas, I.; Adelman, Z.N.; Blair, C.D.; Beaty, B.J.; James, A.A.; Olson, K.E. Engineering RNA interference-based resistance to dengue virus type 2 in genetically modified Aedes aegypti. Proc. Natl. Acad. Sci. USA2006, 103, 4198–4203.

126. Carter, J.R.; Keith, J.H.; Barde, P.V.; Fraser, T.S.; Fraser, M.J., Jr. Targeting of highly conserved dengue virus sequences with anti-dengue virus trans-splicing group I introns. BMC Mol. Biol. 2010, 11, 84–97.

127. Carter, J.R.; Keith, J.H.; Fraser, T.S.; Dawson, J.L.; Kucharski, C.A.; Horne, K.M.; Higgs, S.; Fraser, M.J., Jr. Effective suppression of dengue virus using a novel group-I intron that induces apoptotic cell death upon infection through conditional expression of the Bax C-terminal domain. Virol. J. 2014, 11, 111–128.

128. Nawtaisong, P.; Keith, J.; Fraser, T.; Balaraman, V.; Kolokoltsov, A.; Davey, R.A.; Higgs, S.; Mohammed, A.; Rongsriyam, Y.; Komalamisra, N.; et al. Effective suppression of dengue fever virus in mosquito cell cultures using retroviral transduction of hammerhead ribozymes targeting the viral genome. Virol. J. 2009, 6, 73–89.

129. Isaacs, A.T.; Jasinskiene, N.; Tretiakov, M.; Thiery, I.; Zettor, A.; Bourgouin, C.; James, A.A. Transgenic Anopheles stephensi coexpressing single-chain antibodies resist Plasmodium falciparum development. Proc. Natl. Acad. Sci. USA2012, 109, 11070–11071.

130. Bowman, D.D. Successful and currently ongoing parasite eradication programs. Vet. Parasitol. 2006, 139, 293–307.

131. Dame, D.A.; Curtis, C.F.; Benedict, M.Q.; Robinson, A.S.; Knols, B.G. Historical applications of induced sterilisation in field populations of mosquitoes. Malar. J. 2009, 8, S2–S11.
132. Andreasen, M.H.; Curtis, C.F. Optimal life stage for radiation sterilization of Anopheles males and their fitness for release. Med. Vet. Entomol. 2005, 19, 238–244.
133. Labrecque, G.C.; Bowman, M.C.; Patterson, R.S.; Seawright, J.A. Persistence of thiotepa and tepa in pupae and adults of Culex pipiens fatigans Wiedemann. Bull. World Health Organ. 1972, 47, 675–676.
134. Phuc, H.K.; Andreasen, M.H.; Burton, R.S.; Vass, C.; Epton, M.J.; Pape, G.; Fu, G.; Condon, K.C.; Scaife, S.; Donnelly, C.A.; et al. Late-acting dominant lethal genetic systems and mosquito control. BMC Biol. 2007, 5, 11–21.
135. Miller, L.H.; Sakai, R.K.; Romans, P.; Gwadz, R.W.; Kantoff, P.; Coon, H.G. Stable integration and expression of a bacterial gene in the mosquito Anopheles gambiae. Science 1987, 237, 779–781.
136. McGrane, V.; Carlson, J.O.; Miller, B.R.; Beaty, B.J. Microinjection of DNA into Aedes triseriatus ova and detection of integration. Am. J. Trop. Med. Hyg. 1988, 39, 502–510.
137. Morris, A.C.; Eggleston, P.; Crampton, J.M. Genetic transformation of the mosquito Aedes aegypti by micro-injection of DNA. Med. Vet. Entomol. 1989, 3, 1–7.
138. Nimmo, D.D.; Alphey, L.; Meredith, J.M.; Eggleston, P. High efficiency site-specific genetic engineering of the mosquito genome. Insect Mol. Biol. 2006, 15, 129–136.
139. Morris, A.C.; Schaub, T.L.; James, A.A. Flp-mediated recombination in the vector mosquito, Aedes aegypti. Nucleic Acids Res. 1991, 19, 5895–5900.
140. Haghighat-Khah, R.E.; Scaife, S.; Martins, S.; St John, O.; Matzen, K.J.; Morrison, N.; Alphey, L. Site-specific cassette exchange systems in the Aedes aegypti mosquito and the Plutella xylostella moth. PLoS ONE 2015, 10, e0121097.
141. Labbe, G.M.; Nimmo, D.D.; Alphey, L. piggybac- and PhiC31-mediated genetic transformation of the Asian tiger mosquito, Aedes albopictus (Skuse). PLoS Negl. Trop. Dis. 2010, 4, e788.
142. Jasinskiene, N.; Coates, C.J.; Benedict, M.Q.; Cornel, A.J.; Rafferty, C.S.; James, A.A.; Collins, F.H. Stable transformation of the yellow fever mosquito, Aedes aegypti, with the Hermes element from the housefly. Proc. Natl. Acad. Sci. USA 1998, 95, 3743–3747.
143. Lobo, N.; Li, X.; Hua-Van, A.; Fraser, M.J., Jr. Mobility of the piggyBac transposon in embryos of the vectors of Dengue fever (Aedes albopictus) and La Crosse encephalitis (Ae. triseriatus). Mol. Genet. Genomics 2001, 265, 66–71.
144. Grossman, G.L.; Rafferty, C.S.; Fraser, M.J.; Benedict, M.Q. The piggybac element is capable of precise excision and transposition in cells and embryos of the mosquito, Anopheles gambiae. Insect Biochem. Mol. Biol. 2000, 30, 909–914.
145. Lobo, N.; Li, X.; Fraser, M.J., Jr. Transposition of the piggyBac element in embryos of Drosophila melanogaster, Aedes aegypti and Trichoplusia ni. Mol. Gen. Genet. 1999, 261, 803–810.
146. Matsubara, T.; Beeman, R.W.; Shike, H.; Besansky, N.J.; Mukabayire, O.; Higgs, S.; James, A.A.; Burns, J.C. Pantropic retroviral vectors integrate and express in cells of the malaria mosquito, Anopheles gambiae. Proc. Natl. Acad. Sci. USA1996, 93, 6181–6185.
147. Aryan, A.; Anderson, M.A.; Myles, K.M.; Adelman, Z.N. Talen-based gene disruption in the dengue vector Aedes aegypti. PLoS ONE 2013, 8, e60082.
148. Kistler, K.E.; Vosshall, L.B.; Matthews, B.J. Genome engineering with CRISPR-Cas9 in the mosquito Aedes aegypti. Cell Rep. 2015, 11, 51–60.
149. Dong, S.; Lin, J.; Held, N.L.; Clem, R.J.; Passarelli, A.L.; Franz, A.W. Heritable CRISPR/Cas9-mediated genome editing in the yellow fever mosquito, Aedes aegypti. PLoS ONE 2015, 10, e0122353.
150. Basu, S.; Aryan, A.; Overcash, J.M.; Samuel, G.H.; Anderson, M.A.; Dahlem, T.J.; Myles, K.M.; Adelman, Z.N. Silencing of end-joining repair for efficient site-specific gene

insertion after TALEN/CRISPR mutagenesis in Aedes aegypti. Proc. Natl. Acad. Sci. USA 2015, 112, 4038–4043.
151. Carlson, J.; Olson, K.; Higgs, S.; Beaty, B. Molecular genetic manipulation of mosquito vectors. Annu. Rev. Entomol.1995, 40, 359–388.
152. Kovach, M.J.; Carlson, J.O.; Beaty, B.J. A Drosophila metallothionein promoter is inducible in mosquito cells. Insect Mol. Biol. 1992, 1, 37–43.
153. Beaty, B.J.; Black, W.C.; Carlson, J.O.; Clements, W.H.; DuTeau, N.; Harrahy, E.; Nuckols, J.; Kenneth, E.; Olson, K.E.; Rayms-Keller, A. Molecular and genetic ecotoxicologic approaches to aquatic environmental bioreporting. Environ. Health Perspect. 1998, 106, 1395–1407.
154. Kokoza, V.; Ahmed, A.; Cho, W.L.; Jasinskiene, N.; James, A.A.; Raikhel, A. Engineering blood meal-activated systemic immunity in the yellow fever mosquito, Aedes aegypti. Proc. Natl. Acad. Sci. USA 2000, 97, 9144–9149.
155. Moreira, L.A.; Edwards, M.J.; Adhami, F.; Jasinskiene, N.; James, A.A.; Jacobs-Lorena, M. Robust gut-specific gene expression in transgenic Aedes aegypti mosquitoes. Proc. Natl. Acad. Sci. USA 2000, 97, 10895–10898.
156. Edwards, M.J.; Moskalyk, L.A.; Donelly-Doman, M.; Vlaskova, M.; Noriega, F.G.; Walker, V.K.; Jacobs-Lorena, M. Characterization of a carboxypeptidase A gene from the mosquito, Aedes aegypti. Insect Mol. Biol. 2000, 9, 33–38.
157. Cornel, A.J.; Benedict, M.Q.; Rafferty, C.S.; Howells, A.J.; Collins, F.H. Transient expression of the Drosophila melanogaster cinnabar gene rescues eye color in the white eye (WE) strain of Aedes aegypti. Insect Biochem. Mol. Biol.1997, 27, 993–997.
158. Pinkerton, A.C.; Michel, K.; O'Brochta, D.A.; Atkinson, P.W. Green fluorescent protein as a genetic marker in transgenic Aedes aegypti. Insect Mol. Biol. 2000, 9, 1–10.
159. Kokoza, V.; Ahmed, A.; Wimmer, E.A.; Raikhel, A.S. Efficient transformation of the yellow fever mosquito Aedes aegypti using the piggyBac transposable element vector pBac [3xP3-EGFP afm]. Insect Biochem. Mol. Biol. 2001, 31, 1137–1143.
160. Sinkins, S.P.; Gould, F. Gene drive systems for insect disease vectors. Nat. Rev. Genet. 2006, 7, 427–435.
161. Diamant, G. Screwworm eradication in southeastern United States. Am. J. public health Nations Health 1963, 53, 22–26.
162. Gantz, V.M.; Bier, E. The mutagenic chain reaction: A method for converting heterozygous to homozygous mutations. Science 2015, 348, 442–444
163. Troupin, A.; Londono-Renteria, B.; Conway, M.J.; Cloherty, E.; Jameson, S.; Higgs, S.; Vanlandingham, D.L.; Fikrig, E.; Colpitts, T.M. A novel mosquito ubiquitin targets viral envelope protein for degradation and reduces virion production during dengue virus infection. Biochim. Biophys. Acta 2016, 1860, 1898–1909.
164. Cheng, G.; Liu, L.; Wang, P.; Zhang, Y.; Zhao, Y.O.; Colpitts, T.M.; Feitosa, F.; Anderson, J.F.; Fikrig, E. An in vivo transfection approach elucidates a role for Aedes aegypti thioester-containing proteins in flaviviral infection. PLoS ONE 2011, 6, e22786.
165. Powers, A.M.; Kamrud, K.I.; Olson, K.E.; Higgs, S.; Carlson, J.O.; Beaty, B.J. Molecularly engineered resistance to California serogroup virus replication in mosquito cells and mosquitoes. Proc. Natl. Acad. Sci. USA 1996, 93, 4187–4191.
166. Olson, K.E.; Higgs, S.; Gaines, P.J.; Powers, A.M.; Davis, B.S.; Kamrud, K.I.; Carlson, J.O.; Blair, C.D.; Beaty, B.J. Genetically engineered resistance to dengue-2 virus transmission in mosquitoes. Science 1996, 272, 884–886.
167. Higgs, S.; Rayner, J.O.; Olson, K.E.; Davis, B.S.; Beaty, B.J.; Blair, C.D. Engineered resistance in Aedes aegypti to a West African and a South American strain of yellow fever virus. Am. J. Trop. Med. Hyg. 1998, 58, 663–670.
168. Franz, A.W.; Sanchez-Vargas, I.; Piper, J.; Smith, M.R.; Khoo, C.C.; James, A.A.; Olson, K.E. Stability and loss of a virus resistance phenotype over time in transgenic mosquitoes harbouring an antiviral effector gene. Insect Mol. Biol.2009, 18, 661–672.
169. Franz, A.W.; Sanchez-Vargas, I.; Raban, R.R.; Black, W.C.T.; James, A.A.; Olson, K.E. Fitness impact and stability of a transgene conferring resistance to dengue-2 virus

following introgression into a genetically diverse Aedes aegypti strain. PLoS Negl. Trop. Dis. 2014, 8, e2833.
170. Mathur, G.; Sanchez-Vargas, I.; Alvarez, D.; Olson, K.E.; Marinotti, O.; James, A.A. Transgene-mediated suppression of dengue viruses in the salivary glands of the yellow fever mosquito, Aedes aegypti. Insect Mol. Biol. 2010, 19, 753–763.
171. Carter, J.R.; Taylor, S.; Fraser, T.S.; Kucharski, C.A.; Dawson, J.L.; Fraser, M.J., Jr. Suppression of the arboviruses dengue and chikungunya using a dual-acting group-I intron coupled with conditional expression of the Bax C-terminal domain. PLoS ONE 2015, 10, e0139899.
172. Thomas, D.D.; Donnelly, C.A.; Wood, R.J.; Alphey, L.S. Insect population control using a dominant, repressible, lethal genetic system. Science 2000, 287, 2474–2476.
173. Gossen, M.; Bujard, H. Tight control of gene expression in mammalian cells by tetracycline-responsive promoters. Proc. Natl. Acad. Sci. USA 1992, 89, 5547–5551.
174. Gong, P.; Epton, M.J.; Fu, G.; Scaife, S.; Hiscox, A.; Condon, K.C.; Condon, G.C.; Morrison, N.I.; Kelly, D.W.; Dafa'alla, T.; et al. A dominant lethal genetic system for autocidal control of the Mediterranean fruitfly. Nat. Biotechnol. 2005, 23, 453–456.
175. Bargielowski, I.; Nimmo, D.; Alphey, L.; Koella, J.C. Comparison of life history characteristics of the genetically modified OX513A line and a wild type strain of Aedes aegypti. PLoS ONE 2011, 6, e20699.
176. Bargielowski, I.; Alphey, L.; Koella, J.C. Cost of mating and insemination capacity of a genetically modified mosquito Aedes aegypti OX513A compared to its wild type counterpart. PLoS ONE 2011, 6, e26086.
177. Bargielowski, I.; Kaufmann, C.; Alphey, L.; Reiter, P.; Koella, J. Flight performance and teneral energy reserves of two genetically-modified and one wild-type strain of the yellow fever mosquito Aedes aegypti. Vector Borne Zoonotic. Dis.2012, 12, 1053–1058.
178. Massonnet-Bruneel, B.; Corre-Catelin, N.; Lacroix, R.; Lees, R.S.; Hoang, K.P.; Nimmo, D.; Alphey, L.; Reiter, P. Fitness of transgenic mosquito Aedes aegypti males carrying a dominant lethal genetic system. PLoS ONE 2013, 8, e62711.
179. Lacroix, R.; McKemey, A.R.; Raduan, N.; Kwee Wee, L.; Hong Ming, W.; Guat Ney, T.; Rahidah, A.A.S.; Salman, S.; Subramaniam, S.; Nordin, O.; et al. Open field release of genetically engineered sterile male Aedes aegypti in Malaysia. PLoS ONE 2012, 7, e42771.
180. Harris, A.F.; Nimmo, D.; McKemey, A.R.; Kelly, N.; Scaife, S.; Donnelly, C.A.; Beech, C.; Petrie, W.D.; Alphey, L. Field performance of engineered male mosquitoes. Nat. Biotechnol. 2011, 29, 1034–1037.
181. Carvalho, D.O.; McKemey, A.R.; Garziera, L.; Lacroix, R.; Donnelly, C.A.; Alphey, L.; Malavasi, A.; Capurro, M.L. Suppression of a field population of Aedes aegypti in Brazil by sustained release of transgenic male mosquitoes. PLoS Negl. Trop. Dis. 2015, 9, e0003864.
182. Carvalho, D.O.; Nimmo, D.; Naish, N.; McKemey, A.R.; Gray, P.; Wilke, A.B.; Marrelli, M.T.; Virginio, J.F.; Alphey, L.; Capurro, M.L. Mass production of genetically modified Aedes aegypti for field releases in Brazil. J. Vis. Exp. 2014, 83, e3579.
183. Gorman, K.; Young, J.; Pineda, L.; Marquez, R.; Sosa, N.; Bernal, D.; Torres, R.; Soto, Y.; Lacroix, R.; Naish, N.; et al. Short-term suppression of Aedes aegypti using genetic control does not facilitate Aedes albopictus. Pest Manag. Sci.2016, 72, 618–628.
184. Munoz, D.; Jimenez, A.; Marinotti, O.; James, A.A. The AeAct-4 gene is expressed in the developing flight muscles of female Aedes aegypti. Insect Mol. Biol. 2004, 13, 563–568.
185. Fu, G.; Lees, R.S.; Nimmo, D.; Aw, D.; Jin, L.; Gray, P.; Berendonk, T.U.; White-Cooper, H.; Scaife, S.; Kim Phuc, H.; et al. Female-specific flightless phenotype for mosquito control. Proc. Natl. Acad. Sci. USA 2010, 107, 4550–4554.
186. Wise De Valdez, M.R.; Nimmo, D.; Betz, J.; Gong, H.F.; James, A.A.; Alphey, L.; Black, W.C.T. Genetic elimination of dengue vector mosquitoes. Proc. Natl. Acad. Sci. USA 2011, 108, 4772–4775.

187. Labbe, G.M.; Scaife, S.; Morgan, S.A.; Curtis, Z.H.; Alphey, L. Female-specific flightless (fsRIDL) phenotype for control of Aedes albopictus. PLoS Negl. Trop. Dis. 2012, 6, e1724.

CHAPTER 15

Observations on the Biology and Anatomy of Myerslopiidae (Hemiptera, Membracoidea)

Roman Rakitov

Paleontological Institute, Profsoyuznaya Street 123, Moscow 117647, Russia

ABSTRACT

Adults and nymphs of Mapuchea chilensis (Nielson), from the poorly known family Myerslopiidae, were collected from the litter horizon of temperate forests and shrub bogs in southern Chile. The species apparently feeds on roots and creeping stems of angiosperms. Salivary sheaths of captive specimens terminated in vascular bundles. Indirect evidence suggests feeding on phloem sap. Both nymphs and adults are strong jumpers and both actively disperse, as evidenced by their capture in pan traps. The Malpighian tubules of this species produce no brochosomes and, unlike in most previously studied Membracoidea, comprise no specialized secretory segment. Each tubule comprises secretory cells scattered among excretory ones, a condition not previously known among Hemiptera.

1. INTRODUCTION

The family Myerslopiidae includes three genera, two in New Zealand and one in Chile, altogether containing 19 morphologically similar species [1]. Adult myerslopiids are stout, strongly sclerotized insects, 3 to 7.5 mm in length, with elytra-like, medially sealed tegmina, and without traces of hind wings underneath. Both adults and nymphs dwell cryptically on or near the ground, and their integument, sculptured with ridges and protuberances, is usually camouflaged with hardened soil (Figures 1(a)–1(h)). Members of the family have formerly been classified as a tribe within the leafhopper subfamily Ulopinae (Cicadellidae) but more recently were recognized as a sister lineage to the rest of Membracoidea (leafhoppers + treehoppers), which shares some plesiomorphic traits with cicadoids and cercopoids [2]. The disjunct range of the group has been interpreted as evidence of its considerable age and association with Gondwanaland [1, 2].

Until now the main sources of information on the myerslopiids biology were collecting records of the New Zealand species, the majority of which described the habitat as simply "moss" or "litter" [3]. This led Hamilton to assume that myerslopiids, like peloridiids, are "associated with moss rather than with higher plants." Bryophagy has not been known in Cicadomorpha and is exceptionally rare among other Auchenorrhyncha [4]. Other authors referred to myerslopiids as possible fungivores [1] or even predators [5]. The life history of myerslopiids thus merits clarification.

Also of interest is the activity of the Malpighian tubules of myerslopiids. Secretions of these organs play important roles in the biology of different groups of Cicadomorpha [6–8]. For example, cicadellids actively coat their integuments with water-repellent sheaths of brochosomes, proteinaceous products of specialized Malpighian tubule segments [9–12]. It has been argued that the production of brochosomes had evolved in the common ancestor of Membracoidea [13]. Because myerslopiids have been identified as being among the few leafhopper taxa lacking brochosomes [14], data on their Malpighian tubules are much needed.

An expedition to temperate Chile during the austral summer of 2013-2014 gave the author an opportunity to clarify the microhabitat of myerslopiids, observe behaviors of captive specimens, and preserve material for anatomical studies. A note on the myerslopiids karyotype and tests has already been published [15]. Here, I present ecological, behavioral, and anatomical observations to demonstrate that myerslopiids present no significant departures from cicadellids in their life history traits, but their Malpighian tubules have a different organization, not previously known among Hemiptera.

Figure 1: Mapuchea chilensis. ((a), (b), (c), and (d)) Specimens from the Huerquehue NP. (a) Scanning electron image of a female molted in captivity and not camouflaged with soil; the left tegmen is mangled. (b) The same specimen live, showing color pattern. (c) Freshly molted female. (d) Camouflaged specimen. (e) Specimen from the Puyehue NP. (f) Soil camouflage in combination with disruptive coloration of legs makes this specimen hard to see; note that the eyes remain clean. (g) 5th instar nymph. (h) 3rd instar nymph. All photos of live specimens were taken in captivity.

2. METHODS

Collecting was done mostly by manual sorting of detritus on a white plastic sheet. At the Senda Darwin biological station, additional specimens were collected in water-filled white or yellow disposable plastic dishes placed on the ground as pan traps. Most of the collected specimens were kept live in ventilated plastic food containers on plant cuttings, moss, rotten wood, and detritus from the source microhabitat. Their behaviors were intermittently observed under a stereomicroscope.

Adults and nymphs collected at Senda Darwin were caged on cuttings of Griselinia racemosa roots, identified as a likely food source. To examine salivary sheaths, roots segments where the insects had been observed feeding were marked with a pen marker and then excised, fixed in 4% formaldehyde for 24 hours, and stored in 70% ethanol. These root chunks were subsequently embedded into paraffin, cut into 5 μm sections, and stained with hematoxylin and eosin.

The ultrastructure of the Malpighian tubules was examined in two adult females, one adult male, and two 3rd-instar nymphs. The specimens were preserved in 2.5% glutaraldehyde in PBS buffer, pH 7.4. The tubules were dissected out, stained with osmium tetroxide, embedded in Epon, and sectioned into 60 nm sections. The sections were poststained with uranyl acetate and lead citrate and examined on a JEM-1011 transmission electron microscope (JEOL Ltd., Japan).

3. RESULTS

3.1. Collected Material

Chile: Huerquehue National Park, N of Lake Tinquilco, head of the Lagos trail, 39.155°S, 71.716°W, mature Nothofagus forest with Chusquea understory (Figure 2(a)), 20–25 Dec., 2013, 10 adults and 5 nymphs from leaf litter with Hydrangea serratifolia, Dendroligotrichum dendroides, and small ferns (Figure 2(b)); same site, 9-10 Feb., 2014, 11 adults and 1 nymph (2nd instar). Puyehue National Park, sector Anticura, near park entrance, 200 m N of Road 215, 40.667°S, 72.174°W, 15–17 Jan., 2014, >25 adults and 5 nymphs (5th instar) from ca. 3 $m^2$of leaf litter under a tree, with abundant Hydrangea serratifolia and some herbaceous plants (Figures 2(c)–2(e)); trail Princesa, 40.665°S, 72.172°W, 15 Jan., 2014, 3 nymphs (5th instar) from forest leaf litter along trail; trail Chile, 40.669°S, 72.170°W, 18 Jan., 2014, 2 adults and 1 nymph (5th instar) from forest leaf

Figure 2: Mapuchea chilensis habitats (collecting sites). (a) Nothofagus forest N of the lake Tinquilco in Huerquehue NP. (b) The same forest floor with ferns and Hydrangea sp. (c) Puyehue NP near Anticura campground, leaf litter near a tree base, with abundant Hydrangea serratifolia. (d) The same, detailed view; note the absence of mosses and ferns. (e) The same, with litter removed to expose creeping stems of H. serratifolia; also note the abundant volcanic ash deposited upon the 2011 Puyehue-Cordón Caulle eruption. (f) Shrub bog at the biological station Senda Darwin on Chiloé. (g) The same, showing pan traps. (h) Shrub bog in Chiloe NP near Cucao.

litter with Blechnum, Hydrangea, and Dendroligotrichum; trail Salto del Indio, mature Nothofagus forest, 19 Jan., 2014, 4 adults from leaf litter. Alerce Andino National Park, sector Sargazo, near park entrance, 41.508°S, 72.621°W, open slope near forest edge, 6 Jan., 2014, 1 nymph (5th instar) from short grass and moss under Drimys; the same site, Drimys forest with Chusquea, 8 Jan., 2014, 2 adults from litter with Dendroligotrichum and creeping vines. Chiloé, Senda Darwin biological station, forest, postfire bog with abundant shrubs (Baccharis, Tepualia, and Gaultheria), large ferns (Blechnum spp., Gleichenia squamulosa), and trees (Figures 2(f) and 2(g)), 41.88°S, 73.67°W, 11 Jan., 2014, 3 nymphs (2nd and 3rd instars) from inside the spongy mass of decomposing old fronds around "trunks" of large Blechnum magellanicum; the same site, 27 Dec., 2013–4 Jan., 2014, 6 nymphs (5th instar) from accumulation of litter and peat under Podocarpus and large ferns, with abundant Griselinia racemosa roots; the same site, 13 Jan., 2014, and 1–7 Feb., 2014, 5 adults and 6 nymphs (4th and 5th instars), captured in plastic pan traps (Figure 2(g)), mostly on rainy nights. Chiloé National Park, sector Cucao, shrub bog along the Lahuan trail (Figure 2(h)), 42.621°S, 74.107°W, 13 Jan., 2014, 1 nymph (5th instar) from peat with roots; the same site, 3 Feb., 2014, 2 adults from litter and detritus under large ferns and trees; sector Rancho Grande, 42.572°S, 74.075°W, 3 Feb., 2014, 1 nymph from moss, coll. V. Hartung.

The specimens varied in size, sculpturing, and coloration between and within the collecting sites but were all assigned to Mapuchea chilensis (Nielson). The type locality of that first South American myerslopiid to be described is in the Anticura sector of the Puyehue NP [5]. Szwedo [16] made it the type of his genus Mapuchea, in which he described two additional species, each based on a single specimen: M. burckhardti, from Bosque de Fray Jorge NP, which is the northernmost outpost of Valdivian temperate rain forests, and M. hamiltoni from the Huerquehue NP. The specimens I collected in the Huerquehue NP did not consistently differ from the specimens collected elsewhere.

3.2. Observations of Caged Insects

Due to their cryptic habitat and camouflage, the insects could not be observed in nature. Caged adults and nymphs survived indefinitely until killed. They walked slowly but most of the time clung motionlessly to plant stems or roots, occasionally expelling droplets of clear excrement, which was interpreted as a sign of feeding. The insects were not observed feeding or staying long on green plant parts but twice were observed feeding on damp rotten wood. When disturbed, both nymphs and adults jumped strongly. In one adult, a jumping distance of 50 cm was recorded. The jumps ended by crash-landing followed by brief immobility. Two captive 5th instar nymphs molted to the adult stage. One of these (Figure 1(c)) was closely watched for 9 hours following the molt, during which it remained inactive. No attempts to coat the integument with soil or plant debris were noticed in either of molted specimens during the days following the molt. Grooming, courting, and mating were not observed.

3.3. Salivary Sheaths

Seven salivary sheaths were tracked, all of them terminating in the vascular bundle of the rootlet. Among these, two penetrated into the xylem (Figures 3(a) and 3(b)), while the rest were difficult to interpret, terminating equivocally between the phloem and the xylem (Figures 3(c) and 3(d)) or being poorly preserved. One branched sheath was observed (Figure 3(a)).

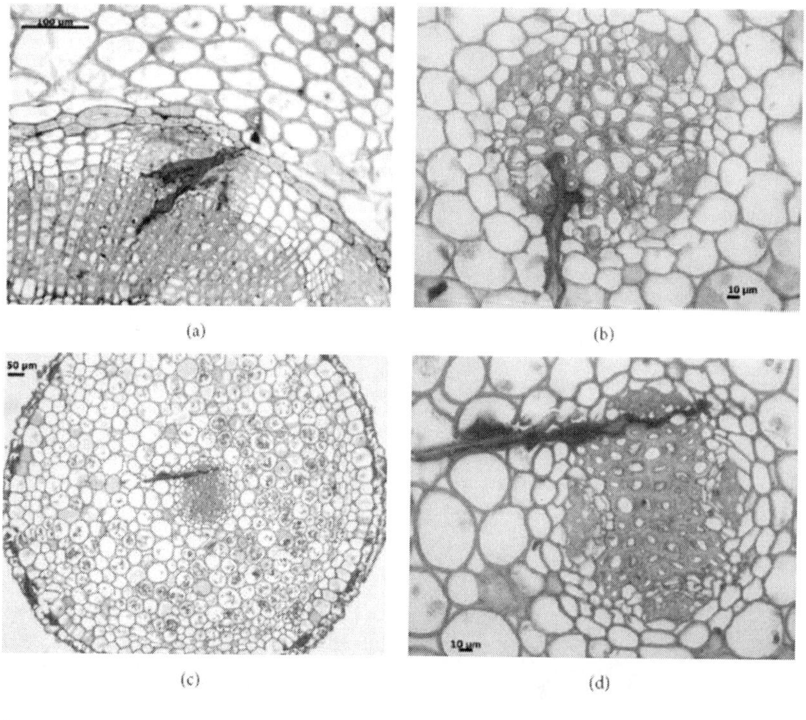

Figure 3: Histological sections of Griselinia racemosa roots with salivary sheaths of Mapuchea chilensis. (a) Branched sheath, with one branch penetrating into the xylem. (b) Sheath terminating in the outermost layer of

the xylem. ((c) and (d)) Sheath terminating equivocally, between the phloem and the xylem.

3.4. Malpighian Tubules

The general arrangement of the digestive tract and four Malpighian tubules is as typical for Cicadomorpha (Figure 4). The shape of the Malpighian tubules is identical in adults and nymphs. Each tubule gradually widens towards the base but becomes constricted again for a short length before entering the filter chamber. The exposed length of the tubule is convoluted and nodulose, resembling twisted rope, distally less so (Figure 4). The tubules form two pairs, the termini of each pair united in a loop attached to the rectum.

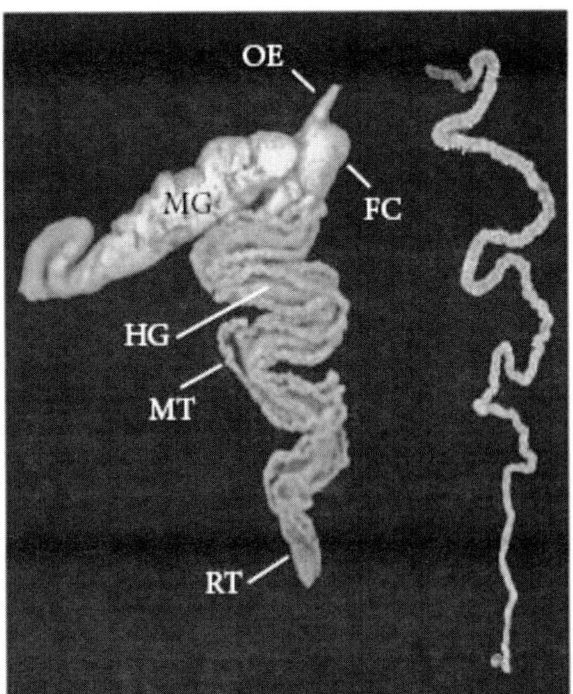

Figure 4: Mapuchea chilensis digestive tract with the Malpighian tubules and an isolated tubule (omitting the basal part entering the filter chamber).

OE: oesophagus; FC: filter chamber; MG: midgut; MT: Malpighian tubules; and RT: rectum.

In both nymphs and adults, the exposed parts of the tubules comprise two types of cells, excretory and secretory. In the thicker, more basal part of the tubule, secretory cells are interspersed among more numerous excretory cells. As the tubule narrows distally, the secretory cells gradually disappear; the most distal part seemingly comprises only excretory cells.

The excretory cells have an extensively developed labyrinth of basal membrane infoldings (Figures 5(a) and 5(b)). Their apical membrane bears a brush of long, simple, or lamellar microvilli, containing mitochondria, and forms intracellular canaliculi, some of them branching or forming flask-shaped cavities (Figures 5(b) and 5(c)). These cavities are also lined with microvilli or lamellae (Figure 5(d); compare with Figure in [17]). The cytoplasm contains numerous conspicuous mitochondria and, in some cells, deposits of glycogen granules (Figure 5(d); compare with [18]).

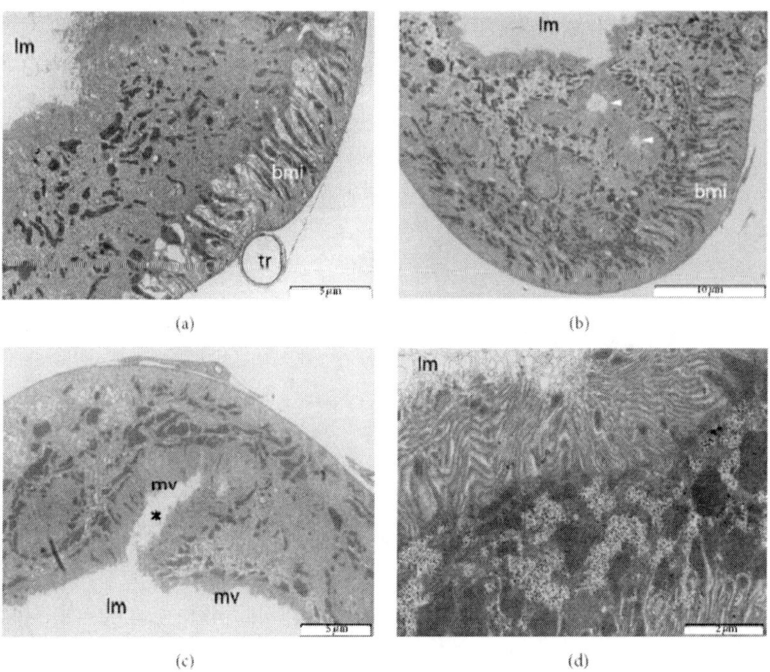

Figure 5: Excretory cells of the Malpighian tubules of Mapuchea chilensis. ((a)–(c)) Adult insects, general structure of the cells. (d) 3rd instar nymph, an excretory cell showing deposits of granular glycogen and lamellar microvilli bordering an intracellular canaliculus. lm: tubule lumen;

tr: trachea; and bmi: basal membrane infoldings. Intracellular canaliculi are shown with arrowheads in (b) and an asterisk in (c).

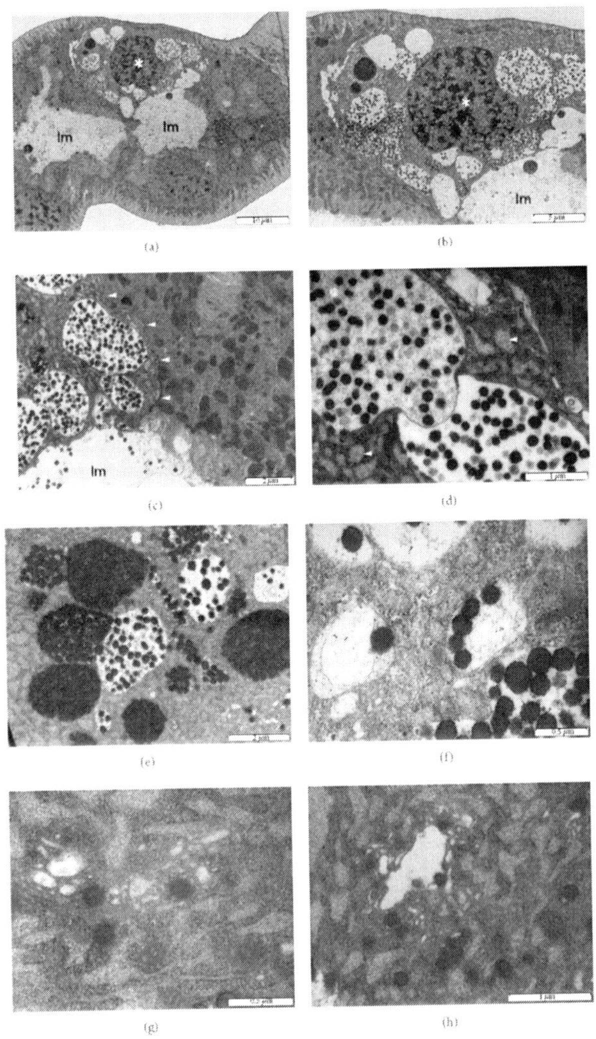

Figure 6: Secretory cells in the Malpighian tubules of Mapuchea chilensis. ((a)–(d)) Cells in the tubules of a 3rd instar nymph. (a) A single secretory cell (its nucleus marked by an asterisk) among excretory cells. (b) The same cell, closeup. (c) Contact between a secretory and an excretory cell; arrowheads point at the junction. Note secretory granules in the tubule lumen. (d) Secretory vacuoles with granules and additional particulate material. Arrowheads point at cisternae of endoplasmic reticulum. ((e)–(h))

Cells in the tubules of an adult. (e), (f) Secretory vacuoles variably loaded with granules. (g), (h) Golgi regions with developing granules. lm: tubule lumen.

The secretory cells (Figures 6(a)–6(h)) are smaller and lack apical microvilli and conspicuous mitochondria. They contain numerous secretory vacuoles filled with spherical or somewhat irregular dense granules, ca. 0.2 μm in diameter, with coarsely granulose surface (Figure 6(f)), and, occasionally, smaller particulate material. The 0.2 μm granules are also present in the tubule lumen (Figure 6(c)). The granules are formed in Golgi cisternae (Figures 6(g) and 6(h)). Most of the remaining cytoplasm is filled with cisternae of endoplasmic reticulum containing homogeneous material (Figure 6(d), arrowheads).

4. DISCUSSION

4.1. Habitats and Hosts

M. chilensis was collected in temperate Chile from forests and shrub bogs. This is similar to the New Zealand myerslopiids, which also occur in both forested and open habitats [3]. M. chilensis dwells cryptically on the ground or within decomposing plant material. On bogs, where the litter horizon is not well defined, the insect can be found at considerable depths within the peat. Most of the microhabitats where M. chilensis was found contained roots or stems of angiosperms, which apparently served as the food source. At several forest locations in the Huerquehue, Puyehue, and Alerce Andino parks, the insects were found between creeping stems of the vine Hydrangea serratifolia. It grew abundantly at the site in Anticura where the largest numbers of M. chilensis, 30 specimens, were collected (Figures 2(c)–2(e)). In at least two sites on the bog at Senda Darwin (Figure 2(f)), M. chilensis appeared to be associated with surface roots of the shrub Griselinia racemosa, abundantly ramifying within the litter and peat. Interestingly, like myerslopiids, the genus Griselinia has a disjunct distribution between southern South America and New Zealand. Caged insects were observed feeding on both G. racemosa roots (Figures 3(a)–3(d)) and H. serratifolia stems. No single plant species was present at every collecting site, indicating that M. chilensis is not associated with one host, but all such sites contained angiosperms. Most of the previous records did not associate myerslopiids with angiosperms, although three New Zealand species have been collected, among other places, from Celmisia and mats of Raoulia, both of Asteraceae [3]. Ferns were present at most of the collecting sites (Figures 2(b) and 2(g)) but were absent from the Anticura site (Figures 2(c)–2(e)). Curiously, three nymphs were found within dead plant material coating the "trunks" of large ferns that microhabitat could potentially contain epiphytes. A single record of a New Zealand myerslopiid from a fern has been listed [3].

Previous records consistently associated myerslopiids with mosses [3]; yet mosses were completely absent from the Mapuchea-rich site in Anticura (Figures 2(c)–2(e)). The mosses on the bogs of Chiloé (Figures 2(f)–2(h)) were intensively sampled, yielding abundant peloridiids ([19]; V. Hartung, personal communication), but only occasional individuals of M. chilensis. The latter species was consistently found on those same bogs in moss-free detritus accumulated under shrubs, large ferns, and trees.

4.2. Feeding Mode

Salivary sheaths revealed that the insects fed by tapping into the vascular tissue. This feeding mode is consistent with the presence of a filter chamber and general structure of the gut and Malpighian tubules, similar to those in previously studied Cicadomorpha (Figure 4). However, the exact source of the ingested sap, phloem or xylem, could not be identified based on the sheaths because of the small size of the sample. Salivary sheaths mark not only feeding, but also probing behavior, thus requiring many observations in order to make inferences. For example, among the 389 examined salivary sheaths of Philaenus spumarius (L.) examined by Horsefield, only 60% terminated in the xylem [20].

Although some of the observed salivary sheaths terminated in the xylem (Figures 3(a) and 3(b)), the external morphology of myerslopiids indicates xylem-feeding to be unlikely. The low nutritional value of xylem sap requires xylem-feeders to intensify ingestion by increasing the volume of cibarial pump muscles [21]. Externally this is manifested in the frontoclypeus being disproportionally enlarged in all stages, but especially so in young instars. The head proportions of myerslopiids follow the opposite developmental trajectory, with the frontoclypeus becoming progressively more inflated towards the adult stage (Figures 1(a) and 1(h); also mentioned in [16]). Therefore, the most likely food source is phloem sap.

4.3. Locomotion and Other Behaviors

While adults jump in all the Cicadomorpha except cicadas, nymphal jumping is thought to have evolved in the ancestral membracoids and been subsequently lost in the lineage that includes several leafhopper subfamilies and treehoppers [13]. The evidence of jumping by both adult and immature Mapuchea is, therefore, not surprising. It is worth mentioning that the leafhopper subfamily Ulopinae, to which myerslopiids have been assigned in the past, contains Ulopa species with nonjumping nymphs.

The fact that adults and nymphs of M. chilensis were consistently captured in pan traps (Figure 2(g)) suggested that they emerged to the surface in order to disperse. The insects quickly drown because their integument is not water-repellent. Climbing or accidentally falling from vegetation into these traps was unlikely, suggesting that the trapped insects had been moving around by hopping.

The origin of the characteristic soil crust of myerslopiids remains a mystery. Its consistent presence on the majority of collected specimens (Figures 1(d)–1(h)) suggests that it may be actively produced, but my limited observations did not discover any such behavior. Similar crusts have been reported for the leafhopper genera Evansiola China and Paulianiana Evans, formerly allied with myerslopiines within Ulopinae [22].

4.4. Malpighian Tubules

It has long been known that the Malpighian tubules of Membracoidea are unique among Hemiptera in having, in their middle or somewhat more distally, a glandular segment [23]. It is usually inflated throughout the life cycle, but in adults of some leafhoppers it becomes secondarily deflated. This segment has been observed in all the previously examined Cicadellidae (e.g., [23, 24]), Membracidae ([23, 24], Rakitov unpublished), and Melizoderidae (Rakitov, unpublished), but not in Aetalionidae (Rakitov, unpublished). The glandular segment consists entirely of specialized secretory cells and is flanked on either side with narrow, nodulose tubule segments consisting entirely of excretory cells [17]. In most adult cicadellids, the secretory cells produce brochosomes, proteinaceous particles of curious structure, which are released after molts and applied in special behaviors onto the integuments (reviewed in [11]). In some cicadellid subfamilies, the glandular segments secrete brochosomes starting from the 1st nymphal instar, while in others they produce other diverse products prior to switching, during the last nymphal instar, to production of brochosomes. These Malpighian tubule secretions of leafhopper nymphs include granules, tubules, or colloids [25]. Their chemistry and function are unknown, but some leafhopper nymphs apply these secretions onto their bodies [26]. Similar products have been found in the glandular regions of the immature and adult Membracidae and Melizoderidae (Rakitov, unpublished). It has been argued that treehoppers have evolved from within leafhoppers and quit secreting brochosomes secondarily [13].

Unlike in most previously studied Membracoidea, the Malpighian tubules of M. chilensis have no glandular segment composed of secretory cells (Figure 4). Instead, the tubules of both nymphs and adults have secretory cells interspersed among excretory cells along the length of tubule, except in the terminal region. The secretory cells produce simple granules, ca. 0.2 μm in diameter, which accumulate in large numbers in secretory vacuoles prior to being released into the lumen (Figures 6(a)–6(f)). This resembles the granular secretions produced in the Malpighian tubules of some cicadellid nymphs prior to switching to the production of brochosomes [25]. In both cases, the granules form in Golgi cisternae (Figures 6(g) and 6(h)), suggesting their proteinaceous nature. In the absence of chemical data, it is impossible to conclude whether the secretory products of M. chilensis are related to those of cicadellid nymphs, but this seems to be a possibility, given their structural similarity and the fact that Myerslopiidae and Cicadellidae are close relatives. The tubule excretory cells of M. chilensis (Figures 5(a)–5(d)) are identical to those of cicadellids [17] and represent a generic type common among insects. Therefore, the essential difference between M. chilensis and the previously studied Membracoidea concerns how the secretory cells are situated along the tubule: interspersed in the first case and forming a compact zone in the second.

Interspersed secretory cells, of unknown function, have been observed in the Malpighian tubules in other insect orders [27] but, to the best of my knowledge, not in Hemiptera. The closest living relatives of Membracoidea are spittlebugs (Cercopoidea) and cicadas (Cicadoidea), the three groups forming together the infraorder Cicadomorpha. During the nymphal stage, the Malpighian tubules of spittlebugs and cicadas produce secretions, spittle components in the first group [6, 7, 28–30] and the so-called anal liquid, of uncertain function, in the second [8]. In both groups, the tubules have a similar organization, representing a third type, different from that of either M. chilensis or other Membracoidea. Each has a short inflated glandular basal segment, composed of secretory cells producing mucosubstances, and a long nodulose distal segment, composed of essentially excretory cells, with characteristic apical microvilli and intracellular canaliculi, which additionally produce proteinaceous granules [8, 28–31].

The basal parts of the tubules show no secretory modification in Membracoidea, including M. chilensis, but their more distal parts secrete proteinaceous substances in all Cicadomorpha, representing three ways (Figures 7(a)–7(c)) to distribute the excretory and secretory functions along the tubule: (1) cells homogenous, each combining the secretory and excretory functions (spittlebug and cicada nymphs); (2) secretory cells interspersed among excretory cells (M. chilensis); and (3) excretory and secretory cells segregated into separate segments (Cicadellidae and Membracidae).

Like in M. chilensis, the Malpighian tubules of Aetalionidae have no morphologically distinct glandular region (Rakitov, unpublished), but their cell composition and ultrastructure have not yet been examined.

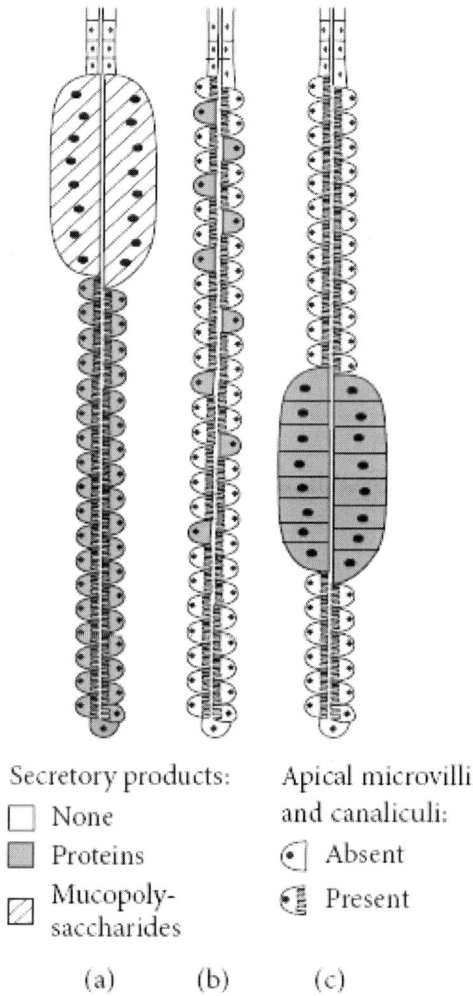

Figure 7: General organization of the Malpighian tubules in Cicadomorpha. The most basal parts, hidden inside the filter chamber, are not shown. (a) Immature Cercopoidea and Cicadoidea. (b) Mapuchea chilensis. (c) Cicadellidae and Membracidae.

I have argued in the past that the secretory specializations of the Malpighian tubules together with the habit of applying their secretions onto the body, displayed by spittlebugs, cicadas, and membracoids, have a common evolutionary origin [8]. The fate of the tubule secretory products of Mapuchea is unknown; no postmolt behaviors similar to anointing of leafhoppers and treehoppers [26] was observed, but the paucity of observations makes conclusions premature. Myerslopiidae merit further study in this regard, as they may represent the only recent group of Membracoidea which had separated prior to and thus may shed light on the origin of brochosomes and related behaviors of leafhoppers.

CONFLICT OF INTERESTS

The author declares that there is no conflict of interests regarding the publication of this paper.

ACKNOWLEDGMENTS

The author thanks Dmitri Shcherbakov for organizing the expedition to Chile and taking him along, Mario Elgueta for making their field work in Chile possible, Alexander Emeljanov, Kirill Eskov, Elena Lukashevich, Elena Rodionova, Daniela Takiya, Chris Dietrich, and Victor Hartung for being supportive teammates, Yuri Zuñiga and Wladimir Silva for assistance during their stay at the biological station Senda Darwin on Chiloé, Margaret Thayer for sharing her own material and records on Mapuchea, which greatly helped planning the collecting trip, Dmitri Tishechkin for attempting to record vibrational signals of Mapuchea, and Elena Kriuchkova for preparing histological sections of roots. A permit to collect insects in the national parks was granted by the CONAF. The study was supported in part by the Russian Foundation for Basic Research Grants 13-04-01839 to Dmitri Shcherbakov and 14-04-00322 to Roman Rakitov.

REFERENCES

1. J. Szwedo, "An annotated checklist of Myerslopiidae with notes on the distribution and origin of the group (Hemiptera: Cicadomorpha)," Zootaxa, vol. 425, pp. 1–15, 2004.
2. K. G. A. Hamilton, "The ground-dwelling leafhoppers Myerslopiidae, new family, and Sagmatiini, new tribe (Homoptera: Membracoidea)," Invertebrate Taxonomy, vol. 13, no. 2, pp. 207–235, 1999.

3. W. J. Knight, "Ulopinae of New Zealand (Homoptera: Cicadellidae)," New Zealand Journal of Science, vol. 16, pp. 971–1007, 1973.
4. A. G. Wheeler Jr., "Bryophagy in the Auchenorrhyncha: seasonal history and habits of a moss specialist, Javesella opaca (Beamer) (Fulgoroidea: Delphacidae)," Proceedings of the Entomological Society of Washington, vol. 105, no. 3, pp. 599–610, 2003.
5. M. W. Nielson, "A new species of Myerslopia from Chile (Homoptera: Cicadellidae)," Entomological News, vol. 107, no. 5, pp. 322–326, 1996.
6. P. Pesson, "Sécretion d'une mucoprotéine par les tubes de Malpighi des larves de Cercopides. Son rôle dans la formation de l'abri spumeux," Bollettino del Laboratorio di Zoologia Generale e Agraria della Facoltà Agraria in Portici, vol. 33, pp. 341–349, 1956.
7. K. Kato, "The origin and composition of the cuckoo spit," The Science Reports of the Saitama University Series B, vol. 3, pp. 33–53, 1958.
8. R. A. Rakitov, "Structure and function of the Malpighian tubules, and related behaviors in juvenile Cicadas: evidence of homology with spittlebugs (Hemiptera: Cicadoidea & Cercopoidea)," Zoologischer Anzeiger, vol. 241, no. 2, pp. 117–130, 2002.
9. M. F. Day and M. Briggs, "The origin and structure of brochosomes," Journal of Ultrasructure Research, vol. 2, no. 2, pp. 239–244, 1958.
10. J. Gouranton and P.-L. Maillet, "Origine et structure des brochosomes," Journal de Microscopie, vol. 6, pp. 53–64, 1967.
11. R. A. Rakitov, "Brochosomal coatings of the integument of leafhoppers (Hemiptera, Cicadellidae)," in Functional Surfaces in Biology, S. N. Gorb, Ed., vol. 1, pp. 113–137, Springer Netherlands, Dordrecht, The Netherlands, 2009.
12. R. Rakitov and S. N. Gorb, "Brochosomes protect leafhoppers (Insecta, Hemiptera, Cicadellidae) from sticky exudates," Journal of the Royal Society Interface, vol. 10, no. 87, 2013.
13. C. H. Dietrich, R. A. Rakitov, J. L. Holmes, and W. C. Black, "Phylogeny of the major lineages of Membracoidea (Insecta: Hemiptera: Cicadomorpha) based on 28S rDNA sequences," Molecular Phylogenetics and Evolution, vol. 18, no. 2, pp. 293–305, 2001.
14. R. A. Rakitov, "On differentiation of cicadellid leg chaetotaxy (Homoptera: Auchenorrhyncha: Membracoidea)," Russian Entomological Journal, vol. 6, no. 3-4, pp. 7–27, 1997.
15. N. Golub, V. Kuznetsova, and R. Rakitov, "First karyotype data on the family Myerslopiidae (Hemiptera, Auchenorrhyncha, Cicadomorpha)," Comparative Cytogenetics, vol. 8, no. 4, pp. 293–300, 2014.
16. J. Szwedo, "A new genus and six new species of ground-dwelling leafhoppers from Chile and New Zealand (Hemiptera: Cicadomorpha: Myerslopiidae)," Zootaxa, vol. 424, pp. 1–20, 2004.
17. D. S. Smith and V. C. Littau, "Cellular specialization in the excretory epithelia of an insect, Macrosteles fascifrons Stål (Homoptera)," The Journal of Biophysical and Biochemical Cytology, vol. 8, no. 1, pp. 103–133, 1960.
18. M. S. Jarial, "Fine structure of the Malpighian tubules of Chironomus larva in relation to glycogen storage and fate of hemoglobin," Tissue & Cell, vol. 20, no. 3, pp. 355–380, 1988.
19. D. E. Shcherbakov, "A new species of Peloridium (Hemiptera: Coleorrhyncha, Peloridiidae) from Chile," Far Eastern Entomologist, vol. 286, pp. 1–11, 2014.
20. D. Horsfield, "Evidence for xylem feeding by Philaenus spumarius (L.) (Homoptera: Cercopidae)," Entomologia Experimentalis et Applicata, vol. 24, no. 1, pp. 95–99, 1978.
21. V. Novotny and M. R. Wilson, "Why are there no small species among xylem-sucking insects?" Evolutionary Ecology, vol. 11, no. 4, pp. 419–437, 1997.
22. J. W. Evans, "Some relict New Guinea leafhoppers and their significace in relation to the comparative morphology of the head and prothorax of the Homoptera-

Auchenorrhyncha (Homoptera: Cicadellidae: Ulopinae)," Pacific Insects, vol. 10, no. 2, pp. 215–229, 1968.

23. E. Licent, "Recherches d'anatomie et de physiologie comparées sur le tube digestif des Homoptères supérieurs," La Cellule, vol. 28, pp. 6–161, 1912.

24. S. P. Mukharji, "The structure and histology of the alimentary canal of the Homoptera (Hemiptera-Insecta)," Indian Journal of Zootomy, vol. 2, pp. 56–121, 1961.

25. R. A. Rakitov, "Secretory products of the Malpighian tubules of Cicadellidae (Hemiptera, Membracoidea): an ultrastructural study," International Journal of Insect Morphology and Embryology, vol. 28, no. 3, pp. 179–193, 1999.

26. R. A. Rakitov, "Post-moulting behaviour associated with Malpighian tubule secretions in leafhoppers and treehoppers (Auchenorrhyncha: Membracoidea)," European Journal of Entomology, vol. 93, no. 2, pp. 167–184, 1996.

27. M. A. Prado, L. M. Montuenga, A. C. Villaro, J. C. Etayo, J. M. Polak, and M. P. Sesma, "A novel granular cell type of locust Malpighian tubules: ultrastructural and immunocytochemical study," Cell & Tissue Research, vol. 268, no. 1, pp. 123–130, 1992.

28. A. T. Marshall, "Spittle production and tube-building by cercopoid nymphs (Homoptera). I. The cytology of the Malpighian tubules of spittle-bug nymphs," Quarterly Journal of Microscopical Science, vol. 105, pp. 257–262, 1964.

29. A. T. Marshall, "Spittle production and tube-building by cercopoid nymphs (Homoptera). 2. The cytology and function of the granule zone of the Malpighian tubules of tube-building nymphs," Quarterly Journal of Microscopical Science, vol. 105, part 4, pp. 415–422, 1964.

30. A. T. Marshall, "Protein synthesis and secretion by the Malpighian tubules of cercopoid larvae (Homoptera)," Journal of Insect Physiology, vol. 19, no. 12, pp. 2317–2326, 1973.

31. Q. Li, H. Zhong, Y. Zhang, and C. Wei, "Comparative morphology of the distal segments of Malpighian tubules in cicadas and spittlebugs, with reference to their functions and evolutionary indications to Cicadomorpha (Hemiptera: Auchenorrhyncha)," Zoologischer Anzeiger, vol. 258, pp. 54–68, 2015.

CHAPTER 16

Evaluating Alpha and Beta Taxonomy in Ant-Nest Beetles (Coleoptera, Carabidae, Paussini)

Simone Fattorini,[1,2] Emanuela Maurizi,[3] and Andrea Di Giulio[3]

[1]Water Ecology Team, Department of Biotechnology and Biosciences, University of Milano Bicocca, Piazza della Scienza 2, 20126 Milan, Italy
[2]Azorean Biodiversity Group (ABG, CITA-A) and Portuguese Platform for Enhancing Ecological Research & Sustainability (PEERS), University of the Azores, Rua Capitão João de Àvila, 9700-042 Angra do Heroísmo, Portugal
[3]Department of Sciences, University of Roma Tre, Viale G. Marconi 446, 00146 Rome, Italy

ABSTRACT

We evaluated completeness, accuracy, and historical trend of the taxonomic knowledge on the myrmecophilous ground beetle tribe Paussini (Coleoptera, Carabidae, Paussinae). Accumulation curves for valid names and synonyms of species, subgenera, and genera were modelled using logistic functions. Analyses of trends in synonymies suggest that few currently accepted taxa will be recognized to be synonymous in the future. This may indicate that Paussini are a taxonomically relatively stable tribe of carabid beetles. However, this result might also be due to the lack of recent taxonomic work in some biogeographical regions.

1. INTRODUCTION

Arthropods are the most diversified animal group [1, 2]. Although it is widely acknowledged that only a small fraction of the extant arthropod species has been described, the magnitude of the so-called Linnean shortfall (i.e., the discrepancy between the number of described species and the number of living species) is a matter of discussion [2]. Also for relatively well-investigated arthropod groups, there is few information about the quality of the taxonomic knowledge [3, 4]. The most basic question is to establish how complete and accurate the taxonomic status of a given group is. With the word completeness we refer here to the problem whether the species list of a given group can be considered fairly complete or if there are still many species to describe. A completely known group is one for which there is no longer a need of an alpha taxonomic work (the discovering and naming of new species [5]). With accuracy we refer to taxonomic stability. An accurately known group is one for which there is no more need of a beta taxonomic work (the study of the relationships between the already described taxa, through systematic revisional work of higher taxa [5]). Because it is not rare that species are redundantly described under different names (i.e., synonyms), a group is known with accuracy when no relevant taxonomic change is expected.

Although the two aspects tend to be interrelated, they are not necessarily redundant, because revisional works are much rarer than descriptions of new taxa.

In this paper, we evaluated the completeness and accuracy of the taxonomic knowledge about a group of myrmecophilous beetles, the tribe Paussini ("ant-nest beetles") of the family Carabidae (Coleoptera, Adephaga, Paussinae), at a global level.

All Paussini are highly specialized social parasites, depending on ants (mainly associated with Myrmicinae and Formicinae) during any stage of their development [6–9]. Adults prey on ants and their broods without any obvious benefit for the ant colonies [10–15]. Because of their specialised behavioural and morphological adaptations, Paussini have long attracted the interest of entomologists working on myrmecophilous insects [13], and they have been recently into focus because of strong incertitude about their relationships with other Paussinae lineages [6, 7]. These studies have prompted our knowledge of Paussini biology, with emphasis on their immature stages and microscopic morphology, but taxonomical work seems to receive little attention.

In general, for assessing the status of the taxonomical process in a given group, the study should be addressed to describe (1) the growth through time of the cumulative number of valid names to estimate the number of species that remain to be discovered in a given taxonomic group, globally or regionally [4, 16–19], (2) the progression of the cumulative number of invalid names (synonyms), and (3) the temporal trends in the proportion of synonyms [20, 21]. Presence of a plateau is considered evidence that no, or few, species remain to be described, but it can be also due to a stop in taxonomic research [22]. In this paper we present an extension and continuation of a recently published study [22] where we have presented a comprehensive treatment of point 1. In the present paper we will treat the additional aspects of points 2 and 3 taking advantage of the statistical methodologies developed in the former paper.

2. MATERIAL AND METHODS

2.1. Data Collection

We used a computerized database including 572 species and 17 subspecies of the tribe Paussini.

The following information was recorded for each species and subspecies: generic assignment, subgeneric assignment, author, year of description, synonyms, and the biogeographical region of species distribution. We also recorded authorship and year of description of genera (see [22] for details).

2.2. Historical Accumulation Curves of Valid Names and Synonyms

We extracted the year of description of all valid species and subspecies names, as well as the year of description of the names that are currently considered synonyms and grouped years into decades. We plotted the raw number of described taxa, and the raw number of valid taxa, the raw number of synonyms, as well as their cumulative number, against the decade of description.

To model species accumulation curves we used the logistic function , where , , and are estimated parameters, because it gave excellent fits and the first parameter is the upper asymptote, thus providing an immediate estimate of the expected number of taxa. Similar analyses were conducted for genera and subgenera. The use of subgenera in the tribe Paussini is very controversial. For this reason, as in our companion study [22], we used subgenera as currently accepted by most authors [23].

2.3. Trends in Synonymies

Both the historical accumulation of species names and the relationship between valid species and synonyms may provide information about the status of the taxonomical knowledge in a given group [24].

Thus, we modelled synonym accumulation curves and measured the temporal variation in the taxonomical efficiency through time in three ways: (1) as the relationship between the number of synonyms versus the number of total described taxa in each decade, (2) as the proportion of names that are now regarded as synonyms over the total number of taxa described in that decade, (3) as the cumulative proportion of synonyms through decades.

Relationship between the number of synonyms versus the number of total described taxa was substantially linear, and we used an ordinary least squares (OLS) regression to model it. We forced the regression to pass through the origin, because when no taxon is described, the number of synonyms must be zero. We used the coefficient of the regression line as a measure of the number of synonyms introduced—on average—for each species in each decade. We used the 95% confidence limits to identify decades with exceptional number of synonyms.

Proportion of synonyms was used as a measure of the relationship between descriptive (alpha) and revisional (beta) taxonomy. We calculated the proportion of synonymous taxa described in each decade to identify a possible temporal trend in synonym proliferation.

The cumulative proportion of synonyms through decades was used as a rough measure of the quality of currently valid names. Following Baselga et al. [24] we assumed that the more taxonomical revisions are carried out, the higher is the probability for a given species name to be synonymized. Given that the synonyms are assigned to the date when the name was introduced, rather than the date when it was recognized as a synonym, the percentage of synonyms will show a diminishing trend with time, as newly described species will have had less time to be reviewed and eventually synonymized [24]. Irrespective of that, the steepness of the decay of this percentage through time can help us to measure the quality of currently valid names.

3. RESULTS

The rate of species description per decade, when the absolute numbers are considered, is very irregular (Figure 1(a)). Between 1775 (when the first species of Paussini was described by Linnaeus) and 1840 only 31 species were described, and no species was described in the decade 1820–1829. In the latter half of the 19th century species were described at an increasing rate, with two peaks, respectively, in the decades 1870–1879 and 1890–1899, in which a total of 38 Paussini taxa were described. However, the description of species peaked between 1920 and 1929, during which period 131 taxa were described, covering almost 17% of the available names. The low level of species descriptions in the decades 1910–1919 and 1940–1949 may be explained by the effects of the First and Second World Wars. Since the 1950s descriptions decreased progressively. When the cumulative numbers are considered, the increase per decade was low until the 1840s. The cumulative numbers of species/subspecies descriptions have reached a plateau, the estimated asymptotic value for the fitted curve being 768 taxa (Figure 1(b)).

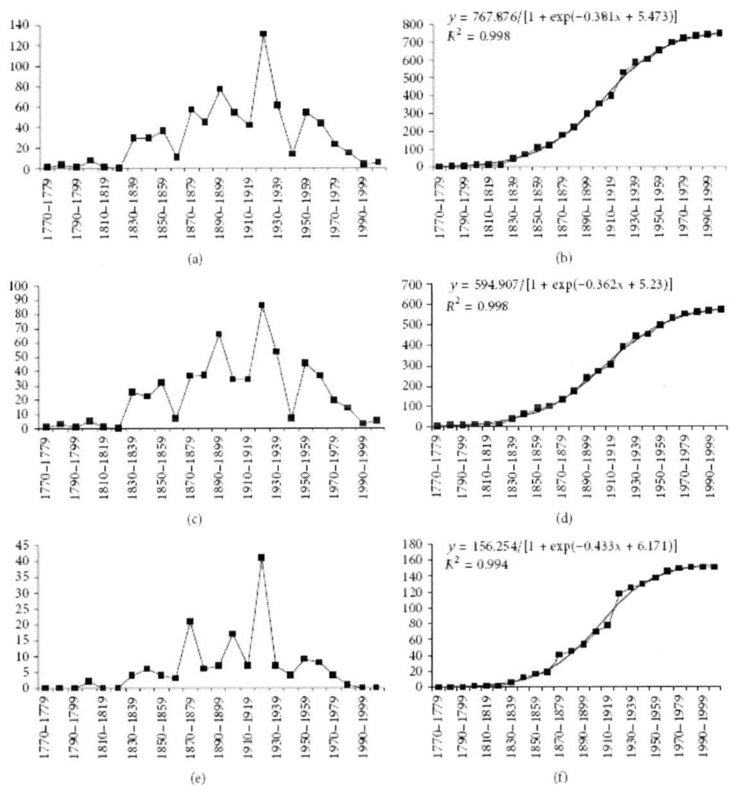

Figure 1: Numbers of total described taxa (a, b), valid species (c, d), and synonyms (e, f) of Paussini by decade. Figures (a), (c), and (e) report the absolute numbers, and Figures (b), (d), and (f) the cumulative numbers along with the equations of the fitted curves.

Analyses omitting synonyms and subspecies revealed similar patterns (Figure 1(c)), with an estimated asymptotic value of species number at 595 (Figure 1(d)). Because the number of currently recognized valid species is 572, the model predicts the existence of 23 undescribed species, with about 96 per cent of the world fauna described.

Patterns in synonyms were also similar to the general trend (Figures 1(e) and 1(f)). The asymptotic value for the number of synonyms is 156 names, very close to the current number of recognized synonyms.

Number of synonyms per decade was directly proportional to the number of described taxa, with a mean rate of one synonym per four taxa each decade (Figure 2). However, the decades 1870–1879, 1900–1909, and 1920–1929 were characterized by an exceptional high number of synonyms. An in-depth analysis of the percentage of synonyms per decade shows a roughly humped trend, with proportion of synonymous taxa increasing from 1830–1849 to 1870–1879, and then decreasing to very low values (Figure 3(a)), which determines a plateau in the accumulation curve of synonymies (Figure 3(b)).

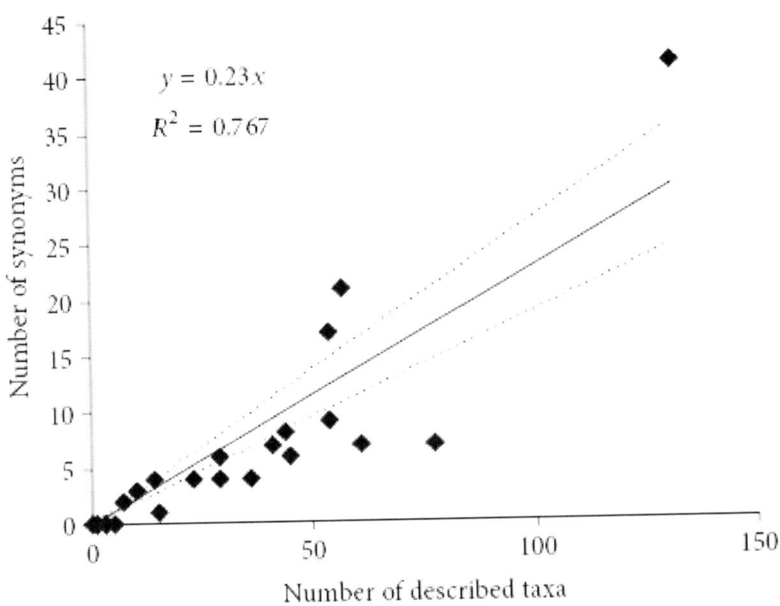

Figure 2: Relationship between number of synonyms and total number of described taxa per decade. Ordinary least square (OLS) regression forced to pass through the origin.

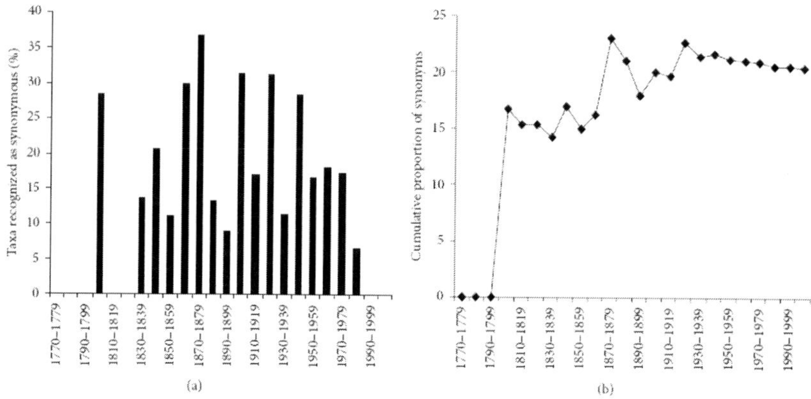

Figure 3: Percentages of synonymous taxa described in each decade (a) and their historical process of accumulation (b).

The historical process of variation in proportion of synonymized names defines the following time spans that correspond to periods of roughly homogenous taxonomical work (Figure 3(a)): (1) the very early stage was obviously characterized by few descriptions (cf. Figure 1) which are still valid species; (2) the relative rate of redescriptions was nearly constant between 1800 and 1870; (3) between 1870 and 1930 we found that at increasing description of species there was also an increasing number of species subsequently found to be synonymous; and (4) finally, from 1930 to present time, the relative rate of descriptions subsequently synonymized diminishes drastically, as less than 20% of the species described during this period have been synonymized (Figure 3(b)).

The largest numbers of described species occur in Africa, followed by the Oriental and Australian regions (Figure 4(a)). This pattern is not paralleled by proportion of synonymies, with the Australian fauna being that with the highest percentage of synonymized taxa (Figure 4(b)).

The distribution of the numbers of authors that have described Paussini taxa is strongly right-skewed (Figure 5). Over 52% of authors have described only one species. The most productive author, Reichensperger, described 88 species, followed by Westwood (79), Wasmann (73), and Luna de Carvalho (71). Altogether, these four authors described more than 40% of known species.

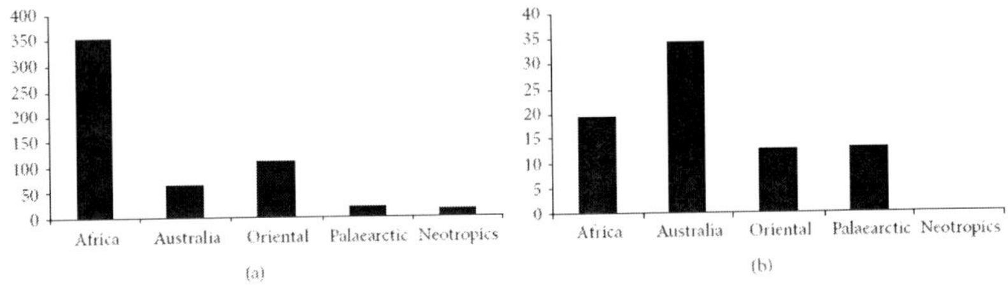

Figure 4: Number of valid species of Paussini per biogeographical region (a) and percentage of synonymous taxa of Paussini per biogeographical region (b).

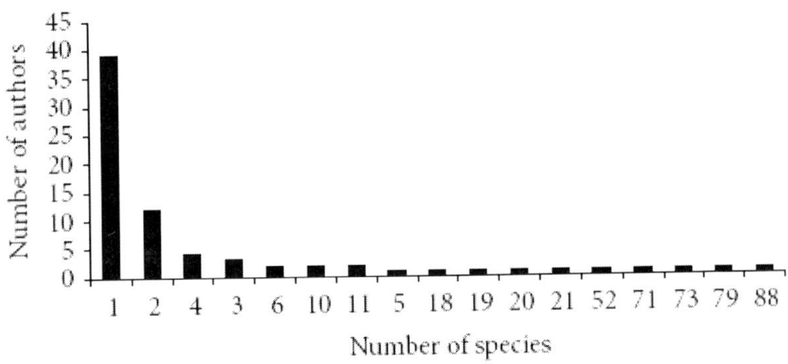

Figure 5: Number of authors in relation to the numbers of species of Paussini that they described.

Reichensperger published his descriptions between 1913 and 1958 (with an average of ca. 2 species per year), covering all biogeographical regions except the Australian. Most of his species (ca. 94%) were described from Africa. Westwood made his descriptions between 1833 and 1874 (with an average of more than 6 species per year) covering all biogeographical regions with a high proportion (ca. 41%) of Oriental taxa. Wasmann also covered all biogeographical regions between 1892 and 1930, with similar proportion of African (49%) and Oriental (42%) taxa and a mean rate of ca. 2 species per year. Finally, Luna de Carvalho described most of his species from Africa (ca. 85%), with a few species from the Oriental and the Palaearctic regions.

Paussini species are currently allocated in 23 genera. The total number of described genera is 51, with 28 synonyms (55%). Most of the genera were described in the decade 1920–1929 (Figure 6(a)). Although some decades were characterized by a high number of descriptions of genera, many were recognized as synonyms (especially among those described in the decade 1920–1929) (Figure 6(e)), so the decade with the highest number of valid genera (4 genera) was 1830–1839 (Figure 6(c)): 75% of the genera described in the decade 1920–1929, and 50% of those described in 1980–1989, were subsequently recognized as synonyms.

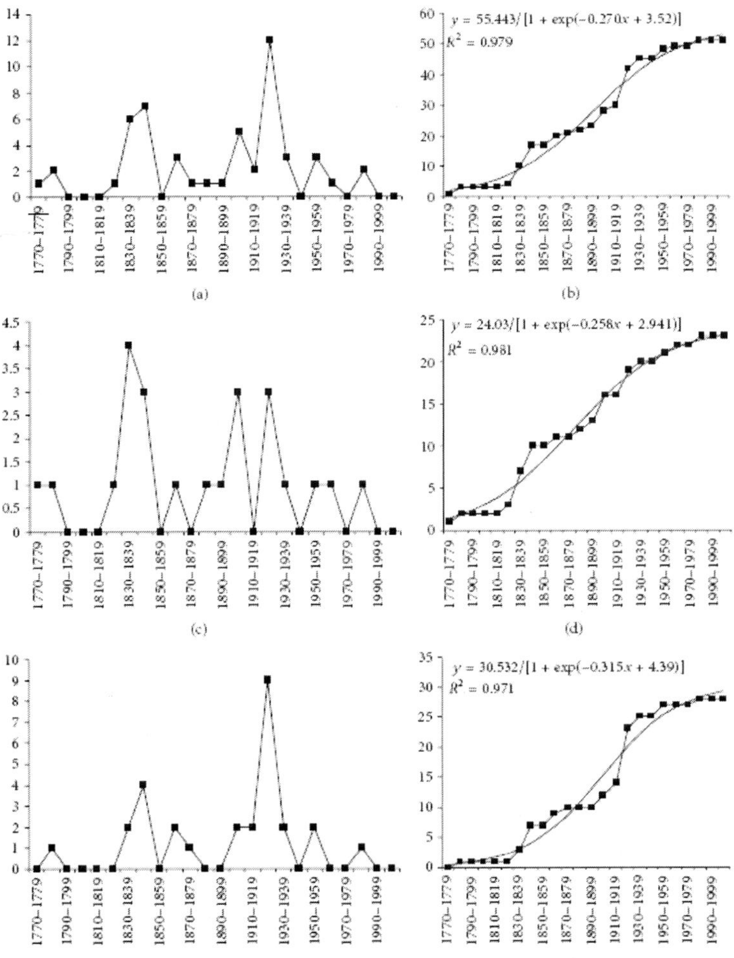

Figure 6: Numbers of total described genera (a, b), valid genera (c, d), and synonyms (e, f) of Paussini by decade. Figures (a), (c), and (e) report the absolute numbers, and Figures (b), (d), and (f) the cumulative numbers along with the equations of the fitted curves.

Patterns of genera accumulation through time indicate a good sigmoid shape for the total number of species (Figure 6(b)), valid genera (Figure 6(d)), and synonyms (Figure 6(f)). In all cases, a plateau has been reached, so virtually no new genus is expected for the future. The historical process of variation in proportion of synonymized genera indicates that after the 1930s there is a substantial stabilization (Figure 7).

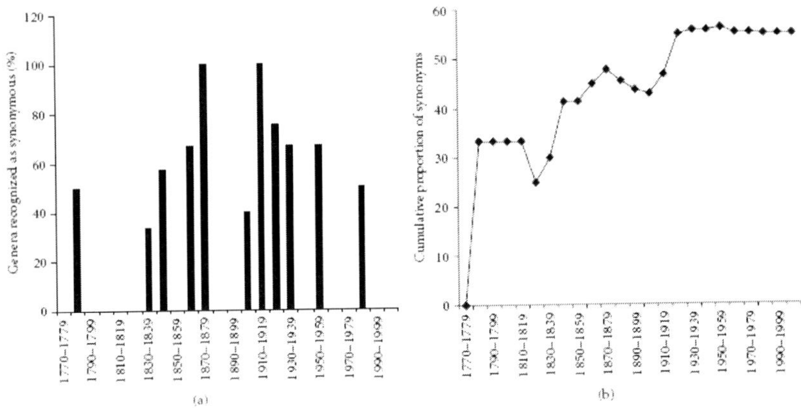

Figure 7: Percentage of synonymous genera described in each decade (a) and their historical process of accumulation (b).

The study of subgenera indicates a proliferation of names in the periods 1920–1929 and 1980–1989 as for the genera (Figure 8). Although these were the two decades which mostly contributed to the current accepted subgenera, these were also the decades in which a large number of synonymous subgenera were described, with proportions of synonyms of more than 54% and 64%, respectively. Accumulation curves showed a stair shape pattern, with apparent plateaus, and were therefore not modelled with fitting curves (Figure 8). Moreover, the historical process of variation in proportion of synonymized genera indicates that there is no substantial stabilization (Figure 9). This was mostly due to the large number of subgenera proposed in a recent time (1980–1989) and subsequently synonymized (Figure 9). These patterns suggest that subdivision into subgenera is not reaching a definitive solution.

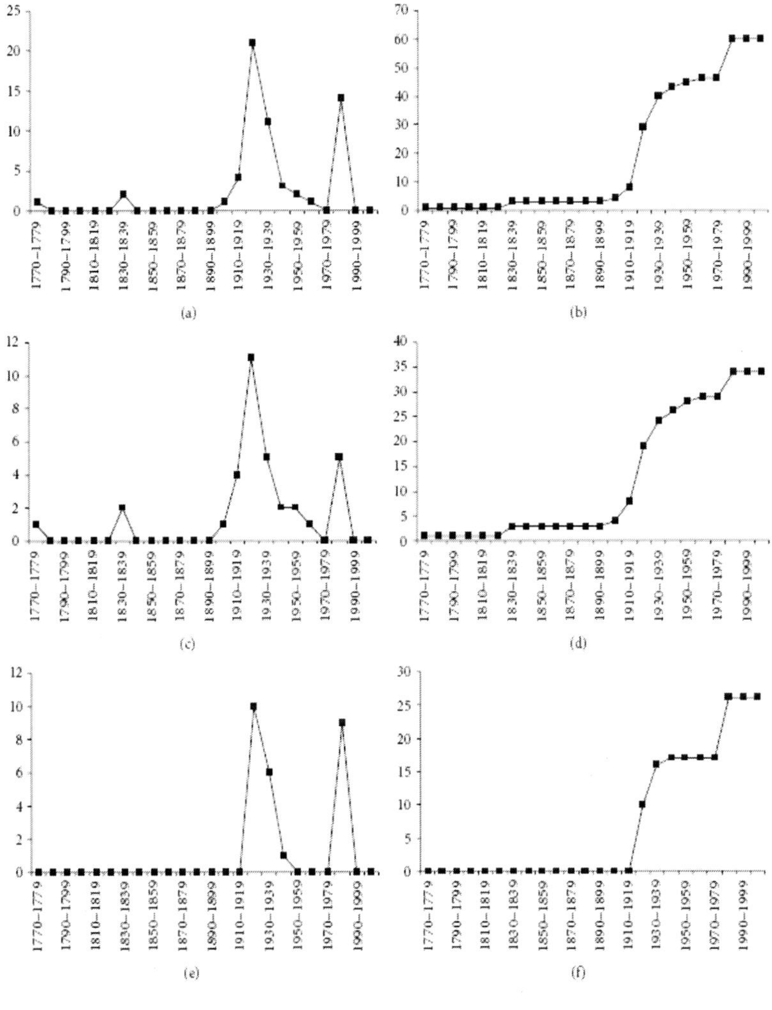

Figure 8: Numbers of total described subgenera (a, b), valid subgenera (c, d), and synonyms (e, f) of Paussini by decades. Figures (a), (c), and (e) report the absolute numbers, and Figures (b), (d), and (f) the cumulative numbers.

Species allocation among genera is strongly dominated by the richest genus (Paussus), with 342 ascribed species (Figure 10) and 25 subgenera. Species distribution among subgenera is also very uneven: the subgenus with the highest number of species is Cochliopaussus (Figure 11).

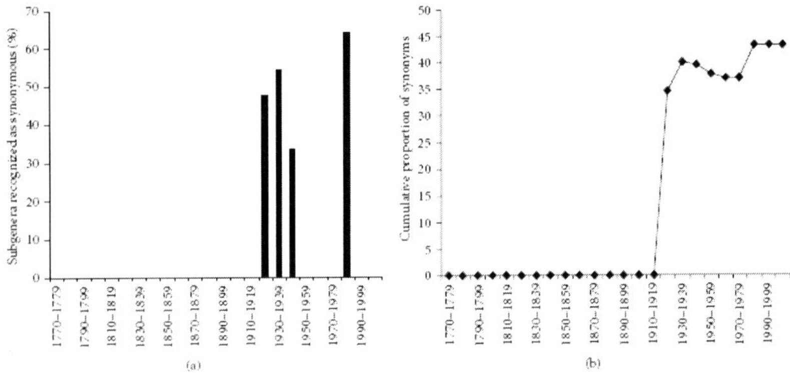

Figure 9: Percentage of synonymous subgenera described in each decade (a) and historical process of accumulation of percentage of synonyms over the total number of names in the Paussini, according to the date of their description (b).

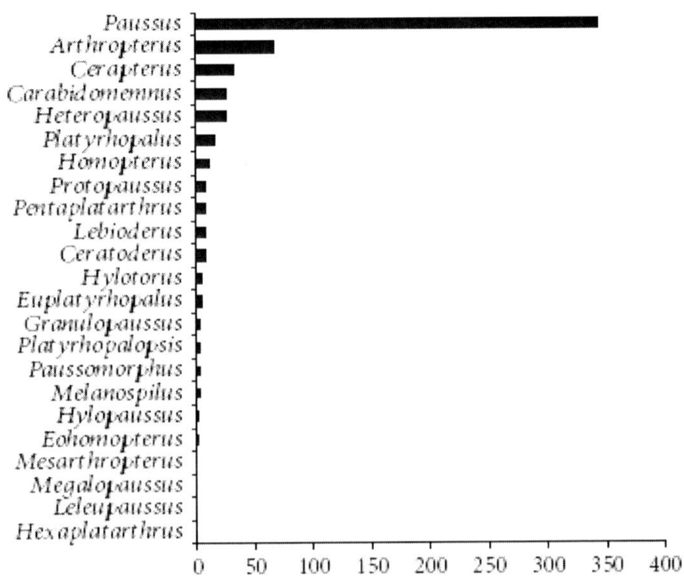

Figure 10: Number of species per genus in the tribe Paussini.

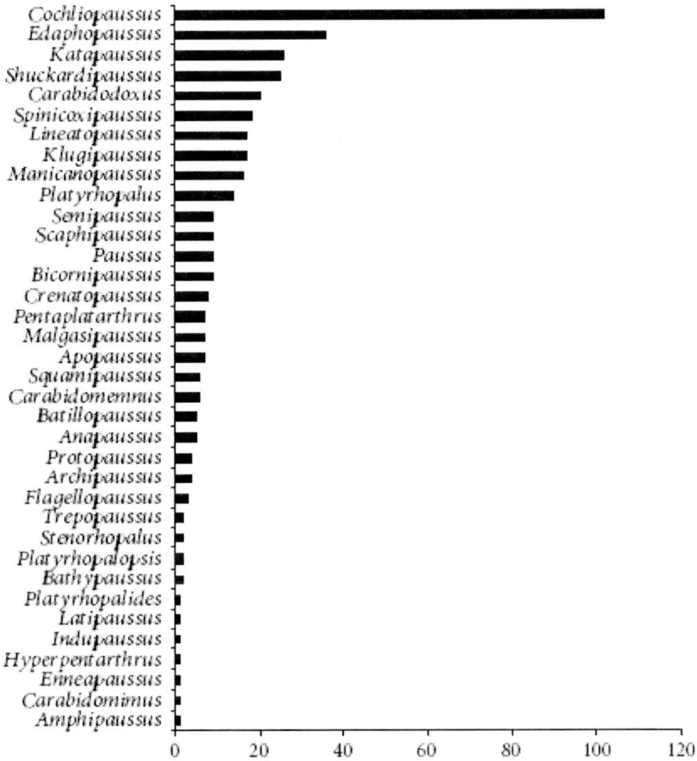

Figure 11: Number of species per subgenus in the tribe Paussini.

4. DISCUSSION

Species accumulation curves of the world Paussini fauna indicate that this tribe of carabid beetle is taxonomically stable but do not prove that knowledge is exhaustive. According to the trends analyzed in the present paper, relatively few species are expected to be described in the future on morphological basis and few currently accepted taxa will be recognized to be synonymous. However, if this situation may reflect a true state of affairs in the best explored regions, it may be an artefact when stabilization is merely due to prolonged taxonomic inactivity.

In general, temporal trends in species descriptions mirror dramatic events in human history. The first peak in African species description occurred in the decade 1880–1899, which can be considered an indirect reflection of the first phase of African explorations that occurred between 1840 and

1870 and especially a direct effect of the German expansion in Africa in the 1880s. The second peak occurred in the decade 1920–1939, which coincides with the third phase of the African colonialism, during which the most influential European states organized and stabilised their territories. The overall trend in species descriptions shows two falls in correspondence with the First and Second World Wars. If taxonomic research was frozen at those dates, we would have a completely false signal of stability. For example, taxonomic knowledge in Australia rested at the 1930s [22]. The lack of recent taxonomic activity, coupled with the low number of described species and the high percentage of synonyms, indicates that the fauna of this region is still poorly known.

Most of taxonomic work on Paussini has been produced by few but very prolific authors. Moreover, the authors that described most species during the 20th century were the same that realised the most comprehensive revisions. This has created a self-referenced system, with an almost complete lack of plurality of views. Therefore, taxonomic stability is largely an effect of the "monopolistic" position of certain taxonomists (e.g., Reichensperger, Westwood, Wasmann, and Luna de Carvalho) for long times. Moreover, each of the most active taxonomists was mostly interested in a different biogeographical region, thus with limited taxonomic overlap.

At global level, the asymptotic value calculated for the synonym curve is very close to the current number of synonyms (151); thus we expect that virtually no taxa will be recognized as synonymous in the next future. This indicates that new species are still being described (alpha taxonomy), albeit at decreasing rate in the best explored regions, whereas virtually no synonyms are currently being described, implying either a lack of beta taxonomy (i.e., redundant descriptions are still considered valid because of the reduced revisional work) or an excellent efficiency of alpha taxonomy (i.e., all new species are valid and none is redundantly described) [24]. We think that failure to recognize synonymies is likely high in the less studied faunas, for which most species have been described from sparse individuals, but this is balanced by the presence of still undescribed species. This may be the case of the Oriental region, which seems to have few species and a moderate percentage of synonymies, but from which so many species are being discovered and no further synonymies established.

Stability in species beta taxonomy indicates that Paussini species are recognized as discrete entities by most researchers. Paussini species were described and are currently recognized on the basis of morphological traits, that is, as groups of phenetically similar individuals that can be separated from other analogous groups by means of phenetic gaps, thus corresponding to a morphological concept of species [25]. Stabilization in synonymies suggests that most taxonomists agree in considering the diagnostic characters presented in species descriptions as gaps sufficiently strong to mark discontinuities among populations. Morphologically defined species do not necessarily correspond to "biological" species (defined as reproductively isolated populations [5]). However, the application of a morphological approach for discriminating species was the practical methodology most frequently used by taxonomists in the past, and the same approach still dominates (and likely will dominate) daily work of the majority of taxonomists. Stability in beta taxonomy of morphological species makes Paussini an ideal candidate for future works using molecular approach to investigate how morphological discontinuities are paralleled by molecular divergences. This would be particularly important to clarify relationships among species. Current taxonomic patterns suggest that most species were allocated into the genus Paussus probably reflecting a real phylogenetic proximity. However, subgeneric divisions appear instable and based on subtle and controversial morphological characters. This suggests that morphological characters are not fully adequate to resolve infrageneric relationships.

ACKNOWLEDGMENTS

The authors are grateful to Wendy Moore (AZ, USA), Augusto Vigna Taglianti (Roma, Italy), and Peter Nagel (Basel, Switzerland) for the continued stimulating discussions about the taxonomy of Paussini and two anonymous referees for their useful comments. Wendy Moore also provided them with an electronic species level database of Paussini and revised the language of the paper. The authors thank Valerio Viglioglia for his help in the data entry and Peter Hlaváč for his suggestions.

REFERENCES

1. K. J. Gaston, "The magnitude of global insect species richness," Conservation Biology, vol. 5, no. 3, pp. 283–296, 1991.
2. R. J. Ladle and R. J. Whittaker, Conservation Biogeography, Springer, West Sussex, UK, 2011.
3. A. Baselga, J. Hortal, A. Jiménez-Valverde, J. F. Gómez, and J. M. Lobo, "Which leaf beetles have not yet been described? Determinants of the description of Western Palaearctic Aphthona species (Coleoptera: Chrysomelidae)," Biodiversity and Conservation, vol. 16, no. 5, pp. 1409–1421, 2007.
4. J. Hortal, A. Jiménez-Valverde, J. F. Gómez, J. M. Lobo, and A. Baselga, "Historical bias in biodiversity inventories affects the observed environmental niche of the species," Oikos, vol. 117, no. 6, pp. 847–858, 2008.
5. E. Mayr, Principles of Systematic Zoology, McGraw-Hill, New York, NY, USA, 1969.
6. A. Di Giulio, S. Fattorini, A. Kaupp, A. V. Taglianti, and P. Nagel, "Review of competing hypotheses of phylogenetic relationships of Paussinae (Coleoptera: Carabidae)

based on larval characters," Systematic Entomology, vol. 28, no. 4, pp. 509–537, 2003.

7. A. Di Giulio and W. Moore, "The first-instar larva of the genus Arthropterus (Coleoptera:Carabidae: Paussinae): implications for evolution of myrmecophily and phylogenetic relationships within the subfamily," Invertebrate Systematics, vol. 18, no. 2, pp. 101–115, 2004.
8. A. Di Giulio, "Fine morphology of the myrmecophilous larva of Paussus kannegieteri (Coleoptera: Carabidae: Paussinae: Paussini)," Zootaxa, no. 1741, pp. 37–50, 2008.
9. W. Moore, X. B. Song, and A. di Giulio, "The larva of Eustra (Coleoptera, Paussinae, Ozaenini): a facultative associate of ants," ZooKeys, vol. 90, pp. 63–82, 2011.
10. C. Escherich, "Zur naturgeschichte von Paussus favieri Fairm," Verhandlungen der Zoologisch-Botanischen Gesellschaft, vol. 49, pp. 278–283, 1899.
11. G. Le Masne, "Recherches sur la biologie des animaux myrmécophiles IV: observations sur le comportement de Paussus favieri Fairm, hôte de la fourmi Pheidole pallidula Nyl.," Annales de la Faculté des Sciences de Marseille, vol. 31, pp. 111–130, 1961.
12. G. Le Masne, "Recherches sur la biologie des animaux myrmécophiles I: L'adoption des Paussus favieriFairm, par une nouvelle societe de Pheidole pallidula Nyl.," Comptes Rendus Hebdomadaires des Scéances de L'Academie des Scéances, vol. 253, pp. 1621–1623, 1961.
13. S. F. Geiselhardt, K. Peschke, and P. Nagel, "A review of myrmecophily in ant nest beetles (Coleoptera: Carabidae: Paussinae): linking early observations with recent findings," Naturwissenschaften, vol. 94, no. 11, pp. 871–894, 2007.
14. A. di Giulio, E. Maurizi, P. Hlaváč, and W. Moore, "The long-awaited first instar larva of Paussus favieri(Coleoptera: Carabidae: Paussini)," European Journal of Entomology, vol. 108, no. 1, pp. 127–138, 2011.
15. E. Maurizi, S. Fattorini, and A. Di Giulio, "Behavior of Paussus favieri (Coleoptera, Carabidae, Paussini), a myrmecophilous beetle associated with Pheidole pallidula (Hymenoptera, Formicidae)," Psyche, vol. 2012, Article ID 940315, 9 pages, 2012.
16. G. C. Steyskal, "Trend curves of the rate of species description in zoology," Science, vol. 149, no. 3686, pp. 880–882, 1965.
17. R. A. Medellín and J. Soberón, "Predictions of mammal diversity on four land masses," Conservation Biology, vol. 13, no. 1, pp. 143–149, 1999.
18. F. J. Cabrero-Sañudo and J. M. Lobo, "Estimating the number of species not yet described and their characteristics: the case of Western Palaearctic dung beetle species (Coleoptera, Scarabaeoidea)," Biodiversity and Conservation, vol. 12, no. 1, pp. 147–166, 2003.
19. A. Baselga and F. Novoa, "Diversity of Chrysomelidae (Coleoptera) in Galicia, Northwest Spain: estimating the completeness of the regional inventory," Biodiversity and Conservation, vol. 15, no. 1, pp. 205–230, 2006.
20. A. R. Solow and W. K. Smith, "On estimating the number of species from the discovery record," Proceedings of the Royal Society B, vol. 272, no. 1560, pp. 285–287, 2005.
21. J. Alroy, "How many named species are valid?" Proceedings of the National Academy of Sciences of the United States of America, vol. 99, no. 6, pp. 3706–3711, 2002.
22. S. Fattorini, E. Maurizi, and A. Di Giulio, "Tackling the taxonomic impediment: a global assessment for ant-nest beetle diversity (Coleoptera: Carabidae: Paussini)," Biological Journal of the Linnean Society, vol. 105, pp. 330–339, 2012.
23. W. Lorenz, A Systematic List of Extant Ground Beetles of the World (Coleoptera "Geadephaga": Trachypachidae and Carabidae Incl. Paussinae, Cicindelinae, Rhysodinae), W. Lorenz, Tutzing, Germany, 2nd edition, 2005.
24. A. Baselga, J. M. Lobo, J. Hortal, A. Jiménez-Valverde, and J. F. Gómez, "Assessing alpha and beta taxonomy in eupelmid wasps: determinants of the probability of describing good species and synonyms," Journal of Zoological Systematics and Evolutionary Research, vol. 48, no. 1, pp. 40–49, 2010.

25. R. R. Sokal and J. Crovello, "The biological species concept: a critical evaluation," The American Naturalist, vol. 104, pp. 127–153, 1970.

CHAPTER 17

Integrative Taxonomy, Phylogeny, and New Species of the Weevil Genus Onyxacalles Stüben (Coleoptera: Curculionidae: Cryptorhynchinae)

Peter E. Stüben[1] and Jonas J. Astrin[2]

[1]*Curculio Institute, Hauweg 62, 41066 Mönchengladbach, Germany*
[2]*Zoologisches Forschungsmuseum Alexander Koenig (ZFMK), Molecular Taxonomy & Biobank, Adenauerallee 160, 53113 Bonn, Germany*

ABSTRACT

A molecular phylogeny of the western Palearctic weevil genus Onyxacalles Stüben, 1999 is presented, combining two mitochondrial genes (COI and 16S) in a Bayesian analysis. Based on molecular data, Onyxacalles pyrenaeus Boheman, 1844 is transferred into the genus Kyklioacalles Stüben 1999 (K. fausti group) and—in an integrative taxonomy framework—the interaction between morphology and molecular analysis is illustrated. The species of Onyxacalles s. str. are assigned to three new species groups, O. henoni, O. luigionii, and O. portusveneris groups. The distribution of the related species in the Mediterranean area is illustrated with values of COI and 16S p-distances. Three new species are described and distinguished from their related species: Onyxacalles nuraghi Stüben sp.n. from Italy (Sardinia), Onyxacalles torre Stüben and Astrin sp. n. from France (Corsica) and Onyxacalles vilae Stüben sp. n. from Croatia (Velebit Mts.). A catalogue of all 20 species of Onyxacalles is given, and a key is finally presented combined with image stacking of the habitus and aedeagus for all species.

1. INTRODUCTION

Together with a number of other genera, the genus Onyxacalles Stüben, 1999 (Curculionidae: Cryptorhynchinae) was separated by Stüben [1] from the formerly excessively broadly circumscribed genus Acalles Schoenherr, 1825 as a group with initially 8 species. Since then, many new species of Onyxacalles have been described, mainly from Spain and North Africa. These discoveries were supported by the morphological finding that the three species from the Canary Islands belong to this genus [1, 2], a thesis that gained support from recent molecular analysis and has contributed to the new subgenus Araneacalles Stüben and Astrin [3]. This closed a "biogeographical gap" (between the Pyrenees and northwestern Africa) as a direct consequence of target-oriented collecting activities and descriptions of many new Onyxacalles species over the past decade [4–9]. Thus, including the new species presented in this work, the genus now comprises 20 valid species.

Most species of Onyxacalles are found in the west Mediterranean area and on the Macaronesian Islands. Only one species, Onyxacalles croaticus (H. Brisout, 1897) [10], reaches Eastern Europe (Carpathians); another species, O. amasyaensis Wolf, 2001, was described from Turkey, but could be a synonym of Onyxacalles denominandus A. and F. Solari, 1907. This species richness in the west Mediterranean is well founded in the ecological preferences of Onyxcalles.

As "nocturnal goblins of the last primeval forests" [7], the species of Onyxacalles are not common in the often disturbed landscapes of the Iberian Peninsula and North Africa [11, 12]. These conditions provide good maps of relictual vegetation and information about the habitats, allowing us to trace these nocturnal Cryptorhynchinae in the dark and humid relicts of natural forests under big oak trees and behind a dense jungle of Smilax aspera L. Therefore, sifting by day is not the ideal method of catching Onyxacalles. With their long legs, Onyxacalles species climb trees by night and were fogged by us from the canopy of the laurisilva on the Canary Islands of Tenerife and La Gomera [13, 14]. The flightless Onyxacalles are highly specialised woodlander inhabitant, and because of their restricted dispersal (above all on the European continent), they are an ideal bioindicator of original forest, highlighting that protection and sustainable development should concern us as entomologists and conservationists.

Moreover, with the study of the cryptic and similar Onyxacalles species, the taxonomist, especially the morphologist, enters a minefield. Due to intraspecific variability, the few external characteristics that are suitable for a differential diagnosis do not guarantee a reliable (re)identification of species: if the tufts of bristles exist or not, if the elytra are short oval or elongated, or if the midgroove of pronotum is more or less distinct—all these are important pointers, but are not conclusive. All Onyxacalles species have the hooked apex of the aedeagus in common, as the genus name implies (cf. Figures 9(a)–9(l)). However, a complex internal sac of the aedeagus (endophallus) does not exist—as it is typical for Kyklioacalles, Dichromacalles, or most genera from Macaronesia. Here, we are quickly stretched to our limits of a descriptive morphology.

With the "Molecular Weevil Identification" Project (see below), we tread a path towards an integrative taxonomy [15–21]. Molecular taxonomy is not only an addition or an accessory "immunisation" to confirm morphological results. The integration of the two approaches must be done with the intention of falsifying and, if so, it can lead to a new way of viewing morphology: interspecific characteristics are discovered that are not regarded as belonging to intraspecific variability, or—more often—substituting the "human eyes," the way we look at things. Instead of "anthropocentric conspicuities" (e.g., forms, colours, sizes), we focus on constitutive characteristics (e.g., apomorphies and homologies) that were previously overlooked in our diagnostic keys. An example is the latest history of the science behind the original species Acalles pyrenaeus Boheman, 1844 (see below).

Another telling example is the present classification of Onyxacalles s. str. into two species groups: the O. luigionii- and O. pyrenaeus-groups, based on the bristles of the elytral intervals (cf. [1, 8]). This hypothesis is no longer tenable, and this has nothing to do with superficial diagnosis of affinities or inaccurate observations. First, molecular analysis of related species reveals new informal species groups (see Figure 1) and makes evidence available to the morphologist, who then looks for new external characteristics. These are inconspicuous paradigm changes with a high impact [22], because the "puzzles," in this case the species, which were previously pressurised within the framework of "normal science" (and more and more frequently caused difficulties) continue, but the emphasis of the characteristics under the new molecular phylogenetic paradigma has changed (see "Catalogue of Onyxacalles"), and new characteristics are discovered (see "Key to the species of Onyxacalles"). The assumption that the morphologist should have seen a characteristic must be abandoned in favour of the question: would the morphologist be able to "see" it?

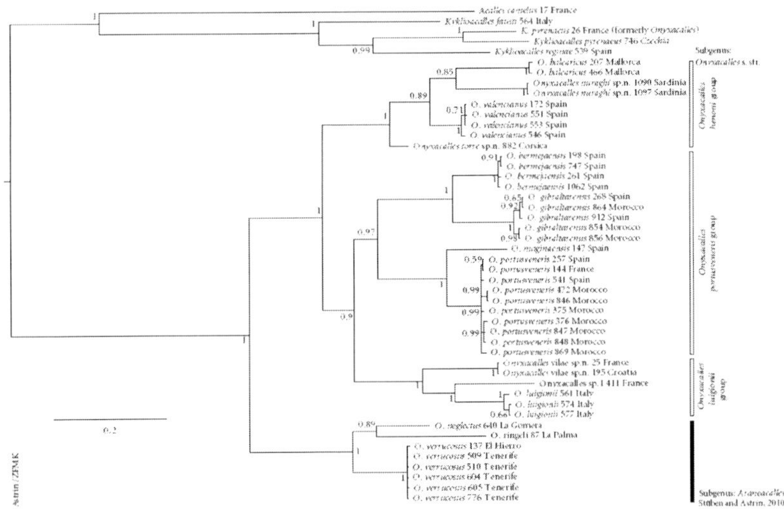

Figure 1: Bayesian consensus tree (50% majority rule) for COI and 16S.

2. CATALOGUE OF ONYXACALLES

Species included in the molecular analysis are printed in bold (l.t. = type locality).

Genus: Onyxacalles Stüben, 1999a.

177 type species Acalles luigionii A. and F. Solari, 1907.

Subgenus: Onyxacalles s. str.

Onyxacalles henoni Groupbalearicus Stüben, 2005: 115, Spain: Majorca (l.t.)croaticus H. Brisout de Barneville, 1867: 62 (Acalles), Croatia (l.t.), Austria, Czech Republic, Germany, Poland, Slovakia, Sloveniahannibali Germann, 2004: 118, Tunisia (l.t.)henoni Bedel, 1888: 36 (Acalles), Algeria: Mt. Edough (l.t.).nuraghi Stüben sp. n., Italy: Sardinia (l.t.).torre Stüben and Astrin sp. n., France: Corsica (l.t.).valencianus Germann, 2005: 104, Western Spain (l.t.).

Onyxacalles portusveneris Groupbermejaensis Stüben, 2001: 145, Spain (l.t.).gibraltarensis Stüben, 2002: 206, Morocco (l.t.), Spain.maginaensis Stüben, 2004: 120, Southern Spain (l.t.).portusveneris Mayet, 1903: 74 (Acalles), France (l.t.), Spain, Moroccoseguraensis Stüben, 2003a: 201, Spain.

Onyxacalles luigionii Groupluigionii A. and F. Solari, 1907: 521 (Acalles), Central (l.t.) and Southern Italy.vilae Stüben sp. n., Croatia (l.t.), France: Isère (perhaps Austria and Slovenia).cf. luigionii, France: Alpes Maritimes.

Incertae Sedisdenominandus A. and F. Solari, 1907: 523 (Acalles), Turkey (l.t.).porcheti Hoffmann, 1935: 162 (Acalles), France: Pyrenees (probably a synonym of A. luigionii, see also Stüben 2007: 149).amasyaensis Wolf, 2001: 150, Turkey (l.t.). (probably a synonym of Acalles denominandus)

Subgenus: Araneacalles Stüben and Astrin, 2010: 78 type species Acalles verrucosus Wollaston, 1863.neglectus Kulbe, 1999: 193, Canary Islands: La Gomera (l.t.), El Hierro.ringeli Kulbe 1999: 196, Canary Islands: La Palma (l.t.).verrucosus Wollaston, 1863: 219 (Acalles), Canary Islands: Tenerife (l.t.), El Hierro.

3. MATERIALS AND METHODS

The molecular analysis is based on 45 (43 after transfer of O. pyrenaeus, see below) individuals in 15 species or putative species of Onyxacalles and on 5 outgroup species (Cryptorhynchinae from 4 other genera; only the two closer genera are shown in the tree for better visualisation, while the 2 more distant ones were removed; see Table 1). Most sequences have been published in Astrin et al. (2012). Collecting and vouchering information as well as GenBank accession numbers are given in Table 1. Voucher specimens and extracted genomic DNA are deposited at the Biobank of the ZFMK (Zoologisches Forschungsmuseum Alexander Koenig, Bonn, Germany).

DNA extraction was carried out on samples preserved in ethanol or on dried material, using Macherey-Nagel Nucleo Spin Tissue kits (Dueren, Germany) or BioSprint 96 kits (Qiagen, Hilden, Germany). We extracted DNA from either 2-3 legs, head and prothorax, or sometimes also the whole weevil, depending on size and conservation of the sample. PCR reaction mixes (50 µL) contained 125 nmol $MgCl_2$, 5 µL 10x PCR-buffer, 25 pmol of forward and reverse primer each, 5 pmol dNTPs, 1.75 units of Taq polymerase, and 5 µL total undiluted DNA template. The lab chemicals were purchased from Sigma-Aldrich (Steinheim, Germany). We used the Qiagen (Hilden, Germany) Multiplex PCR kit in cases where the

Table 1: Collecting data, vouchers, and GenBank accession numbers for the material analysed in this study. All specimens determined by P. E. Stüben, 2010 and 2011. Vouchers (DNA, morphology) are kept at the ZFMK Biobank. Most sequences have been published (or are reviewed) in Astrin et al. 2012 [21]. GenBank accession numbers of new sequences in this study start with "JN…". Taxonomic changes with regard to this publication are printed in brackets (old name).

Taxon	Collecting data	DNA voucher	COI 16S
Acalles camelus (Fabricius 1792)	France: Isère, 2 km SE Lans en Vercors, Montagne de Lans; N45°06′45″ E05°36′21″, 1352 m: Abies, Fagus, Fraxinus, 2005, Stüben	ZFMK-DNA-JJ0017, ZFMK-TIS-cI0026cam	EU286282 EU286447
Acallorneuma doderoi A. and F. Solari 1908	Italy: Sicilia (PA), 6 km SW Godrano, Bosco Ficuzza, Mte. Rocca Busambra; N37°51′38″ E13°23′24″, 1200 m; Quercus, Fraxinus, 2002, Stüben	ZFMK-DNA-JJ0065, ZFMK-TIS-cS0082dod	EU286292 EU286457
Cryptorhynchus lapathi (Linne 1758)	Germany: Bienen bei Rees, Altrheinarm, 2004, Scharf	ZFMK-DNA-JJ0214, ZFMK-TIS-cD0354lap	EU286360 EU286523
Kyklioacalles fausti (Meyer 1896) [25]	Italy: Campania, Cilento, 6 km SE Vallo d. Lucania, M. Sacro o Gelbison; N40°12′41″ E15°19′42″, 1544 m; Fagus, 2008, Stüben	ZFMK-DNA-JJ0564, ZFMK-TIS-cI625fau	GU213776 GU213772
Kyklioacalles pyrenaeus (Boheman 1844) [26] (gen. Onyxacalles)	France: Isère, 14 km N Grenoble, Massif de la Chartreuse, NW Col de Porte; N45°18′40″ E05°45′17″, 1649 m: Abies, Fagus, Fraxinus, 2005, Stüben	ZFMK-DNA-JJ0026, ZFMK-TIS-cI0035pyr	GU988172 GU987762
Kyklioacalles pyrenaeus [26] (gen. *Onyxacalles*)	Czech Republic: W Bohemia (KT), Balkovy, Doubrava Hill (6545), 2008, Kresl	ZFMK-DNA-JJ0764, ZFMK-TIS-cCz798pyr	GU981555 GU981506
Kyklioacalles reginae Stüben 2003	Spain: Teruel, S. Javalambre, Fuente la Risca near Arcos de las Salinas; N39°59′56″ W01°01′21″, 1121 m; Amelanchier ovalis, Acer monspessulanum, Erinacea anthyllis, Ulex, 2008, Stüben	ZFMK-DNA-JJ0539, ZFMK-TIS-cE600reg	GU981544 GU981495
Onyxacalles balearicus Stüben 2005	Spain: Mallorca, 3 km SE Lluc, Sra. de Tramuntana, Sa Maleta; N39°48′47″ E02°53′23″, 571 m: Quercus ilex, 2004, Stüben	ZFMK-DNA-JJ0207, ZFMK-TIS-cE0168bal	EU286357 EU286521
Onyxacalles balearicus Stüben 2005	Spain: Mallorca, 11 km NE Lluc, Sra. de Tramuntana; N39°52′03″ E02°58′20″, 107 m; PT, Smilax aspera, Quercus ilex, 2004, Stüben	ZFMK-DNA-JJ0466, ZFMK-TIS-cE0294bal	GU988348 ———
Onyxacalles bermejaensis Stüben 2001	Spain: Andalucía, 11 km S Ronda, Sierra de las Nieves; N36°39′51″ W05°05′01″, 1047 m; Quercus ilex, 2005, Stüben	ZFMK-DNA-JJ0198, ZFMK-TIS-cE0167ber	EU286350 EU286514
Onyxacalles bermejaensis Stüben 2001	Spain: Málaga, 9 km SE Ubrique, Sierra de Libar; N36°36′52″ W05°23′16″, 663 m; Quercus ilex, Ceratonia, 2007, Stüben	ZFMK-DNA-JJ0261, ZFMK-TIS-cE0194ber	GU988244 GU987827
Onyxacalles bermejaensis Stüben 2001	Spain: Málaga, NW Marbella, Sierra de las Nieves; N36°39′52″ W05°04′57″, 1043 m; Quercus ilex, Echinodera spinosa, 2009, Stüben	ZFMK-DNA-JJ0747, ZFMK-TIS-cE778ber	GU988506 GU988066
Onyxacalles bermejaensis Stüben 2001	Spain: Prov. Malaga, Algatocín, near Opayar; N36°34′39″ W05°18′13″, 576 m, Quercus sp., 17.8.2010, Stüben	ZFMK-DNA-JJ1062, ZFMK-TIS-cES1062	JN121398 ———
Onyxacalles gibraltarensis Stüben 2002	Spain: Cádiz, 10 km SW Algeciras, El Bujeo; N36°04′10″ W05°31′48″, 257 m: Quercus suber, 2007, Stüben	ZFMK-DNA-JJ0268, ZFMK-TIS-cE0206gib	GU988249 GU987832
Onyxacalles gibraltarensis Stüben 2002	Morocco: Rif, SW Oued-Laou, river, O. Laou; N35°17′47″ W05°13′38″, 210 m; Quercus suber, Smilax, Arbutus, 2009, Stüben	ZFMK-DNA-JJ0854, ZFMK-TIS-cE897gib	GU988577 GU988137
Onyxacalles gibraltarensis Stüben 2002	Morocco: W Sebta, vir. Biutz: N35°53′04″ W05°24′08″, 337 m; Quercus suber, Smilax, Arbutus, 2009, Stüben	ZFMK-DNA-JJ0856, ZFMK-TIS-cE899gib	GU988578 GU988138
Onyxacalles gibraharensis Stüben 2002	Morocco: S Ksar-es-Seghir; N35°45′16″ W05°30′49″, 278 m; Pistacia, Quercus suber, 2009, Stüben	ZFMK-DNA-JJ0864, ZFMK-TIS-cE907gib	GU988584 GU988144
Onyxacalles gibraltarensis Stüben 2002	Spain: Cádiz, Los Barrios, Alcornocales N.P., between Facinos, Rio Las Cañas and Mantera Torero; Olea europaea, 2009, Torres	ZFMK-DNA-JJ0912, ZFMK-TIS-cE949gib	GU988608 ———
Onyxacalles luigionii (A. & F. Solari 1907)	Italy: Campania, Cilento, 6 km SE Vallo d. Lucania, M. Sacro o Gelbison; N40°12′41″ E15°19′42″, 1544 m; Fagus, 2008, Stüben	ZFMK-DNA-JJ0561, ZFMK-TIS-cI622lui	GU988407 GU987967

Onyxacalles luigionii (A. & F. Solari 1907)	Italy: Campania, Monti Picentini, 9 km N Acerno, Piano Laceno; N40°48′58″ E15°07′35″, 1210 m; Fagus, 2008, Stüben	ZFMK-DNA-JJ0574, ZFMK-TIS-cI635lui	GU988417 GU987977
Onyxacalles luigionii (A. & F. Solari 1907)	Italy: Basilicata, Monte Pollino, 9 km SE Rotonda, Rif. de Gasperi; N39°54′37″ E16°07′15″, 1486 m; Fagus, 2008, Stüben	ZFMK-DNA-JJ0577, ZFMK-TIS-cI638lui	GU988418 GU987979
Onyxacalles maginaensis Stüben 2004	Spain: Andalucia, 28 km E Jaén, Sierra Magina; N37°43′21″ W03°29′11″, 1600 m; Quercus ilex, 2005, Stüben	ZFMK-DNA-JJ0147, ZFMK-TIS-cE0169mag	EU286327 EU286491
Onyxacalles maginaensis Stüben 2004	Spain: Almeria, 11 km NW Laujar de Andarax, Sierra Nevada, Bayárcal; N37°02′27″ W03°00′12″, 1291 m; Quercus ilex, 2007, Stüben	ZFMK-DNA-JJ0257, ZFMK-TIS-cE0187mag	submitted to GenBank
Onyxacalles maginaensis Stüben 2004	Spain: Teruel, S. Javalambre, Fuente la Risca near Arcos de las Salinas; N39°59′56″ W01°01′21″, 1121 m; Amelanchier ovalis, Acer monspessulanum, Erinacea anthyllis, Ulex, 2008, Stüben	ZFMK-DNA-JJ0541, ZFMK-TIS-cE602mag	GU988390 GU987950
Onyxacalles neglectus Kulbe 1999	Spain: Canary Islands, La Gomera, S Hermigua, El Cedro, Las Mimbreras; N28°07′27″ W17°13′26″, 901 m; laurisilva, 2008, Astrin and Stüben	ZFMK-DNA-JJ0640, ZFMK-TIS-cE713neg	FJ716525 GU988014
Onyxacalles nuraghi sp.n.	Italy: W-Sardinia, E Macomer: above Lei; N40°19′54″ E08°53′49″, 1020 m; Quercus, Acer monspessulanum, 4.10.2010, Stüben	ZFMK-DNA-JJ1090, ZFMK-TIS-cIT1090	JN642097 JN121399
Onyxacalles nuraghi sp.n.	Italy: W-Sardinia, E Macomer: above Lei; N40°19′17″ E08°53′52″, 586 m; Quercus ilex, 7.10.2010, Stüben	ZFMK-DNA-JJ1097, ZFMK-TIS-cIT1097	JN642098 JN121300
Onyxacalles portusveneris (Mayet 1903) [27]	France: Gard, 15 km NE Nimes, Pont du Gard, Collias; N43°57′03″ E04°28′59″, 68 m; Quercus ilex, 2006, Stüben	ZFMK-DNA-JJ0144, ZFMK-TIS-cF0166por	EU286326 EU286490
Onyxacalles portusveneris (Mayet 1903) [27]	Morocco: Rif Mts., 10 km W Ketama = Issague; N34°57′40″ W04°40′51″, 1600 m; Prunus lusitanica, 2001, Stüben	ZFMK-DNA-JJ0375, ZFMK-TIS-cM480por	GU988311 ———
Onyxacalles portusveneris (Mayet 1903) [27]	Morocco: M-Atlas, 10 km S Ain-Leuh; N33°13′48″ W05°20′50″, 1700 m; Quercus ilex, Rubus, Cedrus, 2002, Stüben	ZFMK-DNA-JJ0376, ZFMK-TIS-cM481por	GU988312 ———
Onyxacalles portusveneris (Mayet 1903) [27]	Morocco: High Atlas, E Marrakech, N Taddert, (near Tazouguerte); N31°28′07″ W07°24′59″, 1727 m; Quercus, 2009, Stüben	ZFMK-DNA-JJ0846, ZFMK-TIS-cE889port	GU988573 GU988133
Onyxacalles portusveneris (Mayet 1903) [27]	Morocco: Middle Atlas, S Azrou, Àin Leuh; N33°16′50″ W05°20′18″, 1582 m; Quercus ilex, Euphorbia, 2009, Stüben	ZFMK-DNA-JJ0847, ZFMK-TIS-cE890port	GU988574 GU988134
Onyxacalles portusveneris (Mayet 1903) [27]	Morocco: Middle Atlas, S Azrou, S Àin Leuh; N33°14′57″ W05°21′04″, 1715 m; Quercus ilex, 2009, Stüben	ZFMK-DNA-JJ0848, ZFMK-TIS-cE891port	GU988575 GU988135
Onyxacalles portusveneris (Mayet 1903) [27]	Morocco: Rif, 10 km W Ketama; N34°57′40″ W04°40′51″, 1600 m; Cedrus, Prunus, 2009, Stüben	ZFMK-DNA-JJ0869, ZFMK-TIS-cE912port	GU988587 GU988147
Onyxacalles portusveneris (Mayet 1903) [27] (*Onyxacalles* sp.)	Morocco: High Atlas, 59 km SE Marrakech; N31°28′19″ W07°24′22″, 1500 m; Quercus ilex, Quercus suber, 2002, Stüben	ZFMK-DNA-JJ0472, ZFMK-TIS-cM482mag	GU988350 GU987922
Onyxacalles ringeli Kulbe 1999	Spain: Canary Islands, La Palma, Cumbre Nueva, 4,5 km SE El Paso, El Pilar; N28°37′37″ W17°49′45″, 1432 m; laurisilva, 2006, Stüben	ZFMK-DNA-JJ0087, ZFMK-TIS-cC0171rin	EU286300 EU286465
Onyxacalles sp. 1 (*O. luigionii*)	France: Alpes-Maritimes, 3 km W Sospel, Col de Braus; N43°52′34″ E07°24′17″, 1051 m; Quercus pubescens, Ostrya carpinifolia, broom, 2007, Stüben	ZFMK-DNA-JJ0411, ZFMK-TIS-cF440lui	GU988325 GU987897
Onyxacalles torre sp. n. (*O. henoni*)	France: Corsica, Col de Vizzavona, 22 km S Corte; N42°06′45″ E09°06′49″, 1100 m; Fagus, 2001, Stüben	ZFMK-DNA-JJ0882, ZFMK-TIS-cF479hen	GU988592 ———

Onyxacalles valencianus Germann 2005	Spain: Alicante, 7 km SW Alcoi, Sierra de Menechaor, Santurio de la Font Roja: N38°39′34″ W00°32′29″, 1296 m; Quercus ilex, 2007, Stüben	ZFMK-DNA-JJ0172, ZFMK-TIS-cE0180val	EU286331 EU286495
Onyxacalles valencianus Germann 2005	Spain: Castellón, Morella, Barranco de la Bota; N40°33′12″ W00°00′27″, 814 m; Quercus ilex, Hedera helix, 2008, Stüben	ZFMK-DNA-JJ0546, ZFMK-TIS-cE607val	GU988393 GU987953
Onyxacalles valencianus Germann 2005	Spain: Barcelona, above dry river bed, near Vallirana; N41°22′36″ E01°55′02″, 245 m; Quercus ilex, Ficus carica, Smilax aspera, 2008, Stüben	ZFMK-DNA-JJ0551, ZFMK-TIS-cE612val	GU988398 GU987958
Onyxacalles valencianus Germann 2005	spain: Barcelona, S. Montseny, Tordera valley, near St. Marçal; N41°48′01″ E02°25′15″, 1060 m, 2008, Stüben	ZFMK-DNA-JJ0553, ZFMK-TIS-cE614val	GU988400 GU987960
Onyxacalles verrucosus (Wollaston 1863)	Spain: Canary Islands, El Hierro, 7 km W La Frontera, Pista Derrabado; N27°44′29″ W18°03′24″, 895 m; Laurus azorica, 2006, Stüben	ZFMK-DNA-JJ0137, ZFMK-TIS-cC0170ver	EU286324 EU286488
Onyxacalles verrucosus (Wollaston 1863)	Spain: Canary Islands, Tenerife, 6 km N La Laguna, Monte de las Mercedes; N28°31′50″ W16°17′09″, 950 m; laurisilva, 2003, Stüben	ZFMK-DNA-JJ0509, ZFMK-TIS-cE570ver	——— GU987937
Onyxacalles verrucosus (Wollaston 1863)	Spain: Canary Islands, Tenerife, 4 km S Los Silos, Teno Mts., Monte del Aqua; N28°19′20″ W16°49′14″, 700 m; laurisilva, 2003, Stüben	ZFMK-DNA-JJ0510, ZFMK-TIS-cE571ver	GU988373 GU987938
Onyxacalles verrucosus (Wollaston 1863)	Spain: Canary Islands, Tenerife, NE La Laguna, Anaga Mts. near Moquinal; N28°31′55″ W16°17′24″, 840 m; laurisilva, 2008, Astrin and Stüben	ZFMK-DNA-JJ0604, ZFMK-TIS-cE677ver	GU988433 GU987995
Onyxacalles verrucosus (Wollaston 1863)	Spain: Canary Islands, Tenerife, NE La Laguna, Anaga Mts. near Chinobre: N28°33′21″ W16°10′46″, 808 m; Laurus, Ixanthus viscosus, 2008, Astrin and Stüben	ZFMK-DNA-JJ0605, ZFMK-TIS-cE678ver	GU988434 GU987996
Onyxacalles verrucosus (Wollaston 1863)	Spain: Canary Islands, Tenerife, SW Los Silos, Teno Mts., Monte del Agua, Chupadero: N28°19′23″ W16°49′12″, 940 m; Laurus novocanariensis, 2008, Astrin, Stüben, Behne and Floren	ZFMK-DNA-JJ0776, ZFMK-TIS-cE813ver	GU988524 GU988085
Onyxacalles vilae sp.n. (*O. luigionii*)	France: Isère, 2 km SE Lans en Vercors, Montagne de Lans; N45°06′45″ E05°36′21″, 1352 m; Abies, Fagus, Fraxinus, 2005, Stüben	ZFMK-DNA-JJ0025, ZFMK-TIS-cI0027lui	EU286286 EU286451
Onyxacalles vilae sp.n. (*O. luigionii*)	Croatia: Dalmatian, 8 km E Karlobag, Velebit Mts., Stupacinovo; N44°32′41″ E15°09′58″, 1049 m; Fagus, 2007, Stüben	ZFMK-DNA-JJ0195, ZFMK-TIS-cHR0339lui	EU286348 EU286512

regular protocol failed. PCR primers were taken from Astrin and Stüben (2008; COI is based on the Folmer et al. [23] region; 16S is based on the Crandall and Fitzpatrick (1996) region). Primer sequences were as follows: LCOI490-JJ (COI forward, fw) 5′-CHACWAAYCATAAAGATATYGG-3′, HCO2198-JJ (COI reverse, rev) 5′-AWACTTCVGGRTGVCCAAARAATCA-3′; 16S-ar-JJ (16S fw, erroneously as "rev" in [24]) 5′-CRCCTGTTT ATTAAAAACAT-3′, 16S-1472-JJ (16S rev) 5′-AGATAGAAACCRACCTGG-3′. Thermal cycling was performed on blocks of the type GeneAmp PCR System 2700 (Applied Biosystems, Foster City, CA, USA). PCR program for 16S: first cycle set (15 repeats): 35 s denaturation at 94°C, 35 s annealing at 55°C (−1°C per cycle) and 60 s extension at 72°C. Second cycle set (25 repeats): 35 s denaturation at 94°C, 35 s annealing at 40°C, and 50 s extension at 72°C. PCR program for COI: same as for 16S, but annealing temperatures at 70°C and 55°C, with a decrease of 2°C per cycle in the first cycle set. Double-stranded sequencing was carried out by a sequencing facility (Macrogen, South Korea, and Netherlands) using the same primers as in PCR.

DNA sequence alignment was performed manually (COI) or using the MUSCLE ver. 3.6 programme [28] (16S), run with default parameters. Sequence length was 554 bp for 16S (aligned; longest sequence: 544 bp; shortest: 533 bp) and 658 bp for COI, for concatenated sequence data

1212 bp. The 16S alignment comprised 29 positions with gaps. All of these were included into phylogenetic analysis. We implemented the GTR+I+Γ [29] model of nucleotide substitution for both genes in Bayesian MCMC analyses, run in MrBayes ver. 3.1.2 [30]. Only COI was included for the new sequences. We ran two independent replicates for 10 million generations per analysis (each with 1 cold chain and 3 chains of different temperature). Every 1′000th tree was sampled (20′000 trees retained). Negative log-likelihood score stabilisation was determined in a separate visualisation (in MS Excel). Accordingly, we retained 19.800 trees (after discarding burn-in), of which a 50%-majority rule consensus tree was built, with posterior probabilities (Figure 1).

4. RESULTS AND DISCUSSION

4.1. A Gestalt Switch—Changing the Way You See: Kyklioacalles pyrenaeus (Boheman, 1844)

The species of Onyxacalles are characterised by a particularly long and slender rostrum which is at least 3-4 times as long as wide between the insertions of the antennae. A further conspicuous feature (not typical for western Palearctic Cryptorhynchinae—excepting the species of the Macaronesian Islands) is the unusually long and slender (arachnoid) legs. The name of this genus refers to the hook-shaped tip of the aedeagus. No further species of the former accumulative genus Acalles or of other western Palearctic genera of the Cryptorhynchinae exhibit such a characteristic hook-shaped tip of the aedeagus (onyx; greek: hook, hook-shaped tool). Complex-sclerotised structures of the internal sac are absent or reduced to simple line- or bar-like structures. These structures have been significant, for instance, for the partly phylogeny-based classification and determination of the genera and species of Dichromacalles and Kyklioacalles. The strongly sclerotised median lobe exhibits either only unclear structures or none at all (see [31]).

As early as the beginning of the last century, in their ground-breaking revision of the western Palearctic species of Acalles s.l., A. and F. Solari placed the species A. pyrenaeus Boheman, 1844, A. henoni Bedel, 1888, A. croaticus Brisout, 1867, and A. luigionii Solari, 1907—although together with further species—into the same group (see IV. group; [32]). Together with initially 7 further species (among it the above species denominated by A. and F. Solari, as well as 3 further species from the Canary Islands), they were transferred into the new genus Onyxacalles [1, page 186], a genus that currently comprises 20 species.

However, Onyxacalles pyrenaeus is a polymorphic species, with regard to the outline of the aedeagus as well as the more or less ovally rounded elytra [1, page 188]. This is—among others—the reason why we could not ascertain definitively whether the subspecies Acalles pyrenaeus germanicus Letzner, 1882 (= Onyxacalles boehmei Košťál & Holecová, 2001 [33] syn.) is really a junior synonym or not (even if the first author considers it a synonym; cf. [7, page 123]).

In addition to Onyxacalles portusveneris (Mayet, 1903) (see Figure 4), O. pyrenaeus has an exceptionally large distribution area (cf. [31]). This could explain the high genetic distances of the mitochondrial COI and 16S gene (e.g., France: Lans en Vercors—Austria: Merkersdorf, p-distances: COI = 8,5%, 16S = 2,4%). This species can be found from the Pyrenees to the mountains of Western, Central, and Eastern Europe to the Carpathians and can be beaten from the branches of different conifers, especially larch, but can also be sifted under deciduous trees (e.g., Fagus).

This does not coincide with the other Onyxacalles species, which live on different deciduous trees and never prefer conifers. Apart from these ecological conditions, we can establish that the above-mentioned differences to the other genera of the western Palearctic Cryptorhynchinae point to a closer relationship to the species of Onyxacalles, a view with which most authors concur (e.g., [33]).

In any case, the taxonomists did not pay attention to the distinct ecology and—in comparison with the other species of Onyxacalles—the clearly narrower aedeagus (cf. Figure 6(k) versus Figures 9(a)–9(l)). Are these peripheral characteristics? And in which genus of Cryptorhynchinae is it possible to place this species (without the need for a monotypic genus)? The first author had never imagined assigning this species to the genus Kyklioacalles Stüben, 1999. There was no reason for such a review, not even an "initial suspicion" (see below).

However, this presumed "Onyxacalles" species appears deeply nested within the genus Kyklioacalles in the dendrogram [34, Figure 1]. This inclusion within Kyklioacalles is maximally supported. Furthermore, it is obvious that Acalles pyrenaeus forms a clade with K. fausti (Meyer, 1896), K. reginae Stüben, 2003, K. saccoi(Colonnelli, 1973) [35], and K. reinosae (H. Brisout de Barneville, 1867) [10, 34]. Two species, K. reginae and K. reinosae, are distinguished from the other species by the completely different habitus, but chiefly belong and were allocated early on to the K. fausti group "only" on the basis of the endophallus. This allocation has been shown to be justified by molecular data [34].

But what would have happened if Kyklioacalles had not been defined initially on the basis of the cyclical structure of the endophallus, as the name implies [36], and as an immediate consequence, Acalles pyrenaeuswould not have been eliminated? Putting it the other way round: what if the similarity between this species Kyklioacalles saccoi and Kyklioacalles fausti had been considered in a habitus-to-habitus comparison? That is not only a simple, retrospective, and dispensable "what-if" question, because in this case one brings the fact to mind that A. pyrenaeus must now be coercively placed among the Kyklioacalles species, Figures 8(f)–8(h).

The definition especially of a higher taxon is an arbitrary supposition. As taxonomists we always operate with constructs and as morphologists we find characteristics that are prominent to our eyes (and sometimes "we like to see"). It must be admitted that extremely rarely do we look for homologies, which in theory constitute the best criterion (e.g., [37]), but are in practice often difficult to find when dealing with cryptic and similar-looking species.

This change of mind and perception is similar to a Gestalt switch [22], a figure spinning in two directions: the contour line of the species is the same, but the species is not (as in the Gestalt psychology, the vase-face and duck-rabbit illusions). We cannot explain this Gestalt switch based on morphological research alone, and we cannot build and establish it within this framework. But we know the cause for this inconspicuous paradigm change: only extrinsic evidence from DNA analysis has opened the morphologist's eyes in this case—and this in the true sense of the word.

4.2. Integrative Taxonomy: Changing the Way You Look for Species

Integrative taxonomy sounds like an accumulation of different disciplines: morphology, molecular biology, ecology, ethology, and biogeography deliver the ingredients. But this is not invariably the case, and not so simple. It is a more eventful, reciprocal exchange of evidence, which an initial suspicion either confirms or rejects (comparable to an unsolved criminal case).

An initial morphological suspicion was already available when we discovered the new species Onyxacalles nuraghi (O. henoni group) in the humid Quercus/Acer forest in the mountains of Marghine on Sardinia. The differences from the well-known species of this group are obvious and easy to assemble (see below differential diagnosis of O. nuraghi). In the case of Onyxacalles vilae sp. n. (O. luigionii group) from the Velebit Mountains (Croatia), it was morphologically more ambiguous, but in case of the third species Onyxacalles torre sp.n. (O. henoni group) from Corsica, the specimens of the current type series remained completely unnoticed in the collection for years.

But the molecular results in view of these three (resp. four) species are obvious (see also Figures 1–4). With maximal support, O. vilae is widely separated in the tree topology from O. nuraghi and O. torre (O. henonigroup; Figure 1). Together with Onyxacalles sp. 1, O. vilae groups with O. luigionii (Onyxacalles sp. 1 as sister taxon of the latter), but all these species are separated by considerable p-distances (>9% COI and >4% 16S; Figure 3). Interestingly, the O. vilae specimen from Croatia shares the haplotype of the French specimen. O. torre and O. nuraghi are genetically closer to each other (1,8% 16S; Figure 2), but not sister species.

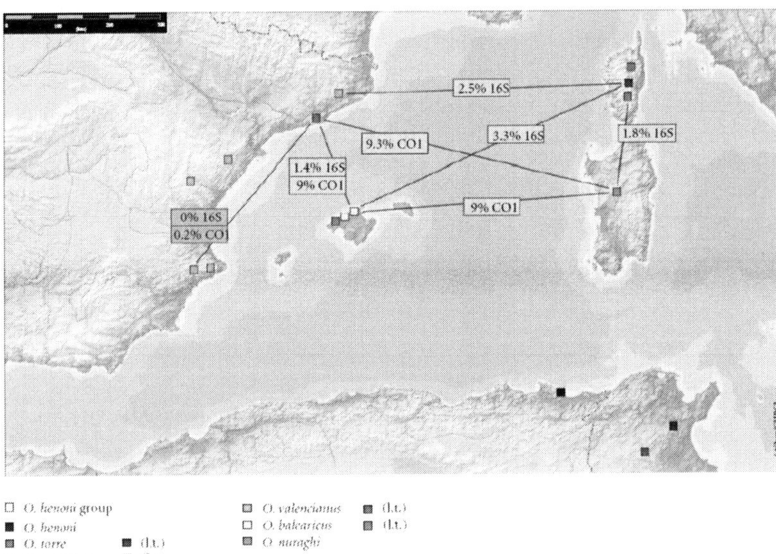

Figure 2: Distribution of related species of the West Mediterranean species of the O. henoni group (without O. croaticus, Eastern Europe) with values of COI and 16S p-distances.

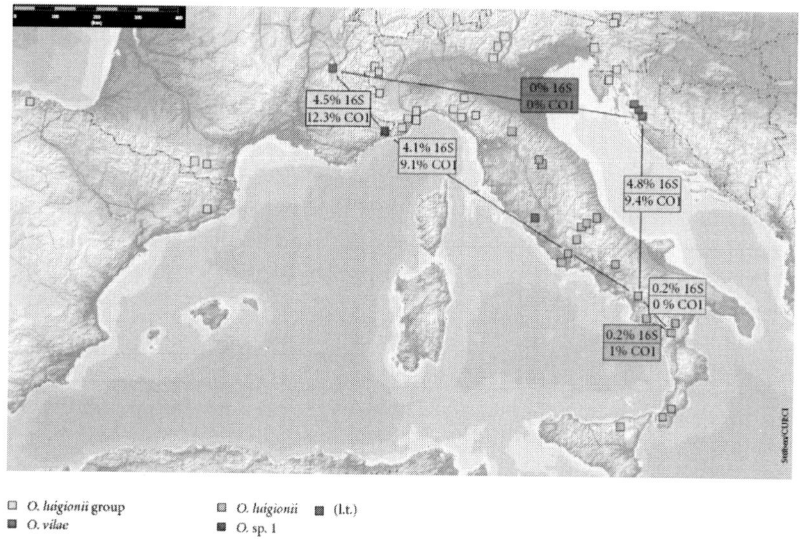

Figure 3: Distribution of related species of the O. luigionii group in the Mediterranean area with values of COI and 16S p-distances.

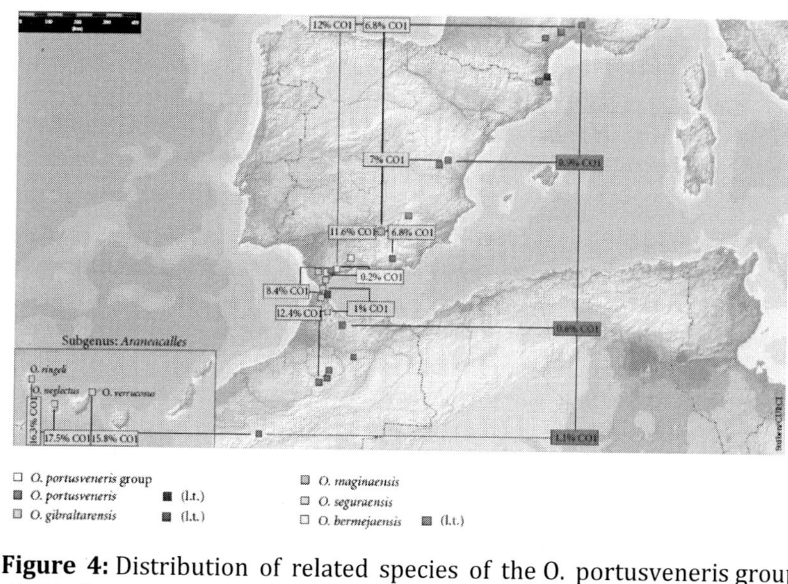

Figure 4: Distribution of related species of the O. portusveneris group in the Mediterranean area with values of COI p-distance (658 bp).

The unambiguous molecular evidence (cf. Figures 2 and 3) prompts the morphological reinvestigation of the material and forces the morphologist to look for new characters (see below "taxonomy")—if they exist!

On the one hand—in the case of Onyxacalles species—we currently give more attention to the apex of the rostrum and its punctures, or of the more or less curved and hook-shaped apex of the aedeagus. Vice versa, we see in the conspicuous and serially placed elytral bristles and its erected tufts on the intervals "seductive anthropomorphisms" and analyses of yesteryears. The perception does not become more precise, and the tuning is not finer adjusted—rather the way we view characters has changed: the perspective regarding criterion and weighting has undergone a shift influenced by extrinsic data.

On the other hand, however, the few specimens in the immediate vicinity of the French village Sospel (Maritime Alps) attest that it does not succeed in all cases (see: Figure 1: Onyxacalles sp. 1). The high p-distances of the mitochondrial COI and 16S genes of these specimens compared to the Middle and South Italian populations of O. luigionii and to O. vilae from the Velebit Mountains and the French Isère indicate the plausibility of a new species (cf. Figure 3). However, morphological differences could not be diagnosed. More findings and larger series of comparison material should establish a clearer picture in future. Therefore, it is premature to postulate a cryptic species and contrast it in a differential diagnosis by nothing more than molecular characters at this stage. Furthermore, this action would have the direct disadvantage that all previous existing records from the French and Italian Maritime Alps could not be allocated to the above-mentioned species. Nevertheless, this could shift in the future, should in depth "DNA barcodes" become available for all applied entomological disciplines.

The first step in this development has already been taken: the Molecular Weevil Identification-Project (ZFMK, CURCI), which is going to establish a molecular (DNA barcodes) and photographic database (stacked images) as well as the highly important associated reference collections for European Curculionoidae (ca. 6000 species). Only by meticulous cross-vouchering can misidentifications be corrected, and molecular results can be linked to the more than 250 years history of entomology. Integrative taxonomy is not just an accumulative or encyclopaedic "furthering of knowledge," but rather—as in this case—an interactive process in an interdisciplinary dialogue.

5. TAXONOMY

Family: Curculionidae Latreille, 1802.

Subfamily: Cryptorhynchinae Schoenherr, 1825.

Genus: Onyxacalles Stüben, 1999, Type species: Acalles luigionii A. and F. Solari, 1907 (L.t.: Central Italy).

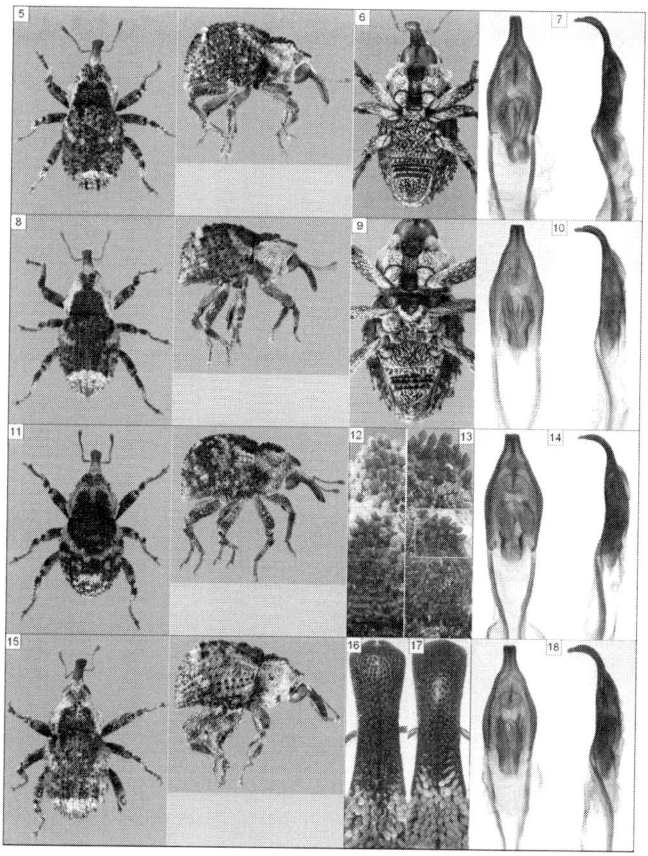

Figure 5: (a)–(c) O. nuraghi sp.n.—habitus (dor./lat./ven.), aedeagus (ven./lat.). (d–f) O. torre sp. n.—habitus (dor./lat./ven.), aedeagus (ven./lat.); (g, j) O. balearicus—habitus (dor./lat.), aedeagus (ven./lat.); (k, n) O. henoni—habitus (dor./lat.), aedeagus (ven./lat.). By comparison, bristle of pronotum—O. balearicus (h) versus O. nuraghi (i); rostrum—O. henoni (l) versus O. torre (m).

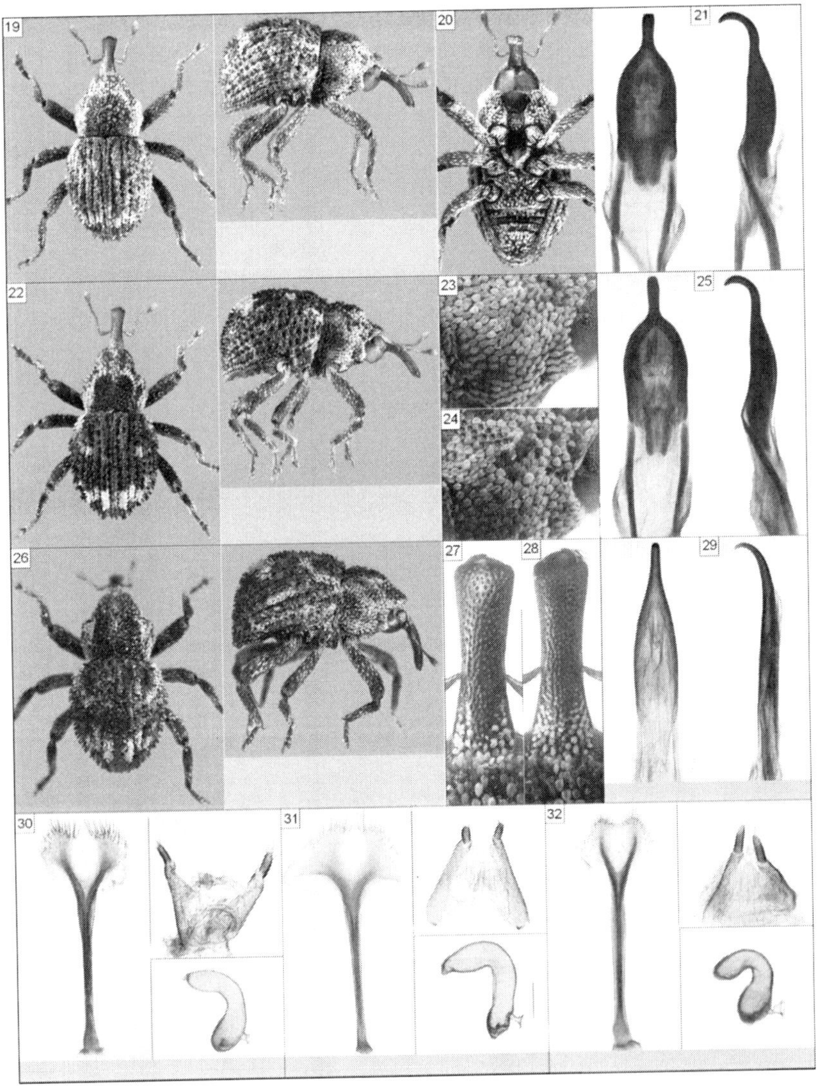

Figure 6: (a–c) O. vilae sp. n.—habitus (dor./lat./ven.), aedeagus (ven./lat.); (d, g) O. luigionii—habitus (dor./lat.), aedeagus (ven./lat.); (h) O. luigionii (lectotype)—habitus (dor./lat.); (k) Kyklioacalles pyrenaeus—aedeagus (ven./lat.); (l–n) female genital (spiculum ventrale, ovipositor, spermatheca) of O. nuraghi sp.n. (l), O. torre (m) and O. vilae (n). By comparison, bristle of pronotum (lat.)—O. vilae (e) versus O. luigionii (f); rostrum—O. vilae (j) versus O. luigionii (j).

5.1. New Species of the Onyxacalles henoni Group

5.1.1. Onyxacalles nuraghi Stüben sp. n.

(Figures 5(a)–5(c), 5(i), 6(l)).

Type Material Holotype (1♂) Italy: Sardinia, Macomer, Lei, N40°19′54″ E08°53′49″, 1020 m, Quercus, Acer monspesulanus, 4.10.2010, leg. Stüben-27-, coll. CURCULIO-Institut, D-Mönchengladbach. Paratypes (1♂) Data as for holotype, coll. Stüben; 4♂, 1♀: Italy: Sardinia, Macomer, Lei, N40°19′17″ E08°53′52″, 586 m, Quercus ilex, 7.10.2010, leg. Stüben-33-, coll. Stüben, CURCULIO-Institut, D-Mönchengladbach (1♀), coll ZFMK, 1♂: ZFMK-DNA-JJ1097, ZFMK-TIS-cIT1097.DNAtype (1♂) Data as for holotype, coll. ZFMK: ZFMK-DNA-JJ1090, ZFMK-TIS-cIT1090; GenBank Acc. no COI: JN121399, 16S: JN642097.

Differential Diagnosis The new species from the south-facing slope of the Chain of Marghine (Italy: Sardinia) belongs—from a morphological and molecular perspective—to the Onyxacalles henoni group and should be compared with the most closely related species from Majorca (Spain): Onyxacalles balearicus Stüben, 2005.

Onyxacalles nuraghi (1)Disc of pronotum with a channel from the base towards the flat sector in front of the fore-margin; with tufts of bristles on both sides of the channel (Figure 5(i)).(2)Bristles on the elytral intervals at least 2x as long as wide; shaping tufts with big gaps; their distances range from 3x the length of bristles.(3)Elytra of male with parallel sides in the middle sector (dorsal view); contour line of elytra forms almost a semicircle in lateral view (Figure 5(a)).(4)Apex ("hook") of the aedeagus (in ventral view) smaller (Figure 5(c)).

Onyxacalles balearicus (1*)Disc of prontoum without a channel and without tufts of bristles (Figure 5(h)).(2*)The free-standing bristles on the intervals shorter, 1.3x as long as wide; placed in a single row, not forming tufts.(3*)Elytra of male broader and stronger (short ovally) rounded (slightly "egg shaped"); contour line of elytra flatter or slightly rounded behind the base in lateral view (Figure 5(g)).(4*)Apex (hook) of the aedeagus (in ventral view) broader (Figure 5(j)).The new species from Sardinia is different from Onyxacalles henoni (Bedel, 1888) [38] (Algeria: Mt. Edough, loc. typ.), with which it has the tufts of bristles on the elytral intervals in common, by (1) darker elytral integument (Figure 5(a) versus 5(k)), (2) finer and longer white bristles on the femora, and (3) longer apex ("hook") of the aedeagus (lateral view, see Figure 5(c) versus 5(n)). It can be distinguished from Onyxacalles valencianus Germann, 2005 from the Spanish mainland

(Barcelona: Villarana, loc. typ.) by (1) elytral tufts of bristles (versus single bristles), (2) deep channel of the pronotum (versus without channel), and (3) longer apex of aedeagus (lateral view, see Figure 5(c) versus 9(h)). For a comparison with all other species see below the "Key to the species, of Onyxacalles Stüben, 1999".

Description Length 2.60–3.40 mm (without rostrum).Head and Rostrum Eyes large; rounded ovally towards front and acuminate towards underside of rostrum; frons between eyes more slender than the base of rostrum; rostrum reddish brown, closely covered with white scales at the base; rostrum of male reaching 3/4 length of pronotum and finely punctuated towards apex; rostrum of females reaching 4/5 length of pronotum, slender, shiny, and even more finely punctuated. The last three funicles of antennae short ovally rounded; the first two funicles elongated; the club clearly separated from funicles.Pronotum Widest at the end of the first third of the pronotum (holotype: 1.17x as wide as long); well rounded laterally towards the fore-margin and the base; with a deep depression at the sides directly behind the fore-margin; disk of pronotum strongly arced, with a channel in the middle from the base towards the flat sector in front of the fore-margin. The integument is rich in contrast consisting of round black scales on the disk and oval, white/brown scales on the flanks of the pronotum. Elongated and black bristles in an upright position in the middle of the disk on both sides of the channel; with a similar, but white tuft of bristles on each side of the pronotum; the deep punctures always covered with scales.Elytra Oblong (holotype: 1.29x as long as wide); widest in the middle and there with nearly parallel sides; only slightly rounded directly in front of the base; short ovally rounded towards the apex. Contour line of elytra strongly arced, almost forming a semi circle. The shiny and predominantly dark brown integument with a beige/white crescent-shaped fascia in front of the base and on the elytral slope. Bristles on the first and third interval (excluding the suture stripe) at most 2x longer than wide, forming flattened tufts, which have big gaps between them (their distances range from three times the length of bristle); bristles sparse on the second and fourth intervals; striae on the disc and at the sides of elytra clearly narrower than intervals, punctures oblong.Legs Long; the marginal front femora reach the base of the rostrum; the hind femora reach the end of the elytral apex. They are covered with predominantly dark brown scales; tibia with long, white, and laterally protruding bristles.Venter The 2nd strip-type sternite only a little bit longer than the 3rd, but not longer than sternites 3 and 4 together. 1st sternite of male with a broad depression (Figure 5(b)).Female Genital See Figure 6(l).Aedeagus Apex ("hook") of the aedeagus (in ventral view) small, see Figure 5(c).Etymology The species name refers to the Nuragic civilization of Sardinia, lasting from the Bronze Age (18th century BC) to the 2nd century AD.Ecology Onyxacalles nuraghi was discovered by the first author on Sardinia near Lei (Macomer) in the mountains of Marghine in 2010. The seven specimens were shifted under Quercus and Acer between 500 and 1000 m above sea level. Distribution This species is so far only known from the Chain of Marghine, Figure 2.

5.1.2. Onyxacalles torre Stüben and Astrin sp. n.

(Figures 5(d)–5(f), 5(m), and 6(m)). Type Material Holotype (1♂) France, Corsica (Haute-Corse): Col de Vizzavona, 22 km S Corte, 1100 m, 8.10.2001, 42°06′45″N 09°06′49″E, Fagus (sift), leg. Stüben-4-, coll. CURCULIO-Institut, D-Mönchengladbach.Paratypes (8♂, 9♀) Data as for holotype, coll. Stüben, CURCULIO-Institut, D-Mönchengladbach (1♀), Zoologisches Forschungsmuseum Alexander Koenig, D-Bonn (1♂, 1♀).DNAtype (1♀) Data as for holotype, coll. ZFMK: ZFMK-DNA-JJ0882, ZFMK-TIS-cF479; GenBank Acc. no 16S: GU988592.Further Material (1♂) France, Corsica (Haute-Corse): Caporalino 10 km N Corte, 350 m, 7.10.2001, 42°23′08″N 09°11′37″E, Alnus, Fraxinus, Quercus (sift), leg. Stüben-2-, coll. Stüben; 2♂: France, Corsica (Haute-Corse): Tattone, 18 km S Corte, 750 m, 8.10.2001, 42°09′21″N 09°09′43″E, Castanea (sift), leg. Stüben-3-, coll. Stüben; 1♂: France, Corsica (Corse-du-Sud): Radicale, 20 km E Ajaccio, 400 m, 9.10.2001, 41°55′31″N 08°58′10″E, 9.X.2001, Quercus ilex (sift), leg. Stüben-9-, coll. Stüben; 2♂: France, Corsica (Corse-du-Sud): Cozzano 2 km NE Zicavo, 750 m, 10.10.2001, 41°55′31″N 09°08′31″E, Castanea (sift), leg. Stüben-12-, coll. Stüben; 4♂, 4♀: France, Corsica (Corse-du-Sud): Coll de La Vaccia N, 9,5 km S Zicavo, 1150 m, 10.10.2001, 41°49′19″N 09°05′04″E, Fagus (sift), leg. Stüben-14-, coll. Stüben.

Differential Diagnosis The new species from Corsica (France) belongs—morphologically and molecularly—to the Onyxacalles henonigroup and is distinguished from Onyxacalles henoni (Bedel, 1888) [38] from Algeria (Mt. Edough, loc. typ.) by the following characteristics.Onyxacalles torre (1)Rostrum finely punctuated towards apex (Figure 5(m)).(2)Scales of the elytra predominantly dark brown or black (Figure 5(d)).(3)Apex ("hook") of the aedeagus broader in ventral view and strongly curved (nearly rectangular) in lateral view (Figure 5(f)).Onyxacalles henoni (1*)Rostrum coarsely and densely punctuated towards apex (Figure 5(l)).(2*)Scales of the elytra predominantly bright: white, beige, or brown (Figure 5(k)).(3*)Apex ("hook") of the aedeagus smaller in ventral view and not so strongly curved in lateral view (Figure 5(n)).The new species from Corsica can be distinguished from Onyxacalles nuraghi from Sardinia (see above) by (1) contour line of elytra behind the base flatter (in lateral view, Figure 5(d) versus 5(a)), (2) elytra more egg-shaped towards the apex (Figure 5(d) versus 5(a)), and (3) apex ("hook") of the aedeagus (in ventral view) shorter and wider (Figure 5(f) versus 5(c)). For a comparison with all other species, see the "Key to the species of Onyxacalles Stüben, 1999" below.

Description Length 3.00-4.00 mm (without rostrum).Head and Rostrum Eyes large; rounded towards front and acuminate towards underside of rostrum; frons between eyes as wide as the base of rostrum; rostrum reddish brown, closely covered with white scales at the base; rostrum of male reaching 2/3 length of pronotum and finely punctuated towards apex (Figure 5(m)); rostrum of females reaching 3/4 length of pronotum, slender, shiny, and even more finely punctuated. The last three funicles of the antennae short ovally rounded, the fourth 1.5x, the third 2x, the second 4.5x, and the first conical funicle 2x longer than wide; the elongated club clearly separated from funicles.Pronotum Widest at the end of the first third of the pronotum (holotype: 1.17x as wide as long); strongly rounded laterally towards the fore-margin and the base; with a depression at the sides directly behind the fore-margin; disk of pronotum arced, in the middle sometimes with a slight channel-like depression in front of the base. The integument is rich in contrast consisting of round black and dark-brown scales on the disk and in front of the fore-margin, and more or less oval, white scales on the flanks of the pronotum. In the middle of the disk on both sides of the flat depression with elongated, studded, and black bristles in an upright position; with similar placed, but shorter and white bristles on each side of the pronotum; the punctures always covered with scales.Elytra Oblong (holotype: 1.31x as long as wide); widest in front of the middle, here with parallel sides or slightly egg-shaped towards the apex; strongly curved in front of the base. Contour line of elytra flatter behind the base in lateral view, the contour line of the elytral slope forming an arc towards the apex. The shiny and predominantly dark brown or black integument with a beige/white crescent-shaped fascia in front of the base and on the elytral slope; sometimes the whole apex can be light brown. Bristles on first and third interval (excluding the suture stripe) 1.5x longer than wide, flattened, and shaping tufts; their distances range from the 2x length of bristles. Bristles sparse on the second and fourth intervals; striae on the disc and at the sides of elytra are small strips, clearly narrower than the intervals, punctures on the disc oblong, round at the sides.Legs Long; the marginal front femora reach the base of the rostrum, and the hind femora reach the end of the elytral apex. They are covered with predominantly dark brown and elongated scales; tibia with white/brown and laterally protruding bristles forming fasciae.Venter The 2nd strip-type sternite only a little bit longer than the 3rd, but not longer than sternites 3 and 4 together. 1st sternite of male with a broad depression (Figure 5(e)).Female Genital See Figure 6(m).Aedeagus Apex (hook) of the aedeagus broad in ventral view and strongly curved (nearly rectangular) in lateral view; see Figure 5(f).Etymology The species name refers to the Torrean civilization in Corsica during the

second millenium BC. The characteristic building of this culture is the "Torre" (tower), the Corsican counterpart of the Sardinian "Nuraghe."Ecology
Onyxacalles torre was sifted by the first author in the mountains of Corsica and is a nocturnal inhabitant of the dark and shady forests like all other Onyxacalles.Distribution
This species is so far only known from Corsica (France); Figure 2.

5.2. New Species of the Onyxacalles luigionii Group

5.2.1. Onyxacalles vilae Stüben sp. n.

(Figures 6(a)–6(c), 6(e), 6(i), and 6(n)). Type Material Holotype (1♂) Croatia: 20 km S Krasno Polje, Northern Velebit Mts., N44°38′14″ E15°05′13″, 1185 m, limestone: Fagus, 26.7.2004, leg. Stüben-10-, coll. CURCULIO-Institut, D-Mönchengladbach.Paratypes (1♂, 2♀) Data as for holotype, coll. Stüben, CURCULIO-Institut, D-Mönchengladbach (1♀), Zoologisches Forschungsmuseum Alexander Koenig, D-Bonn (1♀); 1♀: Croatia: Krasno Polje, Northern Velebit Mts., N44°49′40″ E15°01′49″, 838 m, limestone: Fagus, Quercus, 25.7.2004 leg. Stüben-7-, coll. Stüben; 1♂: Croatia: 5 km W Krasno Polje, Northern Nord-Velebit Mts., N44°48′49″ E14°58′36″, 1534 m, limestone: Fagus, 26.7.2004, leg. Stüben-8-, coll. Stüben; 1♀: Croatia: 6 km W Krasno Polje, Northern Velebit Mts., N44°48′56″ E14°58′08″, 1494 m, limestone: Fagus, 26.7.2004, leg. Stüben-9-, coll. Stüben; 1♂: Croatia: 12 km S Krasno Polje; Northern Velebit Mts., N44°43′00″ E14°59′42″, 1414 m, limestone: Fagus, 27.7.2004, leg. Stüben-14-, coll. Stüben.DNAtype (1♂) Croatia: 8 km E Karlobag, Velebit Mts., Stupacinovo, N44°32′41″ E15°09′58″, 1049 m, limestone: Fagus, 14.07.2007, leg. Stüben-27-, coll. ZFMK: ZFMK-DNA-JJ0195, ZFMK-TIS-cHR339; GenBank Acc. no COI: EU286512, 16S: EU286348.Further Material (9♂, 7♀) France, Isère, 2 km SE Lans en Vercors, Montagne de Lans, 45°06′40″N 05°36′25″E, 1391 m, 23.7.2011, Kalk: Fagus (beaten), leg. Stüben-2-, coll. Stüben; 3♂, 3♀: France, Isère, NW Lans en Vercors: near Autrars, Parc Regional du Vercors, 45°14′12″N 05°34′58″E, 1370 m, 23.7.2011, Kalk: Fagus (beaten), leg. Stüben-5-, coll. Stüben.

Differential Diagnosis The new species from Croatia (Velebit Mts., loc. typ.) belongs—morphologically and molecularly—to the Onyxacalles luigionii group and is distinguished from Onyxacalles luigionii (A. & F. Solari, 1907) [32] from Central Italy (Monte Fogliano, loc. typ.) by the following characteristics. Onyxacalles vilae (1)Rostrum of the male broader, 2.8x as long as wide (as measured by apex); punctures not so densely packed in front of the apex (separated by flat intervals) Figure 6(i).(2)Bristles of the low-contrast elytra on first and third intervals (excluding the sutural stripe) longer and more slender; their distance is larger (Figure 6(a)).(3)Scales of the white fascia at the sides of the pronotum (behind the base) predominantly oblong (Figure 6(e)).(4)Median lobe of aedeagus smaller, 1.64x as long as wide; apex ("hook") a little bit shorter (in ventral view), flatter, and not so strongly curved in lateral view (Figure 6(c)).Onyxacalles luigionii (1*)Rostrum of the male more slender, 3.1x as long as wide; punctures mainly dense towards the apex (only separated by small ridges) Figure 6(j).(2*)Bristles of high-contrast elytra on the on first and third intervals shorter, broader (towards the apex of bristle), and more dense, clearly visible on the white fascia of the elytral slope (Figures 6(d) and 6(h)).(3*)Scales of the white fascia at the sides of the pronotum (behind the base) predominantly round (Figure 6(f)).(4*)Median lobe of aedeagus broader, 1.93x as long as wide; apex ("hook") of the aedeagus longer (in ventral view), strongly (nearly rectangular) curved in lateral view. Internal structure (endophallus) of the sac different (Figure 6(g)).For a comparison with all other species see below the "Key to the species of Onyxacalles Stüben, 1999."

Description Length 2.40–3.20 mm (without rostrum).Head and Rostrum Eyes large; rounded ovally towards front and acuminate towards underside of rostrum; frons between eyes as wide as the base of rostrum; rostrum reddish brown, closely covered with white and oval scales at the base; rostrum of male 2.8x as long as wide (as measured by apex) and finely punctuated towards apex, here separated by flat intervals (Figure 6(i)); rostrum of female clearly longer, slender, shiny, and even more finely punctuated (without punctures in front of the apex). The last funicles of antennae nearly trapezoidal, the funicles 4–6 short oval, the third funicle 1.3x, the second 3x, and the first conical funicle 1.5x longer than wide; the elongated club not clearly separated from the 7th trapezoid funicle.Pronotum Widest at the end of the first third of the pronotum (holotype: 1.12x as wide as long); well rounded laterally towards the fore-margin and the base; with a slight depression at the sides directly behind the fore-margin; disk of pronotum arced, without a channel or a flat depression in the middle. The integument not so rich in contrast, consisting of round, brown scales on the disk and off-white, predominantly oblong scales at the sides of the pronotum (behind the base, Figure 6(e)). In the middle of the disk with elongated brown bristles in an upright position; with a similar, but white tuft of bristles on each side of the pronotum; the deep and dense punctures covered with scales.Elytra Short oval (holotype: 1.19x as long as wide); widest at the end of the first fourth in front of the elytral base; here laterally strongly rounded directly in front of the base; ovally rounded towards the apex. Contour-line of elytra flatter behind the base in lateral view, the contour line of the elytral slope forming a circular arc towards the apex. Bristles of the low-contrast elytra on first and third intervals (excluding the sutural stripe) slender, 2x-3x as long as wide, their distance reaching the double length of bristle, and forming tufts only in front of the base and on the elytral slope; the uprightly protruding bristles on the second and fourth intervals have wider gaps between them (their distances range from three- to fourfold length of bristle); the scales on the intervals do not cover the underground completely; striae on the disc broad, but not broader than the intervals, reaching the width of the intervals at the sides of elytra (but often covered by scales); punctures deep and rounded.Legs Long; the marginal front femora reach the base of the rostrum, and the hind femora reach the end of the elytral apex. They are covered with predominantly brown scales; tibia with short, white, and laterally protruding bristles.Venter The 2nd strip-type sternite only a little bit longer than the 3rd, but not longer than sternites 3 and 4 together. 1st sternite of male with a broad depression (Figure 6(b)).Female Genital See Figure 6(n).Aedeagus Median lobe of aedeagus small, 1.64x as long as wide; see Figure 6(c).Etymology The species name refers to a "Vila" (fairy) in the Velebit Mts. This massif has a similar relevance for Croatians to Olympus for Greeks or the Fuijiyama for the Japanese. In Croatia, the mystical Velebit Mts. range is famous for its fairies, the most celebrated called "Vila Velebita" (The Fairy of Velebit).Ecology Onyxacalles vilae was sifted by the first author in the Velebit mountains of Croatia under Fagus and Quercusbetween 800 and 1600 m above sea level.Distribution A complete distribution map will be given in a separate faunistic study in the future, but this species was also sifted by the author in the Montagne de Lans near Lans en Vercors (France: Isère)—on limestone and under Fagus, too (N45°06′45″ E05°36′21″, 1352 m). Working hypothesis: It could be possible that all specimens of the "Alpine Arc"—between Grenoble and the Velebit mountains—belong to this new species and can be separated from the Central Italian populations of Onyxacalles luigionii (cf. Figure 3).

6. KEY TO THE SPECIES OF ONYXACALLES STÜBEN, 1999

(1)Smaller species with an ovally rounded habitus, legs shorter, femora reach the base of the rostrum; rostrum shorter and broader; if tufts of bristles exist on the elytral slope, these are only densely placed (not tapered). Distribution: continent of Western Palaearctic.

Subgenus: Onyxacalles s. str .... 2(1*)Larger species with a more "elliptical" habitus and with long legs, femora reach the insertions of the antennae; rostrum very long and slender; the tapered tufts of bristles on the elytral slope strongly protruding (Figures 8(b)–8(d)). Distribution: Western Canary Islands.

Subgenus: Araneacalles Stüben and Astrin, 2010……… 16(2)Apex of the aedeagus regularly rounded in lateral view: Figures 6(c) and 6(g) (both species without distinctive tufts of bristles on the uneven elytral intervals, only with densely placed bristles in one or two rows).

Luigionii group ……… 2(2*)Apex of the aedeagus with a second, separated peak in lateral view (cf. Figures 9(a)–9(l)) (most species with more or less characteristic tufts of bristles on the elytra).

Luigionii group --(3)Rostrum of the male more slender, 3.1x as long as wide; punctures mainly dense towards the apex (only separated by small ridges) (Figure 6(j)). Bristles of high-contrast elytra on the 1st and 3rd intervals shorter, broader (towards the apex of bristle), and more dense (Figures 6(d) and 6(h)). Median lobus of aedeagus broader, 1.93x as long as wide; apex ("hook") of the aedeagus longer (in ventral view), strongly (nearly rectangular) curved in lateral view. Internal structure (endophallus) of the sac different (Figure 6(g)). Distribution: Central and Southern Italy (Figure 3).

Onyxacalles luigionii(A. & F. Solari, 1907) [32]= ? Onyxacalles porcheti [39] (Figure 8(e), Pyrenees)(3*)Rostrum of the male broader, 2.8x as long as wide (as measured by apex); punctures not so densely packed in front of the apex (separated by flat intervals) (Figure 6(i)). Bristles of the low-contrast elytra on the 1st and 3rd intervals (excluding the sutural stripe) longer and more slender; their distance larger (Figure 6(a)). Median lobus of aedeagus smaller, 1.64x as long as wide; apex ("hook") a little bit shorter (in ventral view), flatter, not so strongly curved in lateral view (Figure 6(c)). Distribution: Croatia (l.t.), France: Isère (Figure 3).

Onyxacalles vilaeStüben sp. n.(4)Pronotum levelled, "triangular" and stubby, strongly broadened just behind the base; the base 2x longer than the fore-margin (this characteristic is not so pronounced in O. gibraltarensis; however, this species can be clearly separated from all other Onyxacalles s. str. by the completely rounded sides of the aedeagus (Figure 9(a)), see digit 5). Distribution: Southern France, Iberian Peninsula, Morocco.

Portusvenerisgroup ……… 5(4*)Pronotum more arched and marginally broader than long, widest at the end of the first third; the base at most 1.5x longer than the fore-margin. Distribution: Algeria, Tunisia, Western Spain, West Mediterranean Islands, Southeastern Europe, Turkey.

....

9 Portusveneris group --(5)Pronotum more slender, clearly separated from elytra; body outline broadly similar to the species of the henoni group (see digit 4*), but easy to distinguish from these species by the completely rounded sides of the aedeagus. Habitus (Figure 7(a)). Aedeagus (Figure 9(a)). Distribution: Southern Spain, Northern Morocco (Figure 4).

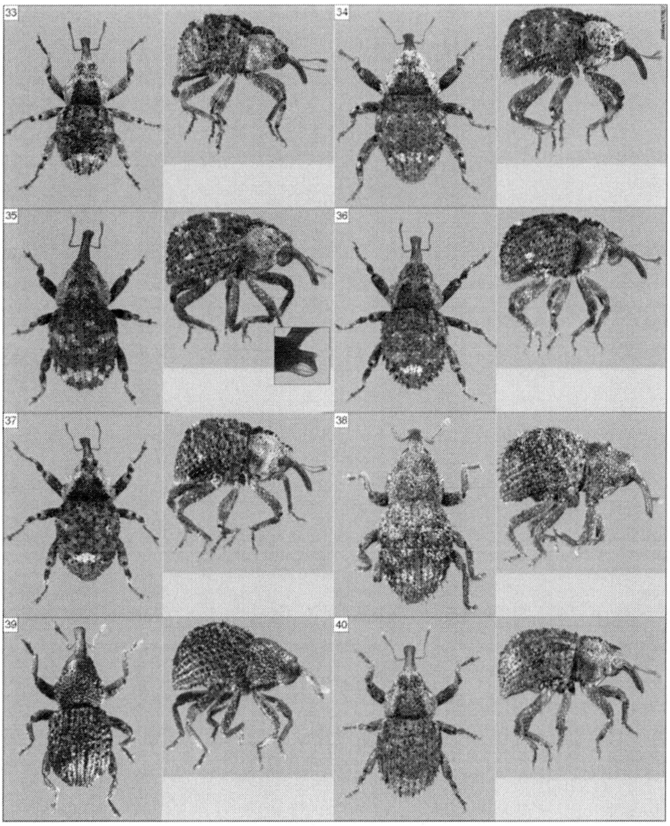

Figure 7: (a–e): Portusveneris group—O. gibraltarensis (a), O. portusveneris (b), O. seguraensis (c), O. maginaensis (d), and O. bermejaensis (e); (f): Incertae sedis—O. denominandus; (g-h): Henoni group (see also next figures)—O. croaticus (g) and O. valencianus (h); all habitus (dor./lat.).

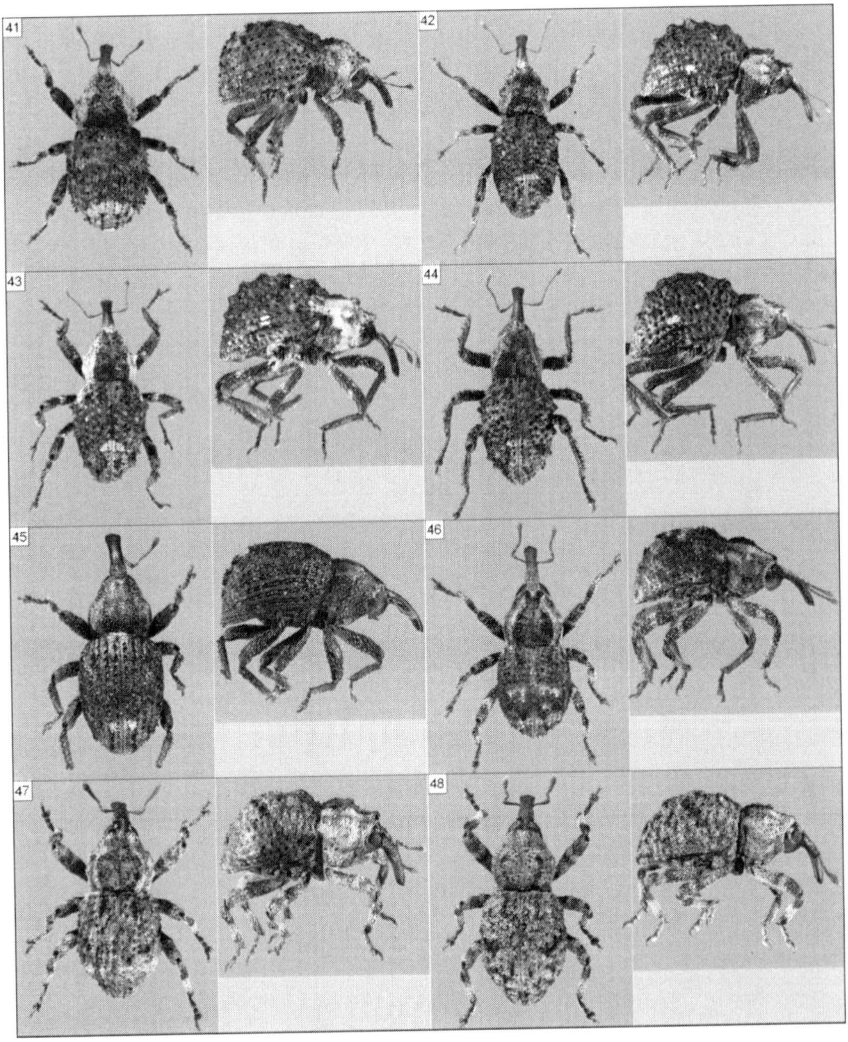

Figure 8: (a): Henoni group—O. hannibali (see also the other species of this group: Figures 5(a), 5(d), 5(g), 5(k)); (b–d): subgenus: Araneacalles—O. neglectus (b), O. verrucosus (c), and O. ringeli (d); (e): O. porcheti (perhaps O. luigionii); (f): Kyklioacalles pyrenaeus; (g) K. fausti; (h) K. saccoi; all habitus (dor./lat.).

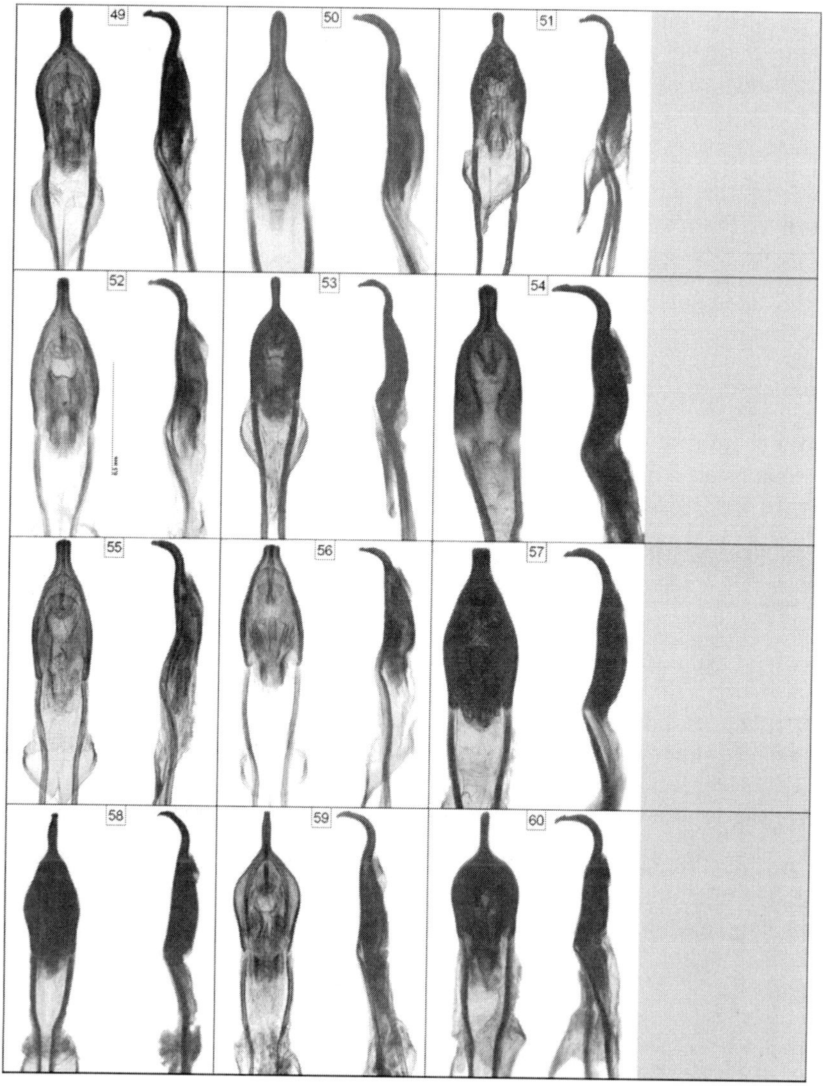

Figure 9: (a–d): Portusveneris group—O. gibraltarensis (a), O. portusveneris (b), O. seguraensis (c), O. maginaensis (d), and O. bermejaensis (e); (f): Incertae sedis—O. amasyaensis; (g–i): Henoni group—O. croaticus (g), O. valencianus (h), O. hannibali (i) (see also the other species of this group: Figures 5(c), 5(f), 5(j), 5(n)); (j–l): subgenus: Araneacalles—O. neglectus (j), O. verrucosus (k), and O. ringeli (l).

Onyxacalles gibraltarensis Stüben, 2002(5*)Pronotum widest directly behind the base and elytra widest directly in front of the base; therefore, pronotum and elytra do not seem separated (Figures 7(b)–7(e)).

.... 6(6)Elytra egg-shaped towards the apex (Figure 7(b)); Aedeagus (Figure 9(b)). Distribution: Southern France, Iberian Peninsula, Morocco (Figure 4).

Onyxacalles portusveneris (Mayet, 1903) [27](6*)Elytra oval or with more or less parallel sides (Figures 7(c)–7(e)).

.... 7(7)Male with a split midtibia spine at the apex; uneven elytral intervals without tufts of bristles. Habitus (Figure 7(c)). Aedeagus (Figure 9(c)). Distribution: Southern Spain (Figure 4).

Onyxacalles seguraensisStüben, 2003(7*)Male without a split midtibia spine at the apex; uneven elytral intervals with tufts of bristles.

.... 8(8)Punctures at the sides of elytra fine and slender; bristles on the first four intervals of the elytral slope in single row. Habitus (Figure 7(d)). Aedeagus (Figure 6(d)). Distribution: Southern Spain (Figure 4).

Onyxacalles maginaensisStüben, 2004(8*)Punctures at the sides of elytra broader and deeper; bristles on the third interval of the elytral slope densely placed, forming a pronounced tuft at the level of the white fascia. Habitus (Figure 7(e)). Aedeagus (Figure 9(e)). Distribution: Southern Spain (Figure 4).

Onyxacalles bermejaensisStüben, 2001 — (9)Pronotum with a deep midgroove and with strong concavities on each side; a species from Turkey. Habitus (Figure 7(f)). Aedeagus (Figure 9(f)). Distribution: Turkey.

Onyxacalles denominandus (A. & F. Solari, 1907) [32] =

? Onyxacalles amasyaensisWolf, 2001(9)Pronotum behind the base at most with a flat depression or a hinted channel; mainly West Mediterranean species, only one species from southeastern Europe.

Henonigroup .... 10 Henoni group --(10)Elytra with superelevated and in tubercles dissected intervals; a species from southeastern Europe, which is added to the henoni group preliminary on the basis of a similar form of the aedeagus. Habitus (Figure 7(g)). Aedeagus (Figure 9(g)). Distribution: East and Southeast Europe [31].

Onyxacalles croaticus (H. Brisout de Barneville, 1867) [10](10)Elytra flat, without tubercles.

....

11(11)Elytra and pronotum (almost) without tufts of bristles; these single, beaded bristles placed in a row.

....

12(11*)Elytra on the intervals 1 and 3 and pronotum at the sides with tufts of bristles; these bristles densely placed in 2-3 rows, forming tufts at regular intervals.

....

13(12)Elytral bristles short, shovel-shaped and densely placed (at most 1.3x as long as wide); elytra rich in contrast (colored); apex of aedeagus strongly curved in lateral view. Habitus (Figure 5(g)). Aedeagus (Figure 5(j)). Distribution: Spain, Majorca (Figure 2).

Onyxacalles balearicusStüben, 2005(12*)Elytral bristles more slender, at least 2x as long as wide and their distance large; elytra poor in contrast; apex of aedeagus flatter curved in lateral view. Habitus (Figure 6(n)). Aedeagus (Figure 9(h)). Distribution: Eastern Spain (Figure 2).

Onyxacalles valencianusGermann, 2005(13)Disc of pronotum with a channel from the base towards the flat sector in front of the fore-margin; elytra of male with parallel sides in the middle sector (dorsal view). Habitus (Figure 5(a)). Aedeagus (Figure 5(c)). Distribution: Italy, Sardinia (Figure 2).

Onyxacalles nuraghiStüben sp. n.(13)Disc of pronotum at most with a flat hallow behind the base; elytra of male broader and stronger (short ovally) rounded (slightly "egg-shaped").

....

14(14)Elytral intervals only

with a few bristles and small tufts; elytra poor in contrast. Habitus (Figure 8(a)). Aedeagus (Figure 9(i)). Distribution: Tunisia (Figure 2).

Onyxacalles hannibaliGermann, 2004(14*)Elytral intervals studded with bristles and with numerous distinctive tufts; elytra rich in contrast.

....

15(15)Rostrum coarsely and densely punctuated towards apex (Figure 5(l)); scales of the elytra predominantly bright: white, beige, or brown (Figure 5(k)); Apex ("hook") of the aedeagus smaller in ventral view and not so strongly curved in lateral view (Figure 5(n)). Distribution: Algeria, Mt. Edough (Figure 2).

Onyxacalles henoni [38](15*)Rostrum finely punctuated towards apex (Figure 5(m)); scales of the elytra predominantly dark brown or black (Figure 5(d)); apex ("hook") of the aedeagus broader in ventral view and strongly curved (nearly rectangular) in lateral view (Figure 5(f)). Distribution: France, Corsica (Figure 2).

Onyxacalles torre Stüben and Astrin sp. n. -- Subgenus: Araneacalles --

(16)Punctures of the 1st and 2nd elytral striae rounded, pothole-like, and as wide as the intervals; the underground of elytra in front of the middle on the 8th and 9th intervals with scales, not shiny; fore-margin of pronotum with a curved up collar. Habitus (Figure 8(b)). Aedeagus (Figure 9(j)). Distribution: Canary Is., La Gomera (l.t.), El Hierro (Figure 4).

Onyxacalles neglectusKulbe, 1999(16*)Punctures of the 1st and 2nd elytral striae elongated, clearly smaller than intervals; the underground of elytra in front of the middle on the 8th and 9th intervals without scales, shiny; fore-margin of pronotum without a curved up collar.

....

17(17)Punctures at the extreme striae slender, the intervals broader; elytra widest in front of the middle, egg-shaped. Habitus (Figure 8(c)). Aedeagus (Figure 9(k)). Distribution: Canary Is., Tenerife (l.t.), El Hierro (Figure 4).

Onyxacalles verrucosus (Wollaston, 1863) [40](17*)Punctures at the extreme striae larger and rounded, not broader than intervals; elytra oval, widest in the middle. Habitus (Figure 8(d)). Aedeagus (60). Distribution: Canary Is., La Palma (Figure 4).

Onyxacalles ringeliKulbe, 1999

ACKNOWLEDGMENTS

This paper was prepared by the Molecular Weevil Identification project (MWI) of the CURCULIO Institute (CURCI) and Zoologisches Forschungsmuseum Alexander Koenig (ZFMK). Keith Bensusan (Gibraltar) kindly revised the English text, and Christina Blume performed the lab work.

REFERENCES

1. P. E. Stüben, "Die westpaläarktischen Arten der Gattung Onyxacalles g.n. (Col., Curculionidae: Cryptorhynchinae)," Entomologische Blätter, vol. 95, pp. 175–203, 1999.
2. P. E. Stüben, "Die Arten des Genus Onyxacalles von den Kanarischen Inseln (Curculionidae: Cryptorhynchinae)," Snudebiller, vol. 1, pp. 106–113, 2000.
3. P. E. Stüben and J. J. Astrin, "Molecular phylogeny in endemic weevils: revision of the genera of Macaronesian Cryptorhynchinae (Coleoptera: Curculionidae)," Zoological Journal of the Linnean Society, vol. 160, no. 1, pp. 40–87, 2010.
4. P. E. Stüben and I. Wolf, "Beschreibung neuer westpaläarktischer Onyxacalles-Arten. Mit einem neuen Katalog und Schlüssel der Arten des Genus Onyxacalles. (Col.: Curculionidae: Cryptorhynchinae)," Snudebiller, vol. 2, pp. 143–154, 2001.
5. P. E. Stüben, "Beschreibung neuer Cryptorhynchinae aus Marokko (Coleoptera: Curculionidae)," Snudebiller, vol. 3, pp. 196–225, 2002.
6. P. E. Stüben, "Beschreibung neuer Cryptorhynchinae aus Spanien, Italien und Marokko—mit einem neuen Onyxacalles-und Echinodera-Schlüssel (Coleoptera: Curculionidae)," Snudebiller, vol. 4, pp. 186–224, 2003.
7. Ch. Germann and P. E. Stüben, "Beschreibung neuer Onyxacalles-Arten aus dem mediterranen Raum und Anmerkungen zur Synonymie einiger Arten der Gattung Onyxacalles Stüben 1999. (Coleoptera: Curculionidae: Cryptorhynchinae)," Snudebiller, vol. 5, pp. 116–131, 2004.
8. P. E. Stüben, "Die Cryptorhynchinae-Fauna der Baleareninsel Mallorca (Coleoptera: Curculionidae)," Snudebiller, vol. 6, pp. 114–130, 2005.
9. P. E. Stüben, F. Bahr, Ch. Germann, L. Behne, and Ch. Bayer, "Beschreibung neuer Cryptorhynchinae aus dem mediterranen Raum (Coleoptera: Curculionidae)," Snudebiller, vol. 6, pp. 84–113, 2005.
10. H. Brisout de Barneville, "Nouveaux tableau des Acalles avec la description de deux nouvelles espèces et celle de l'Orchestes quedenfeltii Gerhard," Annales de la Société Entomologique de France, vol. 4, no. 7, pp. 57–64, 1867.
11. P. E. Stüben and V. Thurn, "Die Wälder sind der Menschheit vorausgegangen, die Wüsten folgen ihr," in WüstenErde. Der Kampf Gegen Durst, Dürre und Desertifikation, P. E. Stüben and V. Thurn, Eds., pp. 16–28, Ökozid 7, Gießen, Germany, 1991.
12. V. ThurnP. E. Stüben and V. Thurn, "Die Wüste greift nach Europa. Die Zerstörung der Mittelmeerwälder," in WüstenErde. Der Kampf gegen Durst, Dürre und Desertifikation, pp. 121–136, Ökozid 7, Gießen, Germany, 1991.
13. P. E. Stüben, K. Fabian, and J. J. Astrin, TreeClimbers. Dendroacalles, Silvacalles and Lauriacalles of the Canary Islands (Curculionidae: Cryptorhynchinae), CURCULIO-Institute, Mönchengladbach, Germany, 2009.
14. P. E. Stüben, L. Behne, A. Floren et al., Canopy Fogging in the Canarian laurel forest of Tenerife and La Gomera, CURCULIO-Institute, Mönchengladbach, Germany, 2010.

15. B. Dayrat, "Towards integrative taxonomy," Biological Journal of the Linnean Society, vol. 85, no. 3, pp. 407–415, 2005.
16. K. W. Will, B. D. Mishler, and Q. D. Wheeler, "The perils of DNA barcoding and the need for integrative taxonomy," Systematic Biology, vol. 54, no. 5, pp. 844–851, 2005.
17. A. D. Roe and F. A. H. Sperling, "Population structure and species boundary delimitation of cryptic Dioryctria moths: an integrative approach," Molecular Ecology, vol. 16, no. 17, pp. 3617–3633, 2007.
18. J. M. Padial, S. Castroviejo-Fisher, J. Köhler, C. Vilà, J. C. Chaparro, and I. D. I Riva, "Deciphering the products of evolution at the species level: the need for an integrative taxonomy," Zoologica Scripta, vol. 38, no. 4, pp. 431–447, 2009.
19. M. Heethoff, M. Laumann, G. Weigmann, and G. Raspotnig, "Integrative taxonomy: combining morphological, molecular and chemical data for species delineation in the parthenogenetic Trhypochthonius tectorum complex (Acari, Oribatida, Trhypochthoniidae)," Frontiers in Zoology, vol. 8, article 2, 2011.
20. D. S. H. Tan, Y. Ang, G. S. Lim, M. R. B. Ismail, and R. Meier, "From "cryptic species" to integrative taxonomy: an iterative process involving DNA sequences, morphology, and behaviour leads to the resurrection of Sepsis pyrrhosoma (Sepsidae: Diptera)," Zoologica Scripta, vol. 39, no. 1, pp. 51–61, 2010.
21. J. J. Astrin, P. E. Stüben, B. Misof et al., "Exploring diversity in cryptorhynchine weevils (Coleoptera) using distance-, character- and tree-based species delineation," Molecular Phylogenetics and Evolution, vol. 63, no. 1, pp. 1–14, 2012.
22. Th. S. Kuhn, The Structure of Scientific Revolutions, University of Chicago Press, Chicago, Ill, USA, 1st edition, 1962.
23. O. Folmer, M. Black, W. Hoeh, R. Lutz, and R. Vrijenhoek, "DNA primers for amplification of mitochondrial cytochrome c oxidase subunit I from diverse metazoan invertebrates," Molecular Marine Biology and Biotechnology, vol. 3, no. 5, pp. 294–299, 1994.
24. J. J. Astrin and P. E. Stüben, "Phylogeny in cryptic weevils: molecules, morphology and new genera of Western Palaearctic Cryptorhynchinae (Coleoptera: Curculionidae)," Invertebrate Systematics, vol. 22, no. 5, pp. 503–522, 2008.
25. P. Meyer, "Diagnose sechs neuer Acalles-Arten der palaearctischen Region, nebst einigen synonymischen Bemerkungen über die Gattung. (Gegeben als Vorarbeit zu einer Bestimmungs-Tabelle der Cryptorrhynchiden Europa's und der angrenzenden Länder)," Wiener entomologische Zeitung, vol. 15, pp. 13–16, 1896.
26. C. H. Boheman and J. C. Schoenherr, "new taxa," in Genera et species curculionidum, cum synonymia hujus familiae, vol. 7 of pars I. Supplementum, Roret, Fleischer, Paris, Farnce, 1844.
27. V. Mayet, "Contribution a la faune entomologique des pyrénées orientales," Coléoptères des Albères. Miscelánea Zoológica, vol. 11, pp. 67–76, 1903.
28. R. C. Edgar, "MUSCLE: a multiple sequence alignment method with reduced time and space complexity," BMC Bioinformatics, vol. 5, article 113, 2004.
29. C. Lanave, G. Preparata, C. Saccone, and G. Serio, "A new method for calculating evolutionary substitution rates," Journal of Molecular Evolution, vol. 20, no. 1, pp. 86–93, 1984.
30. F. Ronquist and J. P. Huelsenbeck, "MrBayes 3: bayesian phylogenetic inference under mixed models," Bioinformatics, vol. 19, no. 12, pp. 1572–1574, 2003.
31. P. E. Stüben, "An illustrated up-to-date catalogue of Westpalearctic Cryptorhynchinae (Curculionoidea)," 2012, http://www.curci.de/illustrated_catalogue/westpalearctic_cryptorhynchinae/.
32. A. Solari and F. Solari, "Studi sugli Acalles," Annali del Museo Civico di Storia Naturale "G. Doria" (3), vol. 3, no. 43, pp. 479–551, 1907.
33. M. Košťál and M. Holecová, "Neotype designation of Onyxacalles pyrenaeus and description of its sibling species Onyxacalles boehmei sp. nov. from Central Europe (Coleoptera: Curculionidae)," Entomological Problems, vol. 32, no. 1, pp. 73–78, 2001.

34. P. E. Stüben and J. J. Astrin, "Molecular phylogeny of the weevil genus Kyklioacalles Stüben, with descriptions of a new subgenus Glaberacalles and two new species (Curculionidae: Cryptorhynchinae)," Zootaxa, no. 2662, pp. 28–52, 2010.
35. E. Colonnelli, "Una nuova specie di Acalles Schönherr del Massiccio del Pollino (Coleoptera, Curculionidae)," Bollettino dell 'Assoziazione Romana di Entomologia, vol. 27, no. 1-2, pp. 25–30, 1973.
36. P. E. Stüben, "Taxonomie und Phylogenie der westpaläarktischen Arten der Gattung Kyklioacalles g.n. (Coleoptera, Curculionidae: Cryptorhynchinae)," Stuttgarter Beiträge zur Naturkunde Serie A, vol. 584, p. 38, 1999.
37. J. W. Wägele, Foundations of Phylogenetic Systematics, Pfeil, Munich, Germany, 2005.
38. L. Bedel, "Description d'une Curculionide nouveau d'Algérie," Bulletin de la Société Entomologique de France, vol. 8, article xxxvi, 1888.
39. A. Hoffmann, "Description d'un Acalles nouveau de France (Col., Curculionidae)," Bulletin de la Société Entomologique de France, vol. 40, pp. 162–164, 1935.
40. T. V. Wollaston, "Diagnostic notices of new Canarian Coleoptera," Annals Magazine of Natural History (3), vol. 11, pp. 214–221, 1863.

Index

I

L

M

M

O

P

Q

R

S

T

V